Farming, Famine and Plague

Kathleen Pribyl

Farming, Famine and Plague

The Impact of Climate in Late Medieval England

Springer

Kathleen Pribyl
Climatic Research Unit
University of East Anglia
Norwich, UK

ISBN 978-3-319-55952-0 ISBN 978-3-319-55953-7 (eBook)
DOI 10.1007/978-3-319-55953-7

Library of Congress Control Number: 2017942814

Cover illustration: Luttrell Psalter, Lincolnshire, circa 1325-40 (British Library, Add. MS 42130, f.172v)

Printed on acid-free paper

This Springer imprint is published by Springer Nature
The registered company is Springer International Publishing AG
The registered company address is: Gewerbestrasse 11, 6330 Cham, Switzerland

Acknowledgements

I would firstly like to thank Christian Pfister for his invaluable advice and support throughout the PhD thesis which largely forms the basis of this book. I also wish to thank the EU project 'Millennium' and the OCCR (Oeschger Centre for Climate Research) in Switzerland for initially funding this research and for offering a platform for the exchange of ideas between historians and natural scientists that provided the stimulus for many topics covered in this book. I am also indebted to Jordan Claridge, Richard Cornes, Andrea Kiss, Christian Rohr and Phil Slavin for encouragement and counsel.

The staff at the Norfolk Record Office in Norwich are gratefully acknowledged for their help with accessing the medieval account rolls, and for granting permission to use example images of these *compoti* in this book. Due to the poor condition of a number of these account rolls, they could only be consulted with the assistance of the conservator, Nick Sellwood. I am also grateful to the Bodleian Library in Oxford, the Climatic Research Unit at the University of East Anglia and the National Archives in London for granting me permission to view their archive material.

I would also like to thank researchers who have provided data that made this study possible: Aryan van Engelen, Jan Buisman and Folkert IJnsen for providing the temperature series for the Low Countries in an easily accessible database (www. knmi.nl/klimatologie/daggegevens.html) and Christian Pfister from the University of Bern, Switzerland, for the unpublished central European summer precipitation index as well as Rob Wilson, Richard Cooper and colleagues for both English tree ring series that have been used extensively throughout this book. The Euro-Climhist database hosted by the OCCR and the University of Bern also served as a valuable resource in the preparation of this work (www.euroclimhist.unibe.ch) and is gratefully acknowledged.

Contents

Abbreviations

AM	Assumption of the Blessed Virgin Mary
BLO	Bodleian Library Oxford
CUL	Chicago University Library
CRU	Climatic Research Unit, University of East Anglia
DCN	Dean and Chapter Records
DN	Records of the Estates of the Bishop of Norwich
LEST	Le Strange of Hunstanton
MC	Minor Collections
NCR	Norwich City Records
NRO	Norfolk Record Office
NRS	Norfolk Record Society
SJ	St James
SL	St Laurence
SPC	St Peter in Chains
TNA	The National Archives, London

Chapter 1
The Historical Climatology of Late Medieval England

In the Bodleian Library in Oxford lies under the reference MS Digby 147, fols. 125–138 a curious text: the 'Tractatus de pronosticacione aeris' by William Merle, written around the year 1340. William Merle is mostly known for his detailed weather diary for the years 1337–1344 'Consideraciones temperiei pro 7 annis', one of the earliest weather diaries in Europe. 'De pronosticacione aeris' is not dedicated to the recording, but to the forecasting of weather in England and, unusual for the Middle Ages, to the study of the impact of extreme weather conditions on land and people. The work on weather prediction is not of the astrometeorological character that is so typical for the late medieval and early modern period, but it is based on a mixture of weather lore and the classical authorities. A large part of the traditional knowledge compiled by Merle came from the people most familiar with 'reading the skies' and the signs of nature: sailors, shepherds and agriculturalists. The treatise concludes with a detailed analysis of the influence of specific weather conditions on the agricultural production and is particularly concerned with the meteorological causes for the development of subsistence crises.[1] Merle's interest in the agrarian economy not only led him to a nonbiased attitude towards his sources of information, but also caused him to transfer the subject of meteorology and agricultural production from the Mediterranean – which had been the regional focus of the ancient natural philosophers – to the northwest of Europe, in particular to England with its maritime climate.[2]

Equally important for peoples' livelihood, or rather death, was the occurrence of plague waves, which abound in late medieval Europe, England included. William Merle died in 1347, a year before the Great Pestilence arrived in England. Nonetheless he might have heard of the rumour about an epidemic approaching from the Orient and landing on Europe's Mediterranean shores, but in any case he must have known the prognostication of great mortality, wars and extreme weather from the lunar eclipse and planetary conjunctions in 1345 by his colleague John

[1] Thorndike, History of magic and experimental science, vol. 3, 141–145. Jenks, Astrometeorology, 193–194 and Snedegar, Between scholasticism and folk wisdom, 30–34.

[2] Snedegar, Between scholasticism and folk wisdom, 32–34.

K. Pribyl, *Farming, Famine and Plague*, DOI 10.1007/978-3-319-55953-7_1

Ashenden.[3] Such a prediction could hardly fail in the Late Middle Ages. Among the political and academic elite, at court and at university, astrology and astrometeorology stood in high favour and were to rise in esteem even more during the insecure and volatile times after 1350, which were perceived by the people as a Dance of Death. Ashenden's work, however, resonated particularly deeply with his contemporaries, because the 'Great Mortality' did indeed befall England and Europe when *Yersinia pestis* swept across the continent. Plague was to be, as famine had been already, in the focus for those men engaged in the science of the stars. The occurrence of dearth and epidemic disease was of course open to a variety of interpretations by contemporaries, and these ranged from the religious to the proto-scientific.

As harvest failures are often linked to specific weather conditions, even more so in pre-industrial times, so do climatic factors play a role in the outbreak of some diseases. Seasonal patterns of plague outbreaks across the world are known, and research on the Third Pandemic, which began in the nineteenth century, has revealed the climatic factors raising the risk for a plague outbreak in a variety of regions and climate zones. At the conjunction of the medieval astronomers and physicians yearning to predict plague, and the modern knowledge about the epidemiology of the plague, the involvement of rodents, fleas and the bacterium *Yersinia pestis* together with the role of meteorological conditions in the development of outbreaks of the Third Pandemic, emerges the question if there were weather factors involved in the recurrent plague outbreaks in England once the Second Pandemic had reached the British Isles in 1348. The Late Middle Ages provide an ideal research field for this question since this time pre-dates the implementation of public health measures on the British Isles.

The Late Middle Ages were a time of crisis. The heaviest blows were concentrated in the fourteenth century which saw the blossom of the High Middle Ages wither: the Great Famine 1315–1317 was followed by the Great Pestilence 1347–1353. Europe's population decreased dramatically. Whereas in western Europe the expansion of the High Middle Ages met with its environmental limits in the first half of the fourteenth century, the eastern European societies continued to flourish until about 1400. War became widespread and long-lasting. England competed with France for the hegemony in western Europe, and although the Hundred Years' War was not fought on English soil, English resources were bound up in the strife, and northern England suffered from the conflict between England and Scotland. In the post-1350 period England's demographic development became mortality driven. Profound change to the established order, including the religious and economic realm, went hand-in-hand with the demographic decline. The church was divided by a schism, and as the Lollards challenged ecclesiastical hierarchy and authority, the Peasants' Rising in 1381 questioned the social order, culminating in 'When Adam delved and Eve span, where was then the Gentleman?' Much has been written about the Late Middle Ages and the scourge of mortality crises in the form of famine and plague that induced the late medieval crisis. However, until recently, the influence

[3] Thorndike, History of magic and experimental science, vol. 3, 326–328.

of weather and climate often featured as a mere supplement to such appraisals, even though historians were aware of the changing climate. It is the intention of this book to demonstrate that far from being a footnote to the events, climate had a governing effect upon life and death.

During the Late Middle Ages Europe's climate experienced major shifts. The Medieval Climate Anomaly with its warm conditions was coming to an end in the later thirteenth century. The High Middle Ages had been a time of minimal volcanic activity and very low solar forcing. In the twelfth and thirteenth centuries a sequence of large volcanic eruptions, amongst them the catastrophic eruption of the Rinjani complex in Indonesia in 1257, heralded the end of this quiet period. Solar forcing was becoming more variable from the eleventh century onwards and was declining rapidly in the late thirteenth century.[4] Temperatures began to fall and it followed a period of climatic transition during the fourteenth century. Solar forcing recovered shortly in the decades around 1380, but by 1420 the Spörer Minimum set in. The Little Ice Age began during the fifteenth century. However, regional climate responses to these forcing mechanisms often did not simply mirror the global average conditions, and it is these regional variations in weather and climate that had profound influences on the society of late medieval England. The analysis of this regional signature of climate and its influence on the basis of life of ordinary people is the focus of this book.

For an analysis of the impact of climate on agriculture and disease in England the use of regional climate data is indispensable. However, only two tree-ring based drought reconstructions are available for England and a rainfall proxy from Scotland, so new data are needed to gain further insight into the medieval climate.[5] Proxy data are not only provided by natural archives. Apart from direct weather references, written records also contain information that can be employed to reconstruct long-term climate series as is done with the proxy series based on tree-rings or speleothems. The most famous such proxy found in documentary sources is the date of the vine harvest, which reflects the mean temperature during the vine growing season. For medieval England no *vine* harvest dates are available, but in the records from the medieval East Anglian countryside – so called manorial accounts – the date of the *grain* harvest is found, which functions in a similar manner. Manorial accounts are a documentary source type that is almost limited to England, and the information preserved in those parchment rolls is the delight and foundation of English medieval agrarian and economic history.

The grain harvest date in the manorial accounts from East Anglia allows the reconstruction of the mean temperature during the growing season, spring and early summer, from the second half of the thirteenth to the first half of the fifteenth

[4] For cause and temporal as well as spatial extension of the Medieval Climate Anomaly or Medieval Warm Period, see Goosse et al., The origin of the 'Medieval Warm Period' and Bradley et al., The Medieval Quiet Period, About the eruption of the Rinjani complex in 1257, see Lavigne et al., Source of the great A.D. 1257 mystery eruption, 16742–16747.

[5] Wilson et al., March–July precipitation reconstruction, Cooper et al., Hydroclimate variability, Proctor et al., A thousand year speleothem proxy record.

century; this temperature reconstruction based on proxy data from documentary sources is hence the earliest for Europe.[6] For the British Isles no other reliable contemporary natural proxy for summer half year temperatures is available for this period. Whereas most natural proxies do not represent the high frequency variability well – and indices based on direct weather references in written records are limited in their capacity to reflect the long-term variability – a temperature reconstruction based on the phenological signal of annual plants catches both the interannual as well as the long-term variability. While the variation in global mean temperature over the last 1000 years – including this highly interesting transition period from the Medieval Climate Anomaly to the Little Ice Age – has been studied extensively,[7] considerable uncertainty still exists in the variability at the regional scale.[8] For improving the spatial pattern more regional temperature reconstructions are needed. A reconstruction of late medieval East Anglian temperatures is not only valuable for this region, but is crucial for establishing northwest European conditions. Temperatures in East Anglia achieve high correlations with temperature in other parts of the British Isles[9] and also the Benelux countries given their geographical proximity.

In addition to the temperature reconstruction, a precipitation index for the months shortly before and during the grain harvest is constructed in this book by using the duration of the grain harvest in medieval eastern England as an indicator of regional rainfall levels. Since the grain harvest length is also influenced by other factors including labour supply and bulk of the harvest, and summer rainfall is much more localized than summer temperature, this index has to be used with care, but its extreme values points to summer seasons with very high or very low levels of precipitation. The study of rainfall levels is supported also by evidence both on drought and above average rainfall conditions supplied by the East Anglian manorial accounts themselves.

The majority of the data on the date and duration of the grain harvest come from the manors of Norwich Cathedral Priory. For allowing an evaluation of the data quality and an appropriate source criticism, information on the priory estates and the making of the manorial accounts is needed, and the conditions of medieval East Anglian agriculture must be ascertained. The grain harvest not only contains a valuable phenological signal, but it was first and foremost the climax of the agricultural year, and for the village community it provided the bulk of the food for the coming year. It was also a socio-economic cornerstone of village life, with the harvest work constituting a substantial proportion of the possible wage earnings. Furthermore, the food for the harvest workers at the lord's table was for many of a higher nutritional quality than that which could be afforded throughout most of the year.[10] The

[6] Pribyl et al., Reconstructing medieval April-July mean temperatures.

[7] Jones et al., High-resolution palaeoclimatology of the last millennium, 5–21.

[8] Mann, Global signatures and dynamical origins of the Little Ice Age and the Medieval Climate Anomaly, 1256–1259.

[9] Jones and Hulme, The changing temperature of 'Central England', 179.

[10] Dyer, Everyday life, 77–100.

grain harvest also played an important role in cultural traditions binding the village community together, and its completion was the occasion for festivities. Hence this book also provides an analysis of the conflicting parameters of practicality and custom and their influence on work organisation and communication of the harvest date. While Norfolk became the centre of the agricultural revolution in the eighteenth century, the harvesting methods remained comparable over time, and this offers the possibility of relating the grain harvest dates around 1800 to contemporary instrumental temperature observations and for thus establishing the connection between temperature conditions and harvest date in medieval East Anglia.

Historical climatology has been very active for the early modern and modern period in Europe, for which much climate information and many seasonal climate indices are now available, but the medieval period is still under-researched.[11] Proxy data are not the only type of information on past climate in documentary sources: far more common are direct references to extreme weather conditions. Many of those are found in narrative sources: chronicles, annals and diaries. In England Thomas Short in his work of 1749 'A general chronological history of the air, weather, seasons, meteors, etc.' was the first to achieve the painstaking collection of such weather references from chronicles and annals from Antiquity to his own days. As with Merle and Ashenden, Short was primarily interested in the link between weather conditions and major events interfering with human livelihood, such as dearth and epidemic disease, but by the eighteenth century the astrometeorological approach had largely been abandoned. With the rise of meteorology, weather compilations remained a popular genre throughout the nineteenth century, but these older compilations suffer from misdating, mislocating and misinterpretation of weather events, because they were made without regard to the historical-critical method. The first largely reliable compilation of extreme weather events on the medieval British Isles, 'A meteorological chronology to AD 1450', was produced in 1937 by Britton.[12] For medieval Europe in general 'Le climat en Europe au Moyen Âge' by Alexandre and 'Duizend Jaar Weer, Wind en Water in de Lage Landen' by Buisman are highly valuable.

The direct weather references assembled in these compilations, or more recently in databases such as the Euro-Climhist database held by the Oeschger Centre for Climate Research and the University of Bern (Switzerland), were mostly recorded by individuals, and during the Middle Ages until the fifteenth century often by monks. Care has to be taken to include only reliable data in an analysis of climate, such data are generally contemporary to the lifetime of the author. Chronicles often ended with the death of the author, so the resulting runs of information are no longer than a few decades. To allow comparability and a quantitative analysis, the climate parameters – most importantly temperature and precipitation – have to be indexed and the information from different sources has to be combined in one regional

[11] Brázdil et al., European climate of the past 500 years, 9–34.

[12] Regarding the care that must be taken when using weather compilations for the analysis of the medieval climate, see Bell, Ogilvie, Weather compilations as a source of data, and Brázdil et al., Historical climatology in Europe, 373–375.

series.[13] The indices are founded on bio-physical properties, e.g. the duration of frost in winter. In the UK, historical climatology was revived with the work of Hubert Lamb;[14] for the medieval part of his climate indices for the British Isles Lamb used mostly information supplied by Britton.[15] To ensure that only good quality, original and contemporary documentary information were considered, Ogilvie and Farmer[16] remade those indices and subjected the weather references collected by Britton to the process of source criticism and enlarged the database by including new medieval weather information. These had been compiled from administrative sources. Contrary to narrative sources, administrative sources were created within an institutional framework, often their purpose was financial and they frequently take the form of accounts. Direct weather or proxy information was recorded as a by-product, often because adverse weather was a cost factor. Since the administrative sources were not limited by the lifetime of individuals they can produce long series; some municipalities began drawing up accounts in the Late Middle Ages and maintained them throughout the centuries. The data in administrative records are reliable, since they are always contemporary. This source type also records mainly direct weather references, but proxy data such as the grain harvest date in the manorial accounts of East Anglia are also available.[17]

For the study of the relationship between agricultural productivity and weather a number of agrarian historians in England collected weather references from various sets of manorial accounts. As a by-product of his research on grain yields and the agrarian economy in the thirteenth and fourteenth century, Titow gathered the direct weather references found in the manorial accounts of the Bishopric of Winchester 1208–1448 and Glastonbury Abbey 1305–1345.[18] The influence of weather on yields and crop cultivation in coastal Sussex, as well as the incursions of coastal flooding were studied by Brandon.[19] Not only direct weather references, but also proxy information like the timing and expenses for ploughing, sowing rates and mortality at lambing time were used by Stern for assessing adverse climatic conditions and their impact on output and profitability of the agricultural and pastoral sector on the manor of Kinsbourne in Hertfordshire.[20]

Although not all series of manorial accounts contain information on weather,[21] these documents harbour a great quantity of unused weather information and remain a largely under-researched source for historical climatology. Since the 1980s they

[13] Brázdil et al., European climate of the past 500 years, 17–18.

[14] For the development of historical climatology on the British Isles, see Pribyl, Study of the climate of medieval England.

[15] Lamb, Climate. Past, present and future, vol. 2.

[16] Ogilvie, Farmer, Documenting the medieval climate.

[17] Brázdil et al., Historical climatology in Europe, 375–376.

[18] Titow, Evidence of weather, and idem, Le climat à travers les rôles de comptabilité.

[19] Brandon, Late medieval weather in Sussex.

[20] Stern, A Hertfordshire demesne.

[21] Stern, A Hertfordshire demesne, 21. For example the accounts for Wisbech Barton, Cambridgeshire, are usually mute concerning weather references, Stone, Wisbech Barton, 645.

have been neglected in this respect; only Hallam employed account rolls of Norwich Cathedral Priory for a rough comparison of weather evidence in the chronicles with the quantity of the grain harvest in Norfolk 1264–1328.[22] Recently Campbell has focused again on the relationship between agricultural productivity and climate as well as the influence of climate variability on the great social and demographic changes of the Late Middle Ages, but without using direct or proxy weather information from manorial accounts.[23]

With their interest in the nexus of meteorology, harvest success, dearth and plague, Merle and Ashenden were not only focusing on the most prominent problems of their time, the decades around the arrival of the Black Death, but they were already encapsulating the basic questions guiding this work: how did meteorological conditions in late medieval England influence agricultural production and in which way did they contribute do the development large scale plague outbreaks? Through the systematic analysis of medieval sources, the application of modern statistical techniques and the building on the methods of historical climatology this book sheds new light on the complex interplay of climatic conditions and variability with mortality peaks induced by famine and plague in the late medieval Dance of Death.

---

[22] Hallam, The climate of eastern England.

[23] Campbell, Four famines and a pestilence; idem, Physical shocks; and idem; Nature as a historical protagonist, idem, Great transition.

Chapter 2
The Keeping of Agricultural Records in Late Medieval England

2.1 Late Medieval Agriculture and Manorial Accounts

In the Late Middle Ages demesne farming, the direct management of part of the lord's land instead of its complete leasing to tenants, was considerably more widespread in England than on the continent. For ensuring the honesty of the administrative personnel and for giving information concerning the state of the agricultural and pastoral sectors and for evaluating the profitability of the manor, it was essential that the manorial officers rendered account of the activities on the manor, its income and expenditure to the lord. The resulting records are known as manorial accounts, and they are the source that allows agrarian historians to form a comprehensive picture of the English seigniorial agriculture in the Late Middle Ages, including cropping trends, sowing rates, harvest success, livestock density and labour input. By their very nature these records contain a plethora of direct and indirect information on the environment, and particularly on weather conditions.

The East Anglian countryside differed from the Midlands with respect to village and field layout; the fully nucleated village was not the standard form of habitation and the field layout was less regular than in the Midlands. The parochial and manorial organisation was also marked by differences, villages in East Anglia frequently possessed more than one parish church and multi-manorial vills were common. With regard to the social composition of the village population, the high percentage of freeman in the east was unrivalled in England.[1] The information stored in the manorial records shows that seigniorial agriculture in Norfolk was highly intensive during the Late Middle Ages.

Based on manorial accounts from southern and eastern England and the Midlands – areas where the direct management of the demesne land was common – Campbell has defined farming regions with respect to seigniorial agriculture. The differences between those regions are the varying degree of intensity of the

[1] On those differences and their causes, see Williamson, Explaining regional landscapes.

K. Pribyl, *Farming, Famine and Plague*, DOI 10.1007/978-3-319-55953-7_2

agricultural production and the type of crops.[2] Eastern Norfolk and other areas close to Norfolk's north and northeast coast, stand out by being subjected to the most intensive and productive cropping type in medieval England, which featured wheat and barley and devoted a large percentage of the sown acreage to legumes.[3] Other areas – especially regions on poorer soils like the sandy Breckland in the south-west of the county (with access to the market and port of King's Lynn), the 'Good Sands' in northwest Norfolk (close to the coast and its small ports) and the hinterland of Norwich with its light sandy soils – were managed on a less intensive level and employed the 'rye with barley' regime.[4]

The intensive agricultural regime in eastern Norfolk was able to maintain soil fertility and favourable yields,[5] although the number of livestock units per hundred sown acres was comparatively low and the fallow almost eliminated. This was the result of the large-scale cultivation of legumes (as a fodder crop and for fixing nitrogen to the soil), as well as of the careful management of other resources to maintain soil fertility. The demesne cattle and other animals were often stall-fed, the resulting farmyard manure was spread on the arable; this was a laborious task. Marling, spreading dung from sheep-pens on the fields and keeping the sheep on the fallow over night were other methods employed to supply the soil with nitrogen. Productivity was also raised, because the light soils of eastern Norfolk could be tilled with horses, which could work faster and longer than oxen. Often sowing rates, especially for oats, were high to smother weeds, and fallow ploughing was employed for eliminating thistles. The high population density of medieval East Anglia ensured a cheap labour supply for tasks like weeding and manuring.[6]

Although no comparable sources are available for peasant agriculture, it is likely to have been even more labour-intensive and productive. The peasants had to ensure their families' survival from the produce of the soil. Additionally the location of eastern Norfolk also provided an easy access to the urban markets of Norwich and Great Yarmouth and the maritime trade.[7]

[2] Campbell, Seigniorial agriculture, 249–305.

[3] Campbell, Seigniorial agriculture, 270–271. The medieval agriculture of Norfolk has been thoroughly investigated by Campbell. Of his numerous works on the subject the ones used here primarily are: Campbell, Field systems, idem, Eastern Norfolk; idem, Arable productivity in medieval England, and idem, Overton, Norfolk farming c.1250 – c.1850. Comparably advanced agricultural regimes as in eastern Norfolk were to be found in eastern and northeastern Kent as well as parts of coastal Sussex, Campbell, Seigniorial agriculture, 271–272. The agricultural system of eastern Norfolk matched that of the Low Countries, the first reference to the elimination of the fallow in East Anglia even predates the continental one by more than fifty years, idem, Eastern Norfolk, 41.

[4] Campbell, Seigniorial agriculture, 267–269.

[5] For a comparison of yields in eastern Norfolk with other regions, see Campbell, Land, labour, livestock, and productivity trends, 161.

[6] Campbell, Field systems, 21–22; idem, Eastern Norfolk, 28–39; idem, Seigniorial agriculture, 269–271. On cost, use and timing of weeding and fallow ploughing, see Postles, Cleaning the medieval arable, 133–142. Fallow ploughing was probably more common in East Anglia than in the Midlands, ibid., 142.

[7] Campbell, Eastern Norfolk, 28, 39–41; idem, Seigniorial agriculture, 270.

The inclination to favour the direct management of the demesne over leasing it out, is usually attributed to the prices for agricultural products which began to rise in late twelfth-century England. This was most likely due to population pressure. The trend continued throughout the thirteenth and early fourteenth century, so that by 1330 the price for grain had increased four- or five-fold compared to the price in 1180. However, because of the population growth and the consequent ample supply of labour, wages displayed no similar tendency in the thirteenth century. It was not before the 1330s that wages caught up with the price level.[8] This situation of rising prices and stable wages prompted the great landlords to reconsider their economic strategy after 1200. Up to the end of the twelfth century most of them – like their continental counterparts – had followed a system of leasing their estates, whereby they received a fixed annual rent from the lessee. If they wanted to profit from the new economic circumstances, the direct management of their resources would prove to be much more advantageous. Thus around 1200 landlords began to abandon leases and take their manors into hand.[9] During the first half of the thirteenth century the movement gained ground and embraced eastern England, which at the beginning had showed itself more conservative than the rest of the country.[10] In the north of England demesne farming remained more confined than in the southern part of the country.[11]

For non-resident landlords the direct management of the demesne land was effected with the help of administrative personnel: usually on smaller estates a reeve, sergeant or bailiff would run the manor, on larger estates a reeve or sergeant would manage the day to day business on the manor, under the supervision of a bailiff, who was responsible for a group of manors.[12] To ensure their honesty, to oversee their work and capability, and later on also to check the profitability of the demesne land in hand, accounts had to be created.

The account survival rate increases sharply after 1270.[13] Consequently the indirect information on weather as well as direct references to adverse weather that interfered with farming and raised costs or cut profits, also multiply in the late thirteenth century. Tendencies to abandon the direct management of the demesne land gained ground in the decades after the Great Pestilence and with it the manorial accounts became superfluous. Between c.1380 and 1400 many manors were leased

---

[8] Farmer, Prices and wages, 1042–1350, 718.

[9] Harvey, The adoption of demesne farming, 345, 353; idem (ed.), Manorial records of Cuxham, 13; Britnell, Britain and Ireland, 225–228.

[10] Harvey, The adoption of demesne farming, 354. However, Campbell, Seigniorial agriculture, 27–28, states that early extant lay and ecclesiastical accounts relate to East Anglia.

[11] Britnell, Britain and Ireland, 228; Campbell, Seigniorial Agriculture, 33, 36.

[12] Harvey (ed.), Manorial records of Cuxham, 12–13; Bennett, English manor, 157–158, 162–175 on steward, seneschal, bailiff and reeve and their obligations and responsibilities.

[13] Britnell, Winchester Pipe Rolls, 31, Harvey (ed.), Manorial records of Cuxham, 17. For the treatises on estate management and accounting appearing after the mid-thirteenth century, see Oschinsky, Walter of Henley, 5, 9, 89, 144. Although many copies of the texts are to be found in the archives of Benedictine foundations, it appears that the monasteries rather used texts of more individual character for their own estates, ibid., 56–59.

again and the end of the direct management came about 1430. Only few manorial accounts, apart from those made for home farms, which were directly managed for a longer time, survive in detailed form after this point.

The long and more continuous series of account rolls, that are still available today, come mainly from ecclesiastical landlords. The longevity of those institutions, their high level of education and the advantage of proper muniment rooms ensured a better survival of their archives than those of lay lords.[14]

Manorial accounts were usually drawn up annually after the end of the agricultural year at Michaelmas (29 September). An account was made for one individual manor with information supplied by the responsible manorial officer.[15] They are written in medieval Latin. The front of the parchment rolls (face) records information on receipts (from rents, income from manor courts and sales of manorial produce or customary labour services) and expenses (on the various sectors and activities of the manor, for example ploughing, carting, construction and building maintenance, dairy farming, harvesting). On the back of the roles (dorse) are the grange account (issue and receipts, outgoings with details on quantity sown, sowing density, acreages sown, liveries), the stock account (issue and receipts, loss in death, liveries) and – especially from the fourteenth century onwards – the detailed works account (day- and boon works).[16]

The definition of seasons in the manorial accounts differs from the modern one. The agricultural year started after the harvest with the winter sowing. Winter itself covered the months from October until March or even April. Summer was loosely defined as the months May to July, whereas autumn, *autumpnus*, referred to the harvest season which lasted normally from 1 August to 29 September and marked the end of the agricultural year.[17] Naturally for an agricultural record the information in the manorial accounts in general, and consequently also the information on weather conditions, is concentrated on the growing period and harvest season; autumn and winter are less well represented.

In some collections of manorial accounts the direct weather references are frequent, as in the Pipe Rolls of the Bishopric of Winchester.[18] The references on weather in the manorial accounts invariably take the form of complaints and outline – farmers of all times would feel sympathy – how the weather put stress on the agricultural or pastoral sector. The pastoral sector was particularly vulnerable to summer droughts, which interfered with the growing of grass and hay and consequently endangered the fodder supply of cattle, sheep and horses. For the success of the hay and grain harvest on the other hand, dry weather at harvest time was crucial, as wet weather would increase the drying time for hay and grain; hay and corn would have to be turned more often, corn-sheaves might also have to be unbound,

[14] Britnell, Winchester Pipe Rolls, 34, Campbell, Seigniorial agriculture, 31–36.

[15] Harvey (ed.), Manorial records of Cuxham, 22–23.

[16] Harvey (ed.), Manorial records of Cuxham, 19; Bennett, English Manor, 188.

[17] Titow, Le climat à travers les rôles de comptabilité, 312–313.

[18] Weather references transcribed by Titow, Evidence of weather and idem, Le climat à travers les rôles de comptabilité.

dried and tied again. Stacking wet hay and storing wet corn would have resulted at least in a loss of nutrients or even a total loss of the hay or grain due to spoiling.[19] Consequently, precipitation that hindered the hay- or grain harvest was mentioned in the *compoti*. Causes for concern were also flooding due to excessive precipitation (mostly of meadow or pasture land in the winter half year), or a hard and long winter that required the supply of the manorial livestock with extra-fodder, especially of the draught animals at ploughing time.[20] In addition to those direct weather references, indirect information on weather (proxies) can be gleaned from the accounts; for example late winter and early spring weather conditions are reflected in the mortality rate among lambs and the time of spring ploughing and sowing, the date of the grain harvest can serve as a proxy for the mean temperatures of spring and summer, and the harvest length reflects to a good degree the precipitation frequency and amount at harvest time.[21]

The weather information supplied by the manorial accounts is very reliable: it is contemporary and it was checked by audit. The audit was undertaken by men who knew the manor and land, and the auditing process was aimed at detecting fraud or mismanagement.[22] During this process, information on weather was used to explain the underperformance of a sector of the manorial economy. Occasionally weather information was added to the accounts during the audit in the margin in another hand than the main text. The accounting and auditing process, as well as the possibility to collect parallel evidence from different manors or estates, makes the *compoti* a high quality source for weather related information.

2.2 Norwich Cathedral Priory

2.2.1 Norwich Cathedral Priory and Its Temporalities Until c.1300

In 1095 bishop Herbert de Losinga moved his see from Thetford to the more important and populous Norwich. By 1300 Norwich's inhabitants may have exceeded 15,000 and made it potentially the most populous provincial town in England[23]; by 1330 the town had grown to 25,000 inhabitants.[24] Norwich was an inland port with access to international trade via the rivers Wensum and Yare and the North Sea port

[19] For hay, see Stone, Wisbech Barton, 645; for grain, see Ault, Open-field farming, 27.

[20] Titow, Evidence of weather, 361.

[21] Stern, A Hertfordshire demesne, 29–30, Oliver, Problems of agro-climatic relationships in Wales, 193.On harvest date and mean growing season temperature, see Pfister, Getreide-Erntebeginn und Frühsommertemperaturen, 29 and on harvest duration and precipitation, see Chap. 7.

[22] Harvey (ed.), Manorial records of Cuxham, 51–53; Bennett, English Manor, 175, 188–192; Drew, Accounts of St Swithun's Priory, 15–16.

[23] Campbell, Norwich before 1300, 29.

[24] Campbell, Ecology versus Economics, 80.

at Great Yarmouth.[25] In fact, stone used for the construction of the cathedral arrived from Caen, France, and could be shipped directly into the cathedral precinct on an artificial canal.[26]

When Losinga decided to construct the new cathedral in Norwich he also established a community of about sixty Benedictine monks for its upkeep. The size of the priory placed it in the first league of monastic cathedrals.[27] In addition to the monks there were servants, clerks and visitors populating the precinct, bringing the number of inhabitants that needed to be provisioned with victuals to 250–270.[28]

To guarantee the priory's income and food supply,[29] Losinga, his successors and a few lay men granted the priory temporalities in the form of landed estates and other revenues. These possessions and rights were mostly situated in Norfolk (Fig. 2.1).[30]

The estate of Norwich Cathedral Priory included Martham and Hemsby, these villages lie close to the North Sea coast in the fertile and highly productive Flegg district of eastern Norfolk. In nearby Scratby the revenues of the monks came from the appropriated church. A substantial group of manors lay in the direct vicinity of Norwich: Catton, Monks' Grange (Pockthorpe), Trowse Newton, Lakenham and Eaton. Together with the manor of Heigham owned by St Benet's of Hulme and the nunnery at Carrow these ecclesiastical estates almost encircled Norwich and contributed to the permanent tensions between its citizens and the cathedral monks.[31] These conflicts led to several violent eruptions, the most severe of them took place in 1272.[32] Still close to the town were the manors of Arminghall and Taverham and the lands at Plumstead and Bawburgh. Further west were the possessions stemming from the first see of the bishopric at North Elmham, the manors North Elmham and

[25] Although as pointed out by Edwards, Hindle, Transportation system, 131 the increasing size of the sea-going ships would have cut Norwich off the direct access to the sea. Dunn, Trade, 224–225 states, that in the Late Middle Ages it would have been unlikely, that the sea-going vessels would navigate the Yare. In consequence goods had to the transshipped at Yarmouth, adding to transport costs.

[26] Blake et al., The Norfolk we live in, 39. The canal protruded at Pull's Ferry from the river Wensum towards the cathedral. No local stone was suitable for the surface work of the cathedral. The location of the canal is still visible in today's street layout; it followed Ferry Lane.

[27] Dodwell, Monastic community, 231. Norwich Cathedral Priory was on par with the older foundations of Worcester and Winchester.

[28] Saunders, Obedientiary and manor rolls, 92, 162.

[29] The priory's grain provision has lately been studied by Slavin, Bread and ale.

[30] Saunders, Obedientiary and manor rolls, 35. A map of the property rights is displayed ibid., xii. A complete list of the revenues of Norwich Cathedral Priory is given by Blomefield, History of Norfolk, vol. 4, 556–62. On the reluctance of the lay magnates to endow the priory with property rights, see Virgoe, Estates, 342, who also provides a simplified list of the landed possessions of the priory, ibid., 346. There was a relative paucity of lay investment to Norwich Cathedral Priory, the nobility rather established monastic communities on their own estates, Harper-Bill, Church and the wider world, 302.

[31] Campbell, Norwich before 1300, 9.

[32] Tanner, The cathedral and the city, 259–261.

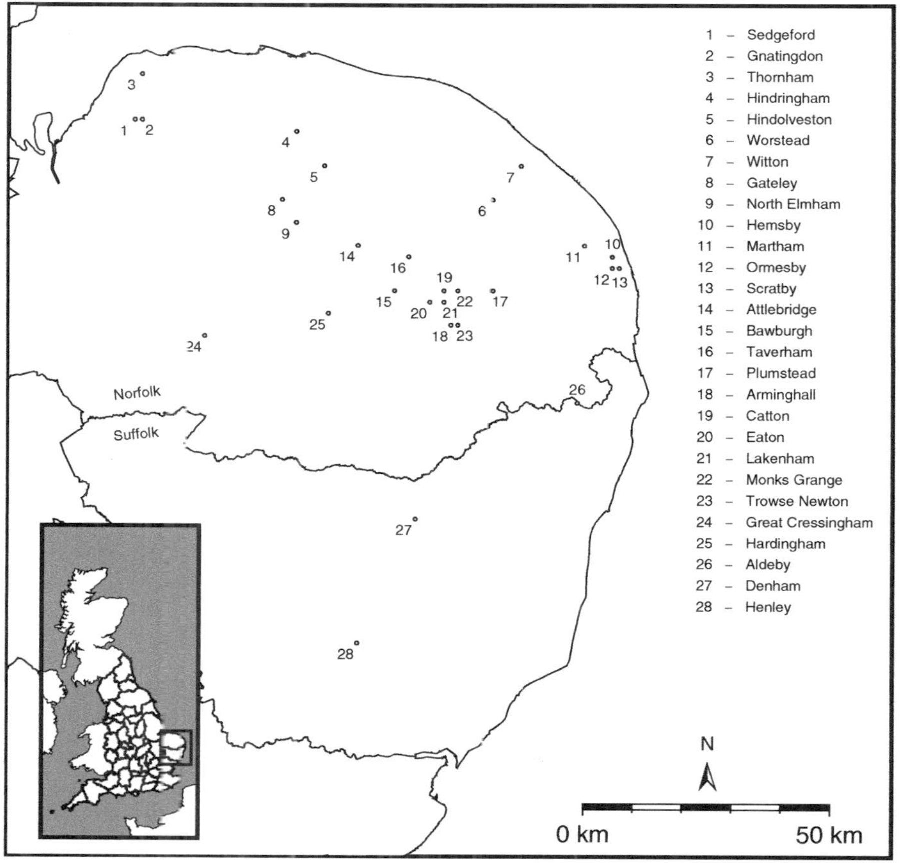

Fig. 2.1 Norwich Cathedral Priory: distribution of manors and other endowments. Shown are only places where accounts detail harvest information. Map with modern county boundaries

Gateley.[33] The monks were also endowed with manors at Hindolveston and Hindringham towards the north coast. A large agglomeration of the priory's land was situated in the northwestern corner of Norfolk, close to the Wash and the North Sea: the complex of the extensive manors of Sedgeford and Gnatingdon and the small unit at Thornham. The priory also owned the manor of Great Cressingham, in the Breckland in western Norfolk,[34] land at Worstead – a village famous for its textile production in the northeast of Norwich – and lands at Denham, Hoxne and Henley in Suffolk and Chalk in Kent.

Until the thirteenth century most of those manors were fee farms, partly leased out for rents in kind. In the thirteenth century the continuing population growth[35] and inflation made the direct exploitation of the land and a market-oriented production of grain more profitable for the landowner. Therefore in this period, Norwich Cathedral Priory, following the national trend, showed a marked tendency towards enlarging its demesne lands. Since donations in the form of landed property were no longer common, the priory had to acquire land by purchase.[36]

At the beginning of the thirteenth century the main temporalities and thus most manors had been apportioned to the *magister celarii*, an office peculiar to Norwich Cathedral Priory, and the cellarer himself. The other obedientiaries held sources of revenues according to their current need and expenses. This allocation stabilized around 1270 so that certain sources of revenues became fixed to certain offices, and the master of the cellar was left in control of most manors and demesne land.[37] Under prior Henry de Lakenham (1289–1310) the policy of acquisition came to a halt, and a move to a centralized policy and to an expansion of the income of the priory through efficient exploitation of its possessions was realized.[38] Hence, when the landed estates and property rights of the cathedral priory stabilized at the end of the thirteenth century, the 16 large estates, know as the prior's manors, were under supervision of the *magister celarii*: Martham, Hemsby, Plumstead, Trowse Newton, Monks' Grange, Catton, Eaton, Taverham, North Elmham, Gateley, Hindolveston,

[33] The see was transferred from North Elmham to Thetford in the early 1070s, Harper-Bill, Church and the wider world, 281. This relocation was not unusual, the Council of London in 1075 authorized the transfer of the bishoprics of Lichfield, Selsey and Sherbourne. North Elmham was a small manor; bishop Herbert de Losinga gave the manor and land to the cathedral priory, Yaxley, North Elmham Park, 517, 562

[34] For the economic development of the Breckland during the later Middle Ages, see Bailey, A marginal economy.

[35] The English population reached a climax around 1300, estimates range from 4.75 million to almost 7 million. They are summed up by Britnell, Economic development, 11–12.

[36] Virgoe, Estates, 343.

[37] Ibid., 348–350.

[38] Virgoe, Estates, 351–352 and Stone, Profit-and-loss accountancy, 41–42. These strategies of prior Henry seem to position him within the wider movement of heads of religious houses being apt administrators aiming at the successful exploitation of their estates at the end of the thirteenth century, Postles, Administrators, 38.

Hindringham, Sedgeford, Gnatingdon, Thornham and Denham (Suffolk)[39]; the cellarer had control over the manors at Great Cressingham and Hopton (Suffolk)[40]; the chamberlain over the manors at Arminghall and Lakenham[41] and the precentor was in charge of land associated with the church in Plumstead.[42]

As in other Benedictine houses the main attention at Norwich Cathedral Priory was turned to cereal farming,[43] nonetheless the extensive landed estates allowed for a certain degree of specialisation on individual manors. Sheep were mainly raised at considerable distance from Norwich: the priory's wool manor was in the northwest of the county, at Sedgeford.[44] Sedgeford's flocks would also graze on Gnatingdon, Thornham and 'Lyng' ground.[45] The Sedgeford-Gnatingdon complex – Gnatingdon would also be referred to as 'East Hall', whereas Sedgeford constituted the 'West Hall' situated in the village itself – was also the largest grain producing unit of the priory[46]: around c.1300 Sedgeford had 430.5 acres under crop, Gnatingdon 423 acres (Fig. 2.3).[47] They were situated on the 'Good sands', well draining soil, and they were less intensely worked than the eastern Norfolk manors.[48] They would retain their status as the largest demesnes managed directly until the end of demesne farming at Norwich Cathedral Priory, although their arable demesne would shrink, especially after the Black Death, until finally in 1417 there would be 276 acres under crop in Sedgeford and 243 acres in Gnatingdon and their sown acreage was to fall further until 1431, when Sedgeford manor, and Gnatingdon most likely, too, were leased out (Fig. 2.3). Yet, the demesne of other manors also dropped or they were leased out altogether. The highly productive eastern Norfolk manors of Martham and Hemsby were also kept in hand by the priory until the 1420s.[49]

The main grain in Norfolk was barley, which also served as bread grain. It was the largest crop on the cathedral priory manors, usually making up about 60% of the demesne produce.[50] Wheat, rye, oats and peas were grown virtually everywhere.

---

[39] Meeres, Records of Norwich Cathedral, 1. Saunders, Obedientiary and manor rolls, 76–77 lists the manors apportioned to the department of the master of the cellar, but does not give Eaton and Trowse Newton, though he gives additionally Aldeby.

[40] Virgoe, Estates, 353.

[41] Saunders, Obedientiary and manor rolls, 114, Virgoe, Estates, 353.

[42] Saunders, Obedientiary and manor rolls, 134.

[43] Smith, Canterbury Cathedral Priory, 128.

[44] Saunders, Obedientiary and manor rolls, 35, on the following page he lists further sheep rearing places. Virgoe, Estates, 352 explains that sheep would mainly be raised on manors on the periphery, because the profitability of sheep farming was lower than that of grain production.

[45] Yaxley, The prior's manor-houses, 21.

[46] Saunders, Obedientiary and manor rolls, 35.

[47] The data on the acreages of the different crops and the total arable demesne of each manor have already been collected by Bruce Campbell. However, since his public access database was neither functioning nor maintained during the work on this part of the thesis, the data have been extracted again by the author.

[48] Stone, Estates, 348, 355.

[49] Virgoe, Estates, 355, Campbell, Seigniorial Agriculture, 235–236.

[50] Virgoe, Estates, 352.

However, the Flegg manors, in fertile eastern Norfolk, like other manors in that region, usually grew no or very little rye. It was partly replaced by mixed corn, *maslin*, a wheat-rye mixture.[51] The manors at Denham and Gateley also hardly ever sowed rye or *maslin*. At Hindolveston and Hindringham a clear preference for wheat can be detected, too.[52] Plumstead usually had more wheat than rye or *maslin* on its fields.[53] Sometimes small estates also omitted the sowing of rye.[54] On the other hand the manors around Norwich, Monks' Grange, Catton, Eaton and Taverham, grew much rye and partly no or very little wheat.[55]

The preference for wheat or rye is partly due to soil conditions, but the rye cultivation at the manors around Norwich is also explained by a close and hungry market in the town; due to the low price of rye, it was not profitable to transport it over a long distance.[56] Whereas the acreages sown with rye in Norfolk dropped after the Great Pestilence 1348–1349, those sown with wheat remained stable.[57] However, in

[51] Campbell, Eastern Norfolk, 31. The last time rye was sown in Hemsby is in 1287–1288, from 1294–1295 onwards mixed corn appears in the accounts and occupies a falling percentage of the sown acreages until the Hemsby rolls end in 1334–1335. The development in Martham mirrors closely the situation in Hemsby. The last time mixed corn is mentioned is in 1349–1350, afterwards only wheat was grown as winter corn in Martham until the 1420s. Scratby usually grew wheat, but never rye and rarely mixed corn.

[52] In Hindringham wheat and rye were sown until 1312–1313 (except 1287–1288, when no rye was sown). However, normally the number of acres sown with wheat was twice that for rye. Rye was replaced by *maslin* 1317–1318 and 1318–1319. From 1320–1321 onwards no rye or *maslin*, was sown at Hindringham. A similar situation prevailed in Hindolveston. Until the early 1300s wheat and rye can be found on the demesne (except in 1272–1273 and 1273–1274, when rye was omitted, and 1287–1288, when it was replaced by *maslin*). After c.1310 rye was apart from individual years (1320–1321, 1395–1396, 1397–1398 and 1398–1399, probably in connection with wet conditions) not sown any more.

[53] The cropping strategy here was variable. Between 1312–1313 and 1331–1331 rye was replaced by *maslin*, after the mid-1330s both were sown. Very little rye or *maslin* were sown after 1354–1355. In the late 1390s the rye acreages increased again. Throughout the 1410s and 1420s more rye and *maslin* were grown in Plumstead, than ever in the second half of the fourteenth century, excepting the late 1390s.

[54] Thornham in northwest Norfolk sowed no rye in 1318–1319, 1322–1323 and 1325–1326. Worstead probably cultivated no rye after the Black Death. In 1330–1331 no wheat was sown.

[55] Campbell, Seigniorial agriculture, 268. Monks' Grange had often wheat in very small parts of its demesne land between 1317–1318 and 1334–1335, but not before. Eaton generally grew no wheat, but exceptions occurred mainly in short phases during the mid-1290s, the late 1310s and early 1320s and between c.1359 and 1370. At Catton some wheat was grown around 1320 and in the early 1340s. Taverham gave over a small part of its land for wheat cultivation until c. the mid-1290s, then no wheat at all was grown until the mid-1330s. At the end of the 1330s wheat reappeared again and its acreages increased after the Black Death, when between the mid-1350s and the mid-1370s wheat occupied even half of the acreage dedicated to rye. Also during the last phase, c.1415–1425, the wheat acreage averaged one third of the rye acreage; only in 1413–1414 and 1420–1421 it was not grown.

[56] Campbell, Seigniorial Agriculture, 219–20, 267–269.

[57] Campbell, Overton, Norfolk farming c.1250 – c.1850, 54. On the consumption of bread made of wheat, rye and/or barley, see Rogers, Was Rye Ever the Ordinary Food of the English?, 121–124 and Stone, Consumption of field crops, 13, 17–23.

a county that mainly grew barley, barley bread was the most important staple food of the urban poor.[58]

Several estates had woods: Hindolveston, Eaton, Thurning, Gateley, Plumstead and Monks' Grange (part of Thorpe wood). Mostly underwood or coppiced wood was sold, but Hindolveston, Eaton, Thurning and Gateley also occasionally sold timber. The most important and valuable wood was at Hindolveston, which was the location of a wood market.[59] It is noticeable, that even when the manor was leased out in the 1380s this wood was kept under direct management by the priory.[60] Occasionally the wood-accounts in Hindolveston mention storm damage (Appendix 1).

In Eaton, Saunders identified a manor specialized on fuel and carrying. The tenants of Eaton performed carting services also for other manors, as well as heath-reaping and washing the priory's sheep.[61]

According to Saunders, Monks' Grange carried no tenants in the usual sense, at least there were no tenants' lands. Most likely the manor was worked by hired labour and by labour from other manors.[62] As such its income would almost entirely depend upon market sales. As long as prices were high the demand by the inhabitants of Norwich would guarantee fine profits, nonetheless Monks' Grange was highly vulnerable to agricultural depression, when profits from prices would fall and there was no secure income from rents. It is no surprise therefore that the last manorial account from this manor comes from the mid-1330s, a decade when many good harvest and an increasing currency shortage led to deflationary tendencies.[63]

Catton was submitted to another unusual process: between 1282–1283 and 1285–1286 its demesne land disappeared, though no evidence for its leasing-out

---

[58] Rutledge, Economic life, 183; Stone, Consumption of field crops, 17.

[59] Saunders, Obedientiary and manor rolls, 77–78.

[60] NRO, DCN 60/18/43A-49, 1382–1383 to 1391–1392.

[61] Saunders, Obedientiary and manor rolls, 36. Interestingly his example for the carrying service of the men of Eaton (Stow MS., f. 26) is a transport of wheat from Sedgeford to Norwich, stopping overnight at Sedgeford and receiving supplies on the way back in North Elmham. Obviously the c.60 km could be covered by half a dozen empty and – on their way back – half-laden carts in one day. A fully laden cart, carrying about 40 bushel of grain, could travel up to 39 km on a single day, Masschaele, Inland Trade, 202–204. For allowing the exceptionally long distance from Sedgeford to Norwich the carts of the Eaton tenants were merely half-laden, with about 18 bushel of wheat. The arrangement was profitable for Norwich Cathedral Priory, because they could rely on the carrying service of the tenants of Eaton; these men received no or very low wages for their work and kept the horses and carts at their own expense. It should be noted that Farmer, Two Wiltshire manors, 5–7 sets the normal distance of grain transports to the market on roads at 16.5 km, so that a return journey could be made within a single day.

[62] Saunders, Obedientiary and manor rolls, 36–37. Two manors of Canterbury Cathedral Priory operated upon the same principle. Between c.1290 and 1330 a large part of their income – around 85% – was constituted by sales, no tenants' rents contributed to it, Smith, Canterbury Cathedral Priory, 173, 182.

[63] On the harvests of the 1330s, see Campbell, Nature as historical protagonist, 299. For information on the currency shortage and deflation starting in the late 1320s with the situation deteriorating considerably during the 1330s, see Mayhew, Numismatic evidence, 7–12 and Britnell, Commercialisation 1000–1500, 179, 182–183. The exportation of coinage in the Hundred Year War contributed to the deflation and crisis, Bridbury, Before the Black Death, 407–410.

appears. It seems, that the demesne was put under the charge of the officials at Monks' Grange, where an increase of the acreage under crop occurred in the last quarter of the thirteenth century.[64]

2.2.2 *The Making of Manorial Accounts and Their Economic Context*

The concentration on the direct management of the demesne land by Norwich Cathedral Priory in the thirteenth century resulted in the need for stricter supervision and administration. The earliest surviving *compoti* date to 1255–1256, but the start of the keeping of written administrative records must pre-date this year. Early *compotus* rolls appear not to have been preserved with the tightest regularity.[65] The last manorial account of a demesne under direct management survives for 1430–1431. In the period between 1256 and 1431 there are 840 surviving accounts from Norwich Cathedral Priory's manors in East Anglia.

The interpretation of the temporal distribution of the surviving accounts is complicated by the lack of a detailed economic and estate history of Norwich Cathedral Priory.[66] The density of the priory's surviving *compoti* over time as displayed in Fig. 2.2 is to a certain degree linked to the development of the demesne farming and thus to economic trends and administrative decisions at the cathedral priory.

After 1260 the number of surviving account rolls increases, although survival is patchy and there remain very many gap years. This coincides with a significant augmentation in the acreage of demesne land under the plough between 1260 and c.1300. During this time the total acreages of demesne land under crop was increased by an eighth from c.2583 acres to 2928 acres. At Sedgeford the increase was much more drastic, although in Gnatingdon the sown acreage had already been large before (Fig. 2.3). Only Catton, Gateley, Hindringham and Hindolveston witnessed a reduction of demesne land.[67] Central Norfolk and eastern manors increased the acreages for wheat.[68] From 1270 to 1320 the price-wage ratio was favourable for

[64] Stone, Estates, 343 and 620, note 27; Saunders, Obedientiary and manor rolls, 36, had assumed that the 'lord [Norwich Cathedral Priory] was bought out at Catton'.

[65] Dodwell, Muniments, 330. On the reasons for drawing up accounts and keeping them properly see King, Estate management, 6–11.

[66] The economic history of Norwich Cathedral Priory has not been as studied as profoundly as that of similar houses, Virgoe, Estates, 340. Virgoe's article itself is a short overview of the subject from the twelfth century until the dissolution of the cathedral priory 1538. Stone, The estates of Norwich Cathedral Priory, 1100–1300, stops in 1300. Campbell's work on agricultural history includes a high quantity of manorial accounts and other material relating to the estates of Norwich Cathedral Priory, but does not focus on this house. Slavin, Bread and ale, focuses on the supply of the cathedral priory with grain.

[67] Campbell, Seigniorial Agriculture, 232.

[68] Virgoe, Estates, 352.

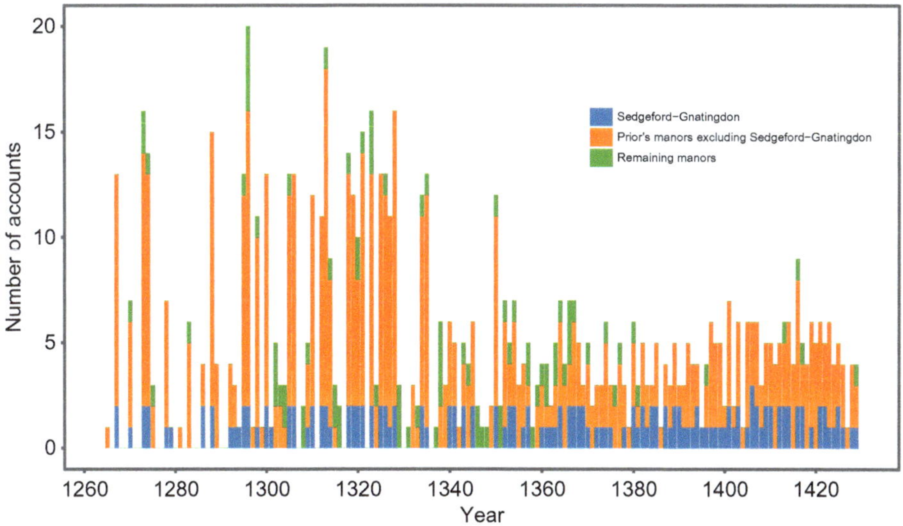

Fig. 2.2 Norwich Cathedral Priory: numbers of accounts per year 1256–1431. The prior's manors supply the most reliable data, and Sedgeford and Gnatingdon form the most complete series amongst the prior's manors

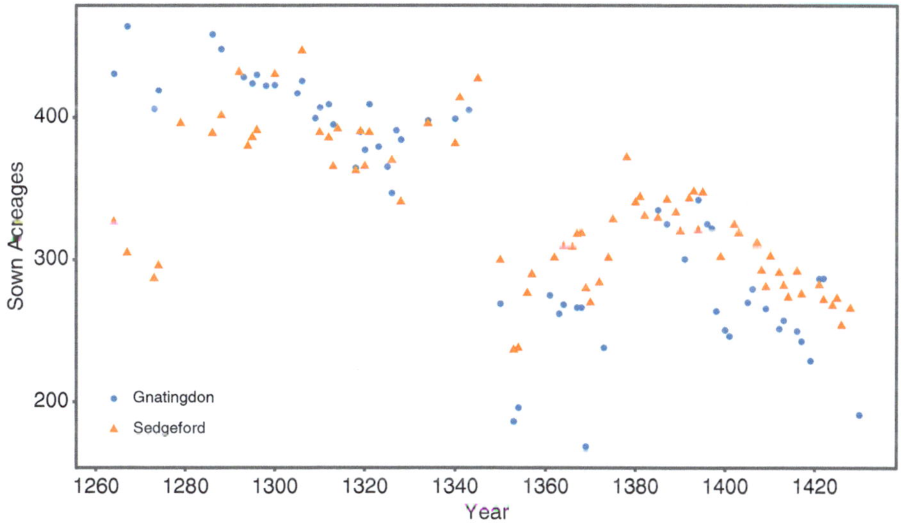

Fig. 2.3 Sedgeford and Gnatingdon: cultivated demesne land 1264–1431. Sedgeford and Gnatingdon possessed the largest demesnes lands of Norwich Cathedral Priory

demesne agriculture, prices were high and wages low[69]; it was the period of 'high farming'.

The very low number of accounts before 1272–1273 and especially in those years immediately preceding 1272 can be explained by the attack of the citizens of Norwich on the cathedral precinct in summer 1272, which caused widespread damage to the priory's buildings by arson and pillaging.[70]

Under prior Henry de Lakenham (1289–1310) a centrally directed process to expand the income of the priory set in. This included a temporary halt of the acquisition of land[71] and a refinement of the accounting process by the introduction of the accounts of *proficua maneriorum*.[72] These records on the profits of the manors appear as separate documents the first time in 1293–1294 and were made until 1341.[73] During the priorates of Henry de Lakenham and Robert de Langley (1310–1326) the interest in the direct exploitation of the demesne combined with enhanced accounting procedures results in a high number of accounts per (surviving) year from the mid-1290s to the late 1320s (Fig. 2.2). Henry de Lakenham's approach towards direct cultivation mirrors the general attitude of his contemporaries, which was marked by a growing concern about profitability and efficiency of demesne farming.[74] This was caused by a temporary drop in grain prices in the latter quarter

---

[69] Stone, Medieval agriculture, 235–236. Stone calculated the ratio between prices and wages by dividing the yearly price of wheat by yearly wages for reaping and binding corn.

[70] Noble, Norwich Cathedral Priory, 16.

[71] Stone, Profit-and-loss accountancy, 41–42. This was a marked change of attitude, as under Prior William de Kirkby demesne farming was seen almost unreflected as positive and was therefore increased, ibid., 41. The halt of enlarging the demesne lasted about a decade, but no major changes took place afterwards. In contrast to the general trend, the small unit at North Elmham was increasing after 1300 and reached its peak around 1320. The land under direct control was then falling in accordance with other manors from c.1327 onwards, Yaxley, North Elmham Park, 570–571. He also indicates a profit maximisation strategy of Norwich Cathedral Priory, ibid., 573. Normally the wheat produced at North Elmham would be sent to the monks, but in years of high wheat prices, it was sold, so 1319–1320 and 1320–1321 60% and 1391–1392 40% were marketed.

[72] Denholm-Young, Seigniorial administration, 129–130 calls the improved accounting methods in the form of the *proficuum* as employed by Norwich Cathedral Priory (and by a few more landlords) 'an attempt to change the bias of the account from an estimate of the liability of the accounting official to an estimate of yearly profit and loss'.

[73] Virgoe, Estates, 351 names 1295 as the first year for the profit of the manors. However, the manorial account Sedgeford NRO, DCN 60/33/10 in 1293–1294 has no profit of the manor entry at the bottom of the face any more (where they were to be found before) and the first *proficua maneriorum* are dated 6 Henry [de Lakenham], which is 1293–1294. Campbell, Seigniorial agriculture, 453–466 and Stone, Profit-and-loss accountancy, 36, too, place the first account of profit of the manors under 1293–1294. In the NRO catalogue they are compiled under DCN 40/13.

[74] Stone, Medieval agriculture, 199–201. Profit calculations were introduced in the accounts of many other ecclesiastical landlords, ibid., 199. According to Harvey, Westminster Abbey, 149, at Westminster Abbey the abbot introduced profit calculations by the end of the thirteenth century, prior and convent followed about a decade later. As a result the less profitable demesnes lands were put at lease.

of the thirteenth century (apart from the famine years in the mid-1290s), rises in wages and a very high taxation level.[75]

Within this period the years 1314–1315 to 1316–1317, the time of the Great Famine, hold a very low number of preserved *compoti* (Fig. 2.2).[76]

The cattle plague 1319–1320 and repeated cattle diseases in the 1320s and 1330s took their toll on the manors of Norwich Cathedral Priory, though its estates and Norfolk in general suffered comparatively light losses in the outbreak of 1319–1320.[77] Nevertheless an impact on the agricultural demesne production is visible in a temporary augmentation of arable at the expense of pasture in the direct aftermath of the cattle plague on some manors of Norwich Cathedral Priory: an increase of about 10% of the sown acreage occurred in Hindringham, Martham and Monks' Granges. This can on the one hand be explained by a depletion of bovine stocks,

[75] Stone, Medieval agriculture, 203–205. The fact that the first group of separate profit of the manors calculations was done at Norwich Cathedral Priory in 1293–1294, introduces another facet. Grain prices were very high in the years following the harvests 1293 and 1294 and still elevated after the harvest 1295, see Munro, Revisions of the Phelps Brown and Hoskins 'Basket of Consumables' commodity price series . For Norwich Cathedral Priory as a landowner this was the opportunity for increased profits through the sale of grain and therefore the establishment of the accounts of *proficua maneriorum* might have been triggered by the desire to evaluate the potential for higher financial gains under these conditions.

[76] For the difficult years of the Great Famine very few manorial accounts survive for Norfolk in general, Campbell, Seigniorial agriculture, 28. Concerning Norwich Cathedral Priory it may that the accounts, especially for the usually well documented prior's manors, were either having difficulties getting through the auditing process, or were indeed not even drawn up. However, the state of the preserved documents during those and bordering years, raises the possibility that the accounts of this period were damaged beyond repair, possibly by humidity or bad storage. The *compoti* of the prior's manors made in the years preceding and following the Great Famine are all in a bad condition. The accounts for 1313–1314 are creased and partly discoloured, their state is often worse on the right edge of the face respectively the left edge of the dorse, than on the rest of the document. In 1317–1318, the right edge of the face of all twelve surviving rolls of the prior's manors is damaged. For the crisis years 1314–1315 to 1315–1316 merely five accounts survive, they are also damaged; only one of them belongs to a prior's manor, BLO, MS Rolls Norfolk, Eaton 25 in 1314–1315. The other accounts are from Scratby, Worstead and Arminghall. The Scratby rolls, for the land attached to the church in this village, for 1314–1316, NRO, DCN 60/30/04–5, are both marked by damage on the right edge of the face respectively the left edge of the dorse. Worstead NRO, DCN 60/39/06 in 1315–1316 also fits the emerging picture. Although Attlebridge NRO, DCN 61/16 for 1314–1315 is not in a good state, either, being partly discoloured and the ink faded, the right edge of the face is in no worse condition than the rest of the account. No accounts are available for 1316–1317. The damage that threatened the *compoti* of the prior's manors for 1313–1314 and 1317–1318 might have destroyed the parchments made during the Great Famine or affected them so badly that they were removed from the collection at a time, when BLO, MS Rolls Norfolk, Eaton 25 had already been separated from the main collection which is today in the NRO. Since most accounts suffered at the same side, the right edge of the face, this side must have been exposed to the cause of damage, maybe while being kept in hutches in the muniment room.

[77] Slavin, Cattle plague, 175. He also notes, that against all odds the estates at Sedgeford (though not Gnatingdon) and North Elmham escaped the cattle plague 1319–1320 and a very low mortality can also be registered for Gateley, close to North Elmham. Manors like Monks' Grange and Eaton lost about half of their bovids. Worst hit were the eastern estates at Martham and Hemsby, Hemsby's loss rate lay above 70%, ibid., 168, 171.

which led to a pasture surplus, and on the other hand by still sufficient numbers of surviving draught-animals to ensure the ploughing of the arable. The additional arable was sown with legumes and oats, to ensure soil fertility and the supply of fodder for stots.[78]

By 1324–1325 the cathedral had begun experimenting in leasing out the dairy production on the prior's manors. In this year Hindringham reports the dairy *ad firmam*.[79] This experiment started, when prices for cheese and butter were falling after a post-cattle plague high in the early 1320s and when a severe drought was impacting on pastures and meadows (Sect. 6.2). Very dry conditions also dominated in the summer half year 1326 and can not have strengthened the faith of the monks in the economic viability of dairy production. In the following years the leasing out of the dairy sector gained ground. In 1326–1327 many cattle and sheep died of pestilence on the Canterbury estates and caused a decline in stock-farming and dairy profits.[80] In the Norwich accounts, too, references to sick or dead cattle appear in the agricultural year 1325–1326, as in Gnatingdon, Hindolveston, Martham, Monks' Grange, North Elmham, Plumstead and Taverham.[81] As a consequence in 1327–1328 the cows in Hemsby, Hindolveston, Hindringham, Martham and North Elmham were farmed out[82]; 1328 was another year of warm and dry weather during the growing season, which must have reduced grass growth. Cattle health was affected, Hindolveston reported some cows not calving, being sterile and being dry or not giving milk; the North Elmham roll mentions two sterile cows. The dairy sector at Plumstead was at farm by 1331–1332 and at Monks' Grange it was leased between this year and 1333–1334[83]; it should be noted that the years 1331–1333 had very warm and dry summer seasons. Sedgeford and Gnatingdon were the only manors that still kept the dairy production under direct management[84] until c.1339–1340, the first Sedgeford account in which the cows are leased is in 1340–1341, in

[78] Slavin, Cattle plague, 174. In Norfolk draught-horses were already often used for ploughing, so this sector of the arable production was less disrupted in Norfolk than elsewhere, Hallam, Eastern England, 298; Langdon, Was England a technological backwater, 282–283. See also Fig. 2.3, which shows the slight increase in the arable land at Sedgeford and Gnatingdon 1318–1319 to 1320–1321, NRO, DCN 60/33/20-22, NRO, DCN 60/14/17-19 as described by Slavin also for other manors. Although Sedgeford escaped the cattle plague, Gnatingdon did not. Both demesne were closely linked and under the supervision of the same sergeant, John de Elmham, so that the loss of livestock in Gnatingdon would have affected both units. Sedgeford in fact send cattle to Gnatingdon and Thornham, Slavin, Cattle plague, 179.

[79] NRO, DCN 60/20/22. The lease started at 2 February 1325.

[80] Smith, Canterbury Cathedral Priory, 109, 165.

[81] Gnatingdon NRO, DCN 60/14/22, Hindolveston NRO, DCN 60/18/25, Martham NRO, DCN 60/23/21, Monks' Grange NRO, DCN 60/26/21, North Elmham NRO, DCN 60/10/21, Plumstead NRO, DCN 60/29/21 and Taverham NRO, DCN 60/35/23.

[82] Hemsby NRO, DCN 60/15/15, Hindolveston NRO, DCN 60/18/28, Hindringham NRO, DCN 60/20/23, Martham NRO, DCN 60/23/22 and North Elmham NRO, DCN 60/10/22.

[83] Plumstead NRO, DCN 60/29/23, Monks' Grange NRO, DCN 60/26/23 and NRO, DCN 62/02.

[84] As already noted, Sedgeford had not been affected by the cattle plague, see, Slavin, Cattle plague, 169.

Gnatingdon the dairy was at farm in 1342–1343.[85] The paragraphs listing the profits of the leased dairy production of the other manors continue to report sterile, dry or non-calving cows as an explanation for low rents paid.[86] Restocking was a slow process and usually the landlords concentrated on rebuilding the numbers of the oxen as plough animals first. In accordance with the national trend the dairy cattle on the estates of Norwich Cathedral Priory were not restocked before the mid-1340s.[87]

During the mid-1320s the prices for agricultural products began to fall, additionally the impact of the cattle plague 1319–1320 was still being felt. These altered macro-economic conditions led to a reassessment of the profitability of demesne farming at Norwich Cathedral Priory under the new prior, William de Claxton (1326–1344): the dairy production was farmed out on many manors, the two vineyards of Norwich Cathedral Priory in Sedgeford and Plumstead leave no further trace in the manorial documents (Sect. 6.7), and there was generally a slight downturn in the acreages of demesne land under plough. In those years the priory reduced its demesne land by about 3% compared to the pre-famine peak.[88] This readjustment mainly hit the northwestern and central Norfolk manors, as well as some estates near Norwich, whereas the demesne at the very productive units at Martham and Hemsby remained stable and the directly managed land at Trowse Newton near Norwich actually increased; the agricultural production of Norwich priory now concentrated on the estates in the region east of Norwich and on some estates near the town itself.[89]

It appears that after 1327–1328 the preservation of the manorial accounts was less strict. Perhaps after the audit and after the processing of the *compoti* data in the rolls for the profits of the manors, the main attention was turned to safe-keeping the *proficua* documents rather than the actual manorial accounts. In any case the *proficua maneriorum* accounts for the late 1320s and the 1330s survive, but the number of preserved manorial accounts is very low. Many account rolls are available only for 1333–1334 and 1334–1335. The *proficua maneriorum* contain indications that by the 1330s Norwich Cathedral Priory was experimenting with piecemeal letting as well as with letting whole manors to farm, as the manors of Hindolveston, Hindringham and Thornham for several years in the 1330s.[90]

---

[85] In 1339–1340 the dairy production was still under direct management in both manors, NRO, LEST/IB 17 and NRO, LEST/IC 06. The following year, 1340–1341, it was leased in Sedgeford NRO, LEST/IB 18 and probably in Gnatingdon, too. However, the next useable Gnatingdon account dates to 1342–1343, NRO, LEST/IB 08; the dairy production was at farm.

[86] Information on infirm, sterile or otherwise unproductive cows is given under the *daeria vendita* section when the dairy production is at lease, since for the non-productive cows less rent would be paid. When the dairy production is directly managed by the cathedral priory, this paragraph displays lists of milk, butter and cheese produced and sold.

[87] Slavin, Cattle plague, 177.

[88] Campbell, Seigniorial Agriculture, 233.

[89] Ibid., 233.

[90] Campbell, Seigniorial agriculture, 233–234, Hindringham and Hindolveston were leased in 1333 and back in hand by 1339, Thornham was at farm by 1334 and again directly managed in the 1340s. Hindolveston's short *compoti* for the mid-1330s indicate leasing, the accounting official is the (rent) *collector*, NRO, DCN 62/02 in 1333–1334 and NRO, DCN 60/18/29 in 1334–1335.

This decline in demesne farming is in accordance with tendencies of other ecclesiastical landowners, such as Canterbury Cathedral Priory, where a downturn is also noticeable.[91] The income of the manors of the bishop of Ely fell by more than half from 1325 to 1333 and sank even further in the years 1333–1346 (compared to 1319–1323), hand-in-hand with this development went piecemeal letting of the demesne.[92] This trend seems to have been caused by a declining population, a heavy fall in prices for agricultural products and impacts of adverse weather, such as sea incursions resulting from storm activity.[93] The grain price recovered shortly in the years around 1330, but due to good harvests, deflationary tendencies and currency shortage it entered a real depression afterwards which lasted until the beginning of the 1350s.[94] As a result of the unfavourable development of the price-wage ratio, landlords turned to less labour-intensive forms of agriculture on demesne land.[95] The shrinkage of market-profits of demesne cultivation had set in before the Black Death, but the epidemic accelerated and aggravated this problem.[96]

The supply of manorial accounts of Norwich Cathedral Priory remains low throughout the 1340s, and the last *proficua maneriorum* accounts were made in 1340.[97] Especially the years preceding the Black Death 1348–1349 are badly accounted for, among them the partly wet summers and bad harvests of 1345–1346 (Sect. 8.2). However, it appears that the return to demesne farming at the central Norfolk manors and at Thornham led to an increase of demesne cultivation on the estates of Norwich Cathedral Priory, so that before the Black Death the area under crop rivalled the extent of the demesne land in the pre-Great Famine years (Fig. 2.3 for Sedgeford and Gnatingdon).[98]

The high number of accounts dating to 1349–1350 can be attributed to the effects of the Great Pestilence. A number of the manors was probably at farm before 1348–

---

[91] Smith, Canterbury Cathedral Priory, 144.

[92] Miller, Ely, 105–106, a consequence was also the reduction of capital investment in the manors by the bishop.

[93] Stone, Medieval agriculture, 45–46 on the role of the indecisive and partly negative population trend, falling prices for agriculture goods, murrains and potential weather impacts. He gives Westminster Abbey and Crowland Abbey as exponents of a policy of leasing some demesne land during the 1330s and 1340s. According to Hybel, Grain trade, 244, the import of Baltic grain, mostly rye and oats for the urban proletariat, contributed to the depression of grain prices in England after 1325–1326. Its most important effect was to smooth the peaks of the grain prices in the months preceding the harvest, when most of those imports entered England, ibid., 235. For the impact on the estates of Canterbury Cathedral Priory respectively on those of the bishopric of Ely, see: Smith, Canterbury Cathedral Priory, 126 and 144, Miller, Ely, 105. Concerning the flooding, see Bailey, *Per impetum maris*, 190–191, 205–207 on the high storm activity c.1275–1350 and the declining profitability of agriculture in coastal areas which resulted in an insufficient upkeep of the sea defences.

[94] Bridbury, Black Death, 579. For literature on the good harvests, deflation and currency shortage, see note 63.

[95] Stone, Medieval agriculture, 236–243.

[96] Britnell, Commercialisation 1000–1500, 190–191.

[97] Virgoe, Estates, 351.

[98] Campbell, Seigniorial agriculture, 234.

1349. Considering the high mortality in this year it seems likely that upon the lessees' death the priory had to take several manors back into direct management.

Although grain prices oscillated at a high level after 1351 for a quarter of a century, and so helped to usher in the 'Indian summer of demesne farming',[99] labour costs were rising[100] and thus narrowing the potential for profit. Hence Norwich Cathedral Priory was reviewing its economic strategy at the beginning of the 1350s.[101] During the spring of 1352 an inventory of the prior's manor houses was drawn up, for assessing their state after the pestilence and probably for helping the monks to form a future strategy.[102] After 1350 it is basically the *compoti* of Catton, Hindolveston, Martham, Plumstead, and Taverham as well as Sedgeford and Gnatingdon that were stored in Norwich Cathedral Priory. Some accounts also survive for Eaton, Gateley, North Elmham and Hindringham. Between 1351 and 1431 the number of preserved accounts per year averages merely four, but the supply is steady and gap years are rare. The demesne land under cultivation of most manors, for which accounts before and after the Black Death are available, drops considerably after 1350 (Fig. 2.3),[103] exceptions are the small units at Taverham, North Elmham and the land attached to the church in Scratby. This reduction in directly managed demesne land and the fact that after 1350 it is a small circle of manors that provides manorial accounts for over 80 years, until the end of the period under consideration here, does imply that the priory made a scrupulous and conscious decision about the manors that were worth keeping in hand under the new economic circumstances and which manors were not sufficiently profitable under such an arrangement and were better off being let at farm.[104] Consequently Hindolveston was leased out and was followed at the end of the decade by the small demesnes at Thornham, North Elmham and probably also Gateley and Catton. In consequence by the mid-1370s the area under direct management had fallen by 70% compared to before 1350. The c. eleven demesne still under direct control also sowed merely three quarters of the land tilled before the Black Death (Fig. 2.3).[105]

Careful decisions and foresight were needed by the monks in the period after the Great Pestilence, because the priory already faced financial problems before that

[99] So termed by Bridbury, Black Death, 584; Mate, Agrarian economy after the Black Death, 345.

[100] Farmer, Prices and wages, 1350–1500, 471, 516–20.

[101] The immediate short term effect of the pestilence on the manors of Norwich Cathedral Priory might have been disastrous; widespread disruption and loss of income was common for landlords in those years, Britnell, Commercialisation 1000–1500, 191. This was also the case on the Canterbury estates, see Smith, Canterbury Cathedral, 144.

[102] Yaxley, The prior's manor houses, 1. The inventory was made for eleven manor houses: Hemsby, Plumstead, Trowse Newton, Eaton, Taverham, North Elmham, Hindolveston, Hindringham, Sedgeford, Gnatingdon and Thornham.

[103] Virgoe, Estates, 354 and Campbell, Seigniorial Agriculture, 235.

[104] As Stone, Medieval agriculture, 83–84, points out, that the 'Indian summer' of demesne farming was severely clouded for some landlords, among them Ramsey Abbey, Battle Abbey, and some demesnes of the bishopric of Winchester. In the 1350s abandoning the direct demesne agriculture became a real option for landlords in England, Britnell, Commercialisation 1000–1500, 188.

[105] Campbell, Seigniorial agriculture, 235.

time. Although the 1330s were generally still good times for Norwich Cathedral Priory the obedientiaries' income was falling since this decade[106] and in 1347 bishop Bateman's injunctions also show his anxiety about the priory's financial state.[107]

In addition to these factors the performance of the administrative personnel for the demesne cultivation might not have been up to the expected standard. After the Great Pestilence landlords faced increasing difficulties in finding capable and reliable managers for their demesne. Complaints about their performance and a high turnover of administrative personnel were not unusual.[108] The estates of Norwich Cathedral Priory, too, were subjected to rapidly changing managers during the early to mid-1350s (Appendix 2), which resulted in the need for closer supervision of the demesne agriculture.

This and the narrowing margin for profit in demesne farming are mirrored in the tightened accounting procedures and careful dating structure introduced under prior Laurence de Leck (1352–1357) in the accounting/agricultural year 1354–1355. From now onwards there appear detailed sections for performed or sold *opera,* and *terra* accounts at the end of the grange account, listing the total acreage sown, the acreage in fallow, the area used as a sheepfold etc. In the grange accounts under each crop the acres sown with the crop in the individual fields appear. All acres and the quantity of seed used are then summed up and the average sowing density is given. For dating, a combination of the prior's years in office with the king's regnal years is used and raises the dating safety.[109] These measures of regaining control during the priorate of Laurence de Leck materialized in Sedgeford and Gnatingdon

---

106 Noble, Norwich Cathedral Priory, 30–32, 39–41. The running of the monastery, its financial status and administration are discussed ibid., 10–91.

107 Cheney, Norwich Cathedral Priory, 94, 97, Dodwell, Monastic community, 253 notes that the annual (obedientiary) accounts were not coming in as they should. Not that all those measures were of much use in the overturned economic and social circumstances after the Black Death: by 1363 the annual deficit recorded in the *status obedientiarorum* for 1347, £173, had trebled, Cheney, Norwich Cathedral Priory, 97.

108 Stone, Medieval agriculture, 216–219.

109 Only in one period after the accounting reform is the dating problematic. In the early part of the reign of Henry IV the prior's years in office are in conflict with the king's regnal years. This irritation was caused by Richard II abdicating on Michaelmas 1399 and the parliament accepting the abdication one day later. Whereas the prior's years of Alexander de Totyngton run smoothly over the 1390s and the first decade of the fifteenth century, the regnal years fall out of step between 1400 and 1408. With the new prior Robert de Burnham (elected 1407) the situation was rectified and on the accounts the regnal year 9–10 of Henry IV is given in two consecutive years, with the first year in office of Robert de Burnham (1407–1408) and with his second year (1408–1409). The confusion is obvious, because the manorial accounts of Norwich Cathedral Priory for this time had in the past been separated into two different collections which are today both in the NRO. The majority of *compoti* are in the Dean and Chapter Archives (DCN), but the accounts for Sedgeford and Gnatingdon are in the Le Strange of Hunstanton collection (LEST). The collections were originally dated independently; whereas the DCN collection was dated over the prior's years, the Le Strange collection was dated over the regnal years and hence is a year misdated in the catalogues and was re-dated for this study.

in the form of a halt to the free fall of the demesne acreage under plough which had begun with the Great Pestilence (Fig. 2.3).[110]

The turbulent 1350s were followed by two decades of relative stability with regard to the manors under direct management and the demesne land under crop. The recurrent waves of plague in 1361, 1369 and 1375 (Chap. 10) and the 'Pokkes' in 1365 (Sect. 8.4), however, brought population numbers down further. Due to high mortality and the dislocation of the workforce in the short run the epidemics had a potentially disruptive influence on the acreage of demesne land sown.[111] This effect is obvious in Sedgeford, Gnatingdon and Taverham in the outbreak 1369, when comparing their demesne under crop between 1368 and 1373 (Fig. 2.3).[112] The information from Hindolveston and Sedgeford in 1375 does not indicate a drop in sown acres (Fig. 2.3),[113] and for the wave of 1361 too few data are available. Generally after the Great Pestilence the productivity of some estates was hindered by an increased use of unwillingly performed, but cheap, customary labour services, instead of expensive hired labour.[114]

In the mid-1370s, however, the favourable conditions, which allowed the continued flourishing of demesne farming in England after the Black Death, came to an end when the bumper harvests of the mid- and late 1370s, which caused the grain prices to fall from 1376 onwards,[115] occurred amidst the decreasing population trend,[116] which lead to rising wages.[117] In consequence the direct exploitation of the agricultural demesne became less and less profitable. At Westminster Abbey – which had been considering putting manors at farm ever since the Black Death – the pendulum swung fully towards a policy of leasing after 1370.[118] In the 1380s more monasteries and cathedral priories decided to return at last to the system of leasing out whole demesnes (apart from the home farm),[119] among them Canterbury

[110] The continued fall of sown acreages at Sedgeford and Gnatingdon in the early 1350s can be due to the shortage of labour as well as to the prevailing dry conditions in at least 1352. Sedgeford and Gnatingdon are very drought sensitive being situated on sandy soil.

[111] Stone, Medieval agriculture, 251–252.

[112] NRO, LEST/IB 30-32, NRO, LEST/IC 17-19, NRO, DCN 60/35/37-40.

[113] NRO, DCN 60/18/41-43 1367–1368 to 1379–1380, NRO, LEST/IB 33-35 1373–1374 to 1377–1378.

[114] The lower productivity of customary labour as compared to hired labour lead to lower hay yields, slower corn harvesting and slower weeding on the manor of Wisbech Barton in Cambridgeshire, Stone, Wisbech Barton, 648–649, 652–654.

[115] Bridbury, Black Death, 584–585, Britnell, Britain and Ireland, 430. In 1378 the price fell further and remained low for the rest of the Middle Ages, Bridbury, Black Death, 579.

[116] In 1377 there were about 2.5 to 3 million people in England, that was a net decline of c.45% since 1348 when population would have numbered around 4.5 to 6 million people, tending rather to the higher estimate (the real peak, however, was around 1300), Hatcher, Plague, 68.

[117] Bridbury, Black Death, 585–586

[118] Harvey, Westminster Abbey, 151.

[119] Bridbury, Black Death, 584, Britnell, Britain and Ireland, 430. Generally, as the price of land fell and that of labour rose, the most intensive farming systems downgraded their degree of intensity, Campbell, Fair field once full of folk, 63. Consequently sheep farming also became more popular

Cathedral Priory.[120] At Norwich Cathedral Priory a similar tendency can be detected. Sedgeford, Gnatingdon and Martham seem to have been kept in hand, but at Sedgeford the sown acreage was reduced around 1380 (Fig. 2.3).[121] After the violent shock of the Peasants' Revolt 1381[122] and with the election of prior Alexander de Totyngton in spring 1382 the economic strategy at the cathedral priory was drastically modified. More manors were put at lease. Plumstead dropped out of direct management after 1383. Hindolveston was farmed out in 1382–1383. Since the large wood at Hindolveston continued to be managed by the priory, accounts for that place continue to be drawn up. Obviously the cathedral priory was not overly hopeful concerning the future prospects of demesne farming in Hindolveston, because the accounts between 1382–1383 and 1391–1392[123] specify that the manor was leased out to the *firmarius* Nicholas Bottes for 40 years starting in 1382–1383. Hindolveston's fate was shared by Taverham, Monks' Grange and probably Eaton. The monks were now cultivating barely half the acreage compared to the time before the Black Death.[124] The economic crisis faced by the monks manifested itself also in the administration of the manors that remained under direct management. At Sedgeford and Gnatingdon the administrative structure was changed (Appendix 2.1). Until 1381–1382 a reeve and a hayward had supported the sergeant in his work in Sedgeford; in Gnatingdon the offices of reeve and hayward had existed likewise. Latest by 1384–1385, however, the office of the reeve was abolished.[125] As soon as the first manorial accounts were drawn up under the new prior Alexander de Totyngton in autumn 1382 the density of weather references in those documents increased substantially (Appendix 1). It may well be that this increased documentation of factors interfering with agriculture or raising costs was linked to the decreased mutual control of the manorial officers due to the reduction in administrative personnel.

After this depressed period, a change in the attitude towards demesne farming came on in East Anglia in the early 1390s. The manor of Hindolveston was back under direct management by the cathedral priory at the latest in 1395–1396.[126] By that time in addition to the *compoti* of Sedgeford, Gnatingdon and Martham, accounts for Plumstead, Eaton and North Elmham are preserved again. It appears

amongst landholders in Norfolk and elsewhere, Britnell, Britain and Ireland, 413–416 and Campbell, Overton, Norfolk Farming c.1250-c.1850, 77–78.

[120] Smith, Canterbury Cathedral Priory, 192.

[121] NRO, LEST/IB 35-38.

[122] The high tide of the Peasants' Revolt swept over East Anglia in June 1381, in Norfolk the manorial records of Abbey of St Benet's of Hulme were burned, but the revolt was over by July; Oman, Great Revolt, 99–134. The uprising was violent and widespread in Suffolk, the Abbey of Bury St Edmunds was ransacked and a number of monks killed, see Bailey, Medieval Suffolk, 186.

[123] NRO, DCN 60/18/43A-49.

[124] Campbell, Seigniorial Agriculture, 235.

[125] Sedgeford NRO, LEST/IB 38-41, Gnatingdon NRO, LEST/IC 19-21.

[126] NRO, DCN 60/18/50.

that the poor harvests 1389 and 1390 and ensuing raised grain prices[127] resonated in Norwich Cathedral Priory and fuelled a desire to revive demesne farming on a larger scale, so that leased demesne land was taken back into hand. Even though another plague wave affected England in the early 1390s, economic conditions during that decade were more favourable for demesne farming than in the 1380s and instead of a loss a slight net profit could be made from the arable sector of the cathedral priory.[128] The changed circumstances where also felt in other eastern areas; at the Westminster Abbey manor of Kinsbourne in Hertfordshire profits were high throughout the 1390s.[129] Hence in the first decade of the fifteenth century the direct cultivation of demesne land of Norwich Cathedral Priory amounted to 64% of the acreage under crop on the eve of the Black Death.[130] Finally, when the North Elmham rolls end in 1410–1411, the bailiff's accounts for Taverham (1413–1414 to 1423–1424) and Great Cressingham (1412–1413 to 1416–1417) reappear. Generally record density is again satisfactory from the mid-1390s onwards, and the 1410s are particularly well documented.

During the 1420s this trend was decisively reversed[131]: Martham was leased by 1424 and the last manors under direct control were Sedgeford and Gnatingdon in 1430–1431 or 1429–1430 respectively.[132] By the early 1430s the era of demesne farming at Norwich Cathedral Priory had passed.[133]

This long insistence upon demesne farming sets Norwich Cathedral Priory apart from many other large ecclesiastical landowners, who, except for the more conser-

[127] Munro, Revisions of the Phelps Brown and Hoskins 'Basket of Consumables' commodity price series. Values advanced by one year. According to Walsingham, Historia Anglicana, vol. 2, 195, 198 the grain harvest was below average in 1389 and bad in 1390, see also Sect. 6.2.

[128] Slavin, Bread and ale, 78–79.

[129] Stern, A Hertfordshire demesne, 154.

[130] Campbell, Seigniorial Agriculture, 235–236.

[131] This development was perhaps helped by the raised mortality amongst the monks throughout the 1420s. Seven monks died in the outbreak of plague in 1420, and then about three monks died every year until 1425. Amongst them were also obedientiaries. Such a mortality in a community of then c. fifty monks resulted in a loss of expertise for managing the monastery as Noble, Norwich Cathedral Priory, 58–61 observes. This would apply to the direct management of the demesne, too.

[132] NRO, LEST/IB 70 and NRO, LEST/IC 42. Sedgeford was at farm in 1432–1433, NRO, LEST/IB 71, probably Gnatingdon was leased out in the same year, there is merely no account of Gnatingdon stating that it is at farm.

[133] As long as the income of Norwich Cathedral Priory was partly made up by demesne farming, the yearly income had varied greatly and enhanced the structural annual deficit. After deciding for a policy of leasing the priory achieved a much greater financial stability over the next fifty years, Virgoe, Estates, 357. However, that does not imply that farming generally was not profitable any more at all. Large, non-resident landowners might have faced particular difficulties during the period c.1410–1430, as pointed out by Stone, Medieval agriculture, 221–228. Apart from a narrowing margin for profit in commercial agriculture, these problems included potentially a decreasing quality of administrative personnel, since apt men could find good chances elsewhere in the agricultural sector, problems to exert labour services and the quality of performed work. It was during the agricultural depression of the 1460s and 1470s, that smaller landowners, like the Pastons in north-east and eastern Norfolk, faced severe difficulties in marketing grain, maintaining the level of rents, finding lessees and receiving arrears from unwilling tenants, Britnell, The Pastons, 137–142.

vative ones, had by 1400 abandoned demesne cultivation aside from their home farms.[134] The movement towards leasing had started in the 1380s and gathered pace in the 1390s[135]: Canterbury Cathedral Priory in a sudden move in the mid-1390s had given up demesne cultivation almost entirely,[136] Westminster Abbey and the bishopric of Worcester turned towards the same direction.[137] Ramsey Abbey followed between 1400 and 1410,[138] and the estates of Durham Cathedral Priory were also generally at lease in the fifteenth century.[139]

Out of the 840 manorial accounts from Norwich Cathedral Priory for the years 1256 to 1431, 561 provide a harvest date. The harvest date supply rate mirrors the survival rate of the accounts on a lower level (Figs. 2.2 and 3.5).The most reliable information comes from the prior's manors listed at the beginning of this chapter. Further information is provided by Aldeby, Great Cressingham, Henley, Plumstead (precentor's property), Scratby and Worstead as well as the much shorter series from Arminghall, Attlebridge, Bawburgh, Hardingham, Hevingham (bishop's manor), Heythe, Lakenham, North Elmham (bishop's manor), Ormesby, Thornage, Wicklewood and Witton.[140]

It can be concluded, that the data density of harvest dates provided by these *compoti* is sufficient for executing a statistical analysis and temperature reconstruction; the periods with low data coverage are the 1250s to 1290, the 1330s and 1340s as well as the 1380s. The information of the 1330s, 1340s and 1380s can be supplemented with data from the manorial records of other landowners.

---

[134] Miller, Introduction: Land and People, 13.

[135] Britnell, Britain and Ireland, 430.

[136] Smith, Canterbury Cathedral Priory, 192.

[137] Britnell, Britain and Ireland, 430.

[138] Raftis, Ramsey Abbey, 266.

[139] Halcrow, Durham Cathedral Priory, 355–356. The movement towards leases began here in the 1380s, Britnell, Britain and Ireland, 430.

[140] These 840 manorial accounts include the listed estates under NRO, DCN 60 (Aldeby, Catton, Denham, Eaton, North Elmham, Gateley, Gnatingdon, Great Cressingham, Hemsby, Henley, Hindolveston, Hindringham, Martham, Melton, Monks' Grange, Trowse Newton, Plumstead, Plumstead (precentor's property), Scratby, Sedgeford, Taverham, Thornham and Worstead), as well as stray accounts from other places: Arminghall NRO, DCN 61/06, Lakenham NRO, DCN 61/10-11, Le Gannoc NRO, DCN 61/14, Attlebridge NRO, DCN 61/15-16, Bawburgh NRO, DCN 61/18-24, Fordham NRO, DCN 61/29, Hardingham NRO, DCN 61/31, Heythe NRO, DCN 61/33-34, NRO, DCN 62/02, Ormesby NRO, DCN 61/39-41, Postwick NRO, DCN 61/42, Wicklewood NRO, DCN 61/45, Wiggenhall NRO, DCN 61/46, Witton NRO, DCN 61/48. Accounts from the episcopal estates, surviving in NRO, DCN 95, are included: Thornegg/Thornage NRO, DCN 95/06, Hevingham NRO, DCN 95/11 and North Elmham NRO, DCN 95/07 and NRO, DCN 95/09. The accounts for Great Cressingham are under NRO, MC/212. Additional accounts of the major estates are also listed under Sedgeford NRO, LEST/IB and Gnatingdon NRO, LEST/IC, Martham NRO, NRS 5889, 20 D1-5920, 20 D3, and BLO, MS Rolls Norfolk, Eaton. The accounts have been checked for harvest information up to 1435. Harvest date information is available for all the manors listed under NRO, DCN 60, except for Melton.

2.2.3 *Archival History of Norwich Cathedral Priory*

The muniments of Norwich Cathedral Priory are mainly held in the Dean and Chapter Archives (DCN) at the NRO in Norwich. Over the course of time some *compoti* have become separated from this main collection: most of the accounts for the manors of Sedgeford and Gnatingdon from 1339–1340 to 1430–1431 are held in the Le Strange of Hunstanton collection (LEST) in the NRO, the manorial accounts of Martham from 1355[141] to 1423–1424 are in the holdings of the Norfolk Record Society (NRS) in the NRO, the Great Cressingham rolls are in the Minor Collections (MC/212), and most rolls of the manor of Eaton from 1358–1359 to 1422–1423 are in the Bodleian Libary in Oxford (BLO) under MS Rolls Norfolk.

Around 1300 Norwich Cathedral Priory witnessed the attempt to install a central storage room for the important documents and *acta*, whose arrangement was also checked and revised. This revision did not extend to the manorial accounts.[142] They would normally have been stored in the office of the obedientiary concerned, probably in hutches.[143] Usually the obedientiaries would also keep their own financial accounts, the so called obedientiary or obediental accounts, at hand for at least a while, although they should have been assembled in the central muniment room. This surely also applies to the less important manorial accounts.[144] Consequently the documents thus kept outside the central storage room were more likely to succumb to fire, riots or simple loss, as is demonstrated by the poor survival rate of manorial accounts in the years preceding the attack on the cathedral and the fire in the precinct in 1272 (Fig. 2.2).

The documents of the prior's manors must have been kept together and in good order throughout the fifteenth century and at least until 1538, the year of the dissolution of the cathedral priory. This continuity is underlined by the fact that the post 1354–1355 accounts for the prior's manors have been updated at some point after the end of the period under consideration here, that is after 1431. *Compoti* of the second half of the fourteenth century and later were often rolled from top to bottom and for identifying an account without unrolling it, information at the lower end, usually the dorse, is necessary. At this place the post mid-fourteenth century accounts were originally just dated with regnal years. In the cathedral priory, however, time was rather reckoned with the help of the prior's years in office, and

[141] NRO, NRS 5891, 20 D1 is an account only for summer 1355.

[142] Dodwell, Muniments, 327.

[143] Ibid., 330.

[144] This outlined storage system of the manorial accounts is supported by the evidence on the rolls themselves. So in 1317–1318, the right edge of the face respectively the left edge of the dorse of all twelve surviving rolls of the prior's manors is damaged. 1-3 cm of the parchment are missing. However, the roll Scratby NRO, DCN 60/30/06, which was made for the land attached to the appropriated church there and which did not belong to the prior, survives in an undisturbed condition with both edges intact. It can be concluded that the accounts of the prior's manors were kept together, perhaps in a hutch or pigeonhole, and all rolls were inserted in the same direction and kept in a yearly order.

consequently at some time after 1431, but probably before the dissolution of the priory, the prior's years in office were added at the lower end of the dorse for all accounts of the prior's manors. The *compoti* of the cellarer's manor of Great Cressingham and the accounts of Scratby and Worstead, which were not held by the prior, do not contain this additional information.

The break in the keeping of the muniments did not occur with the dissolution of the cathedral priory and the establishment of the Dean and Chapter at Norwich Cathedral in 1538, though it is likely the records of the now Dean and Chapter archive were not kept in good order in the later sixteenth century.[145]

The disruption of the continuity for the Dean and Chapter archive came during the Commonwealth in April 1649, when the Dean and Chapter were abolished in England. Their estates were seized for sale and the records of the confiscated lands had to be transported to London. In 1660 Parliament declared the sales of church land void, but the muniments would neither be returned promptly nor in good order to their ecclesiastical home institution. It seems likely that during this time some runs of accounts became detached from the main collection of the Dean and Chapter Archives and are today in minor collections, if not lost. When at Norwich Cathedral Priory a newly appointed prebendary found the records in 1681 in the 'treasury' in the cathedral precincts they were in a sad state and badly preserved.[146] Three hundred years later, in 1975 they were handed over to the NRO.

The Dean and Chapter archives of Norwich Cathedral Priory were first sorted, dated and described by H.W. Saunders.[147] Later Barbara Dodwell also turned her attention towards the history of the muniments of the priory and began sorting and classifying its documents.[148]

2.3 Supplementary Series

Some short and non-continuous series of harvest dates can be gained from account rolls for the holdings of the Abbey of St Benet's of Hulme, St Giles's Hospital (Great Hospital), the Norfolk manors of Hunstanton, Heacham, Ringstead, Fincham and Kempstone, and the Suffolk manors of Hinderclay, Redgrave and Akenham (Fig. 2.4). Many of these series give only short runs of information on the harvest, their value lies in supplementing and verifying the Norwich Cathedral Priory series

145 Dodwell, Muniments, 331–332. The records of the estates of the medieval bishopric did not survive the reformation in a similarly favourable state. In 1535 the estates of the bishopric were almost completely granted to King Henry VIII. Consequently the records of these lands were handed over to the royal administration; the few stray documents left behind in Norwich are in the NRO in the 'Records of the estates of the bishopric of Norwich' under NRO, DN/EST 15 (Records relating to the bishop's estates before the exchange with St Benet's) and in the Dean and Chapter archives under DCN 95/1-23. The muniments sent to the royal administration are lost.

146 Dodwell, Muniments, 332, Meeres, Records of Norwich Cathedral, 2.

147 It should be noted that a number of misdatings occur in the collection.

148 Meeres, Records of Norwich Cathedral, 5.

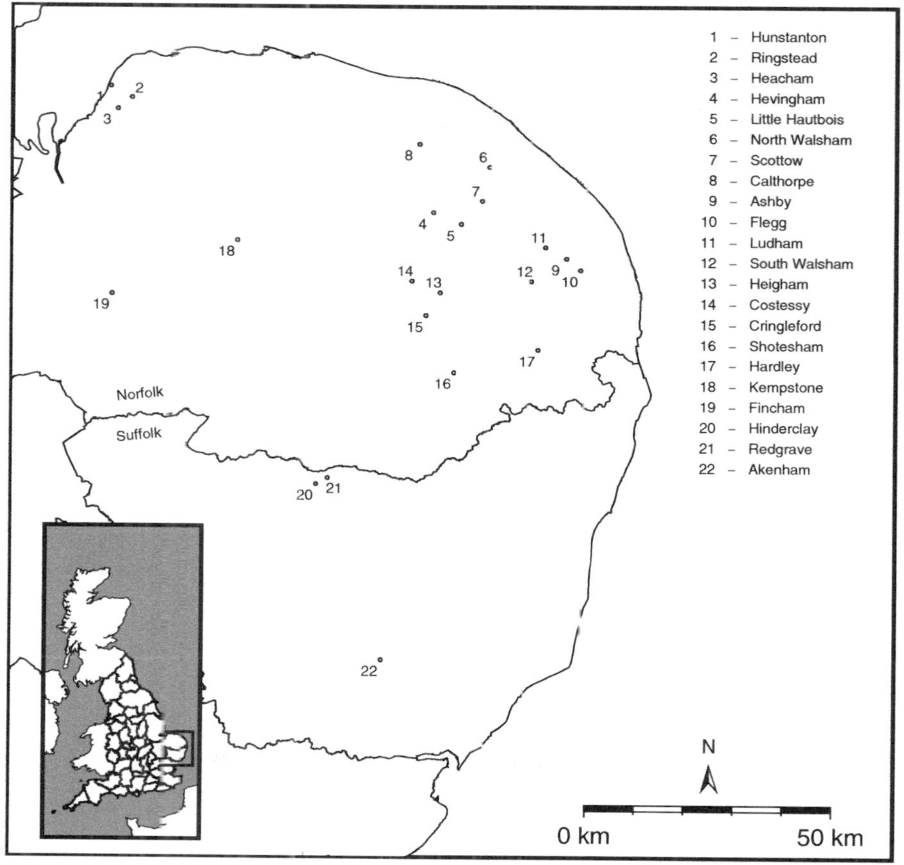

Fig. 2.4 Supplementary series: distribution of manors and other endowments. Shown are only manors, where accounts detail harvest information. Map with modern county boundaries

and filling some gap years. Especially the data density of the 1330s, 1340s and the 1350s, which are periods for which a relatively low number of Norwich Cathedral Priory accounts survives, is raised by integrating the data of those individual series.

A small series of 54 manorial accounts survive for the estates of the Abbey of St Benet's of Hulme in the relevant period 1256–1431. The earliest *compotus* dates 1284–1285 and the last one 1430–1431. They are held amongst the 'Records of the estates of the Bishop of Norwich' at the NRO.

The abbey was situated in eastern Norfolk near the village of Horning in a very isolated location at the junction of the rivers Bure and Thurne. According to legend the first religious community settled there as early as the year 800, but was dispersed by the Danish incursions in 870. The site was rebuilt in the tenth century. In 1019 King Cnut (re-) founded the convent and twenty-five monks were to live under the Rule of St Benedict. The King bestowed on the abbey the manors of Horning, Ludham and Neatishead. The possessions and privileges of the abbey augmented; its extensive property lay entirely in Norfolk. It was an institution of regional importance.[149] In 1536 St Benet's Abbey was annexed to the see of Norwich and whereas the possessions of the bishop were granted to the king, the bishop took over the estates of St Benet's of Hulme.[150]

Consequently the records of St Benet's Abbey were incorporated into the diocesan archive; only a few of its manorial records survive today. During the Peasants' Revolt 1381 many of the abbey's manorial records were lost.[151] Very few account rolls survive for the period before 1350. In the early 1350s and during the 1360s there is a denser supply of rolls. Sporadic accounts are dotted over the 1370s and early 1380s. Towards the end of that decade a period with a relatively good survival rate sets in and continues – in the fifteenth century somewhat reduced – until 1430.

Considering the survival rate of the manorial accounts of St Benet's of Hulme, the substantial number of rolls available in the first part of the 1350s might be explained by changing economic and social conditions. When the new prior William de Haddiscoe was elected in 1349, he was confronted with the social upheaval and changing conditions for demesne farming due to the population decrease caused by the Great Pestilence. Closer supervision and a readjustment were a common answer by ecclesiastical institutions to the new circumstances.[152] Most accounts of the 1350s and 1360s come from Flegg, a fertile area just east of the abbey, where the monks were important landowners.[153] For the late years of William de Haddiscoe

[149] Page (ed.), History of the county of Norfolk, vol. 2, 330. As Pestell, Landscapes of Monastic Foundations, 138–146 points out, there is very little contemporary evidence for the early history of St Benet's of Hulme. For the possessions of the abbey at the time of the Domesday survey, see Blomefield, History of Norfolk, vol. 11, 51–52.

[150] Page (ed.), History of the county of Norfolk, vol. 2, 336. Blomefield, History of Norfolk, vol. 4, 539–542 lists the estates received by the bishopric.

[151] The manor court rolls were destroyed by rebelling peasants in 1381, Page (ed.), History of the county of Norfolk, vol. 2, 334.

[152] For example St Swithun's Priory, Winchester, tried to compensate for a reduced income by tightening its audit process, Drew, Accounts of St Swithun's Priory, 24.

[153] The term probably covers the hundreds of East and West Flegg. In both hundreds St Benet's Abbey possessed considerable estates.

and the first years of his successor, William de Methwold, elected in 1365, a fairly good supply of records survives again. It follows a period of few surviving documents in the 1370s. In England demesne farming was in accelerated decline after 1375 and in many places estates were leased in the 1390s.[154] This development could be reflected in the drying up of the supply of manorial accounts of St Benet's of Hulme over the same time and finally by the years 1379–1380 to 1386–1387, for which no accounts at all survive. Conditions changed towards the end of the 1380s and the early 1390s. The bad harvests 1389–1390 and the plague-ridden early 1390s had changed the socio-economic conditions once more, and Norwich Cathedral Priory showed a renewed interest in the direct management of its estates. For St Benet's manorial accounts are also again available for this time. From 1392–1393 onwards they mainly came from Shotesham, then after 1405 from Flegg.

In seventeen of the manorial accounts from St Benet's Abbey the date of the beginning of the grain harvest is mentioned. Most of those *compoti* were made in the 1350s and 1360s and they were drawn up for Flegg, Little Hautbois (Horning), Ludham, North Walsham, Scottow and Shotesham (Fig. 2.4).[155] The estates of St Benet's were largely situated in eastern Norfolk. Horning and Ludham are villages bordering the abbey to the west and north. North Walsham is a town in northeast Norfolk and whereas Scottow is about 6 km to the south of it, Little Hautbois is c.10 km to its southwest. Shotesham is a village about 10 km south of Norwich. At Little Hautbois and Horning the almoner maintained hospitals.[156] Most *compoti* between 1350 and 1370 record the harvest date.[157] Only two earlier accounts also contain the grain harvest date, they both come from Heigham near Norwich at the beginning of the fourteenth century.[158] The last harvest date is to be found in the roll from Ashby in the hundred of West Flegg in 1377–1378.[159] Generally the manors grew wheat, barley, peas and oats, occasionally also vetches and rye. At the beginning of the fourteenth century Heigham cultivated the latter probably for the Norwich poor, and in 1302 it listed no wheat in the grange account.

---

[154] Mate, Agrarian economy after the Black Death, 354, Stone, Medieval agriculture, 122. Bridbury, Black Death, 586 puts the retreat from demesne farming a decade earlier.

[155] Flegg NRO, DN/EST 09/05-11; Little Hautbois NRO, DN/EST 01/13 summer account; Ludham NRO, DN/EST 01/11; North Walsham NRO, DN/EST 10/02-3; Scottow NRO, DN/EST 11/03; Shotesham NRO, DN/EST 11/05, 10. Concerning the abbey's possessions at the various places see for Little Hautbois and Horning Blomefield, History of Norfolk, vol. 6, 299 and ibid., vol. 11, 56, for Ludham, ibid., vol. 9, 330, for North Walsham ibid., vol. 11, 73–77, for Scottow ibid., vol. 6, 340–341 and for Shotesham ibid., vol. 5, 503, 512–513.

[156] The almoner surveyed the St James's Hospital at Horning as well as the hospital at Little Hautbois, Blomefield, History of Norfolk, vol. 11, 56 on Horning and ibid., vol. 6, 299 on the hospital in Hautbois. According to Blomefield's entry for Hautbois the hospital there would render account together with the hospital at Horning. Consequently the harvest date in Little Hautbois NRO, DN/EST 01/13 might refer to Horning instead of Little Hautbois.

[157] Between 1349–1350 and 1370–1371 eighteen account rolls survive. Fourteen of them give information concerning the onset and duration of the harvest. The other account rolls are either damaged, fragile or almoner's accounts for the hospitals.

[158] Heigham NRO, DN/EST 01/05 harvest 1302 and NRO, DN/EST 01/06. For the lands of St Benet's Abbey at Heigham, see Blomefield, History of Norfolk, vol. 4, 503–505.

[159] Ashby NRO, DN/EST 09/01.

After 1377–1378 until the end of the relevant period accounts from Ashby, Flegg, Heigham an a number of other manors merely list the duration of the harvest. Apart from the data for Flegg,[160] the information is too scarce to be included in any analysis of the harvest duration. On other manors more and more land and activities appear to have been farmed out.[161]

St Giles's Hospital was founded in 1249 and endowed amongst others with churches and land at Costessey, Cringleford, Calthorpe and Hardley and St Mary at South Walsham,[162] but it is in the accounts of those five villages, that usable information on the harvest date can be found (Fig. 2.4). Of those holdings the first two are to the west and southwest of Norwich and today are suburbs of this city. Calthorpe lies about 25 km to the north of Norwich close to the river Bure. South Walsham is situated halfway between Norwich and the east Norfolk coast, and Hardley about 15 km southeast of Norwich.[163] Often several accounts of one place are sewn together; many parchments have suffered in the course of time. Two groups of harvest dates can be distinguished. From Calthorpe and Cringleford come data between 1332 and 1347.[164] These accounts are dated with the help of regnal years and occasionally the dominical year. The information from Hardley is from a methodological point of view more insecure than that of other hospital estates,[165] but since it agrees with parallel data, it has been included. The second group of hospital harvest information ranges from 1392 to 1408 and also includes data from Costessey and South Walsham. The year dating of the rolls is based on regnal years, usually

160 NRO, DN/EST 09/12-15, 17–19.

161 There are 25 rolls in all. One is fragile and was not checked: North Walsham NRO, DN/EST 10/05 another one, Flegg NRO, DN/EST 09/16 is so faded, that it is not clear, if it records the harvest duration. The six accounts made for Shotesham after 1392–1393, NRO, DN/EST 11/06-11, give the impression to come from a manor of which many parts had been at lease. The same applies to Flegg NRO, DN/EST 09/20 and the unidentified manor NRO, DN/EST 01/22. In the almoner accounts NRO, DN/EST 01/17-18 it is stated that Hautbois is *ad firmam*. Barton NRO, DN/EST 01/21 is short and gives no harvest duration.

162 Page (ed.), History of the county of Norfolk, vol. 2, 443–444. For Calthorpe see Blomefield, History of Norfolk, vol. 5, 519–520, for Cringleford, ibid., vol. 5, 35, for Costessey, ibid., vol. 4, 416–417, for Hardley, ibid. vol. 10, 137–141 and for South Walsham, ibid., vol. 11, 138–143. For a description of the holdings and rights of St Giles's Hospital, as well as a map of the estates, see Rawcliffe, Medicine for the soul, 70–83.

163 The hospital records are now part of Records of Norwich in the NRO: NCR Case 24 Hospital Shelf c-i Rolls of various manors.

164 Not all of these available harvest dates stem from the harvest paragraph of the account rolls. The information in the accounts NRO, NCR Case 24 Shelf c Calthorpe 1340–1341, 1345–1346, 1346–1347 is indirect and comes from the *liberatio* list and for the first two rolls also from the *vadium* payments: the harvest date is the date given for the end of the liveries and payments for the *famuli* and the reeve. This would usually coincide with the opening of the lord's table at harvest time. For the different forms of harvest date information in the accounts, see Sect. 3.2.

165 The information does not refer to the harvest or the lord's table itself, but it is found in the *liberatio* section and indicates, when the manorial workforce was stopped being paid in kind in summer (Sect. 3.2). The harvest dates in the Hardley rolls are indirect and also more biased to saints' days, than the dates of the other hospital accounts; for the medieval practice of dating days with references to saints' days, and on the use of saints' days as days for starting the harvest, Sect. 3.5.

listing the year of opening as well as closing of the accounts, and the year runs from Michaelmas to Michaelmas.

Hunstanton, Heacham and Ringstead are situated in the northwestern region of Norfolk,[166] and are almost neighbours of the Norwich Cathedral Priory manors Sedgeford, Gnatingdon and Thornham. The manor of Hunstanton was already in the possession of the Le Strange family in the Late Middle Ages.[167] The manors of Heacham and Ringstead formed part of the *temporalia* of religious houses, Heacham of Lewes Abbey and Ringstead of Ramsey Abbey. After the dissolution the Le Strange family gained control of both manors and received some of their records.[168]

The account rolls of Hunstanton supply harvest data for the period 1331 to 1371. In some respects the Hunstanton *compoti* defy the standard characteristics of manorial accounts: the first three accounts supplying information, for the summers 1331–1333, as well as the account for 1345–1346 cover the whole year, but start and end at 1 August.[169] This has to be taken into consideration, when dating the harvest. The other accounts for the 1330s and early 1340s are half year accounts, usually ranging from late spring to mid-October and mid-October to late spring. The Hunstanton account for 1342–1343 then ranges over a whole year, but starts and ends in mid-October.[170] The rolls around 1370 are also for the full year, but the accounting year now changes at Michaelmas.[171] The accounts date with the help of the regnal years, usually listing the year for the opening as well as for the closing of the account roll.

The account rolls for Heacham offer harvest information for 1296–1297, 1300–1301 and 1303–1304.[172] The Heacham accounts cover the whole agricultural year and run from Michaelmas to Michaelmas. They date the year with the help of the regnal years.

Of the surviving manorial accounts for Ringstead only the *compotus* for 1390–1391 records the harvest date.[173]

For Fincham (New Hall), a lay manor in western Norfolk,[174] a number of account rolls survive between the second half of the thirteenth and second half of the fourteenth century.[175] Some harvests dates are available from early accounts.

---

[166] The manorial records for those three vills are kept in the Le Strange of Hunstanton collection in the NRO: Hunstanton LEST/BG 1-24, Heacham LEST/DG 1-7, Ringstead LEST/EG 1-12.

[167] Blomefield, History of Norfolk, vol. 10, 314–320.

[168] Blomefield, History of Norfolk, vol. 10, 308–9, 340 and Raftis, Ramsey Abbey, 15.

[169] NRO, LEST/BG 2, 4–5, 13.

[170] NRO, LEST/BG 12.

[171] NRO, LEST/BG 15–17.

[172] NRO, LEST/DG 1.1, NRO, LEST/DG 3.3 and NRO, LEST/DG 1.2.

[173] NRO, LEST/EG 03.

[174] Blomefield, History of Norfolk, vol. 7, 348.

[175] Britnell, Winchester Pipe Rolls, 33, NRO, HARE 780–790.

Kempstone belonged to the temporalities of Castle Acre Priory,[176] the manor served as the home farm for this monastic house until 1449, its records give mostly information on harvest length.[177]

The villages of Hinderclay and Redgrave are situated just across the border in the county of Suffolk. Their manors were part of the estates of the Benedictine abbey of Bury St Edmunds and especially for Hinderclay harvest data are available in an impressive series of accounts from the late thirteenth to the early fifteenth century; harvest dates fall between 1296 and 1319 and harvest lengths are supplied by the whole series of account rolls. The material for Redgrave covers a shorter time span from the third decade to the end of the fourteenth century, it contains mostly information on the harvest duration.[178] The *compoti* use regnal years in general, early accounts end in July, later accounts at Michaelmas.

The manor of Akenham is to be found just north of Ipswich in Suffolk. The accounts date to the late thirteenth and to the fourteenth century, but the harvest information clusters in the four decades after 1350; the data are mostly on harvest length and only include a few references to the harvest date.[179]

---

176 Blomefield, History of Norfolk, vol. 9, 523–524.

177 NRO, WIS 02–06, 08–10, 12–15, 17–19, 21–37.

178 Hinderclay CUL, Bacon 405–510, Redgrave, CUL, Bacon 325–374.

179 Raynham Hall, Akenham 1278 to 1397.

Chapter 3
The Medieval Grain Harvest

3.1 Climatological Significance

St Iimes [James] willeth husbandes, get reapers at hande:
the corne being ripe, doe but shead as it stande.
Be sauing and thankfull, for that god hath sent:
he sendeth it thee, for the selfe same entent.[1]

Thomas Tusser's instructions for a successful grain harvest – to assemble the reapers when the maturity of the grain approaches, around St James (25 July, Old Style), and to harvest when the grain is ripe, because otherwise the farmer risks to lose his harvest to shedding – date to mid-sixteenth-century Suffolk. However, they could as well be considered a blue-print for the harvest in the Middle Ages or indeed for the cutting of the grain 300 years later, at the eve of the Industrial Revolution, before the old structures were overturned by the use of machinery. In 1774 Stephen Frost, farmer at Langham in Norfolk, still tried to determine the best timing for the harvest with the object to avoid loss by shedding and remarked in his diary for the week of the 21–27 August:

> Mem: Began to cut my Whate [sic] too late which is in general the case, but for the future begin when the Ear turn Browne [sic] and carnel [sic] tolerable hard and pay no regard to the Straw but shock it well and let it stand some time which will prevent a great loss on the Ground which often happens.[2]

Also Tusser's following verses generally apply to the medieval as well as early modern conditions:

Reape well scatter not, gather cleane that is shorne:
binde fast, shock apace, pay the tenth of thy corne.
Lode saife, carry home, lose no time, being faier:
goife iust, in the barne, it is out of dispaier.

---

[1] Tusser, A hundreth good pointes of husbandrie, point 96. Thomas Tusser lived in Suffolk as a farmer, when he wrote the text in the mid-1550s.

[2] NRO, MC 120/07, 21–27 August 1774.

K. Pribyl, *Farming, Famine and Plague*, DOI 10.1007/978-3-319-55953-7_3

This done, set the pore ouer all for to gleane:
and after thy cattel, to eate it vp cleane.
Then spare it for pasture, till rowen be past:
to lengthen thy dayrey, no better thou hast.
Then welcome thy haruest folke, seruauntes and all:
with murth and good chere, let them furnish th[i]ne hall [...].[3]

Pre-industrial societies could not afford waste and would aim for a clean harvest process, nor could they – especially in England's maritime climate – afford to lose time during harvesting. As long as the corn was not gathered in or well stacked, it was at risk from the changeable weather. Therefore, when Tusser recommends to be 'thankful, for that god hath sent' he merely states the obvious for his contemporaries, medieval predecessors and early modern successors.

As long as the grain was harvested by hand, to reap or mow it at the right time was crucial. In a dry summer, the time window for cutting wheat did not exceed eight to ten days after ripening. During the first half of the nineteenth century a shift occurred from cutting the cereals at the dead-ripe, but not over-ripe, stage, to cutting at the (reap)-ripe stage.[4] If the crops became over-ripe, the kernels would be knocked off the ear when handled and would be lost to the ground.[5] A common threat in medieval as well as early modern times was the loss from shattering and shedding. In 1774 Stephen Frost of Langham in Norfolk still tried to determine the optimum harvest time, for avoiding harvesting too late with the associated shedding. On the other hand, harvesting too early would reduce the nutrition in the grain and lead to spoiling.[6]

Not only was the corn during the harvest time subject to the vagaries of the weather, but the quantity and quality of the harvest itself were largely the result of the prevailing weather conditions, rainfall and temperature, during the growing season. Indeed the main determinant for the time when the grain reached maturity and thus for the onset of the grain harvest, was the mean temperature during the growing season. The decision to start the harvest was based so closely on the phenological development of the corn crop, that the information can be used as a phenological proxy. The reconstruction of temperature in eastern England during the medieval period using this series is described by Pribyl et al. (2012) and is expanded upon in Chap. 5.

Several annually resolved temperature reconstructions for the time after 1450 have been based on the close connection between the phenological phase of the grain, harvesting and the mean growing season temperature. Pfister pioneered the field and used grain harvest-related data from the seventeenth century onwards to reconstruct temperatures in the Swiss Mittelland.[7] Brázdil and Kotyza included fifteenth-century information on the grain harvest in the Louny district in their

[3] Tusser, A hundreth good pointes of husbandrie, point 97–99.

[4] Collins, Harvest technology, 456, 465.

[5] Ault, Open-field farming, 28.

[6] Ault, Open-field farming, 28.

[7] Pfister, Getreide-Erntebeginn und Frühsommertemperaturen.

multi-proxy reconstruction for the Czech Lands.[8] Tarand and Kuiv reconstructed mean summer temperatures from rye harvest dates in the Baltic area 1671–1949[9] and Nordli's reconstruction of nineteenth-century Norwegian May–August temperatures is based on the barley harvest.[10] Možný et al. reconstructed March–June temperatures in the Czech Republic back to 1501 based on the winter wheat harvest date.[11] Work for western Hungary on vine and grain harvest dates was also undertaken by Kiss et al.[12] Wetter and Pfister analysed a series of grain harvest dates dating back to the fifteenth century stemming from the records of the hospital in Basel, which owned estates in the Basel region in Switzerland, in south-western Germany and in the (French) Alsace.[13] The reconstruction presented in this book, is the only reconstruction based on cereal harvest dates for the British Isles and currently the only such reconstruction stretching as far back as the thirteenth century.

3.2 Management and Accounting Practices

In the Middle Ages and early modern times the grain harvest marked the climax of the agricultural year. The people depended largely on regional supplies as foodstuffs for the following year and had almost no technological means at their disposal to rescue a harvest spoiled by weather. Drying ovens, for example, could cope with some quarters of wet grain but never with a whole harvest.

Thus a smooth run of the harvest for maximum efficiency was guaranteed by local custom and village by-laws. Labour was short during harvest time, so every able-bodied adult was obliged to work on the fields. To ensure the labour supply the villagers were forbidden to leave their community during the harvest season. The lord could demand labour services from his customary tenants in the form of day works, *opera*, and additional boon works, *precariae*. The tenants in turn would be entitled to take meals at the lord's table during these harvest works. The lord also had the priority for hiring casual labour in the village. This could cause some difficulty for the customary tenants to reap their own crops.[14] Towards the end of July or the beginning of August usually seven or eight wardens of autumn were chosen, they did not answer to the lord, but ensured a regulated harvest process of the village community.[15] The harvesting of the demesne land, which usually lay intermingled

[8] Brázdil, Kotyza, History of weather and climate (1000–1500), 143–151.

[9] Tarand, Kuiv, The beginning of the rye harvest.

[10] Nordli, Reconstruction of nineteenth century summer temperatures in Norway.

[11] Možný et al., Cereal harvest dates.

[12] Kiss et al., Reconstructed May–July temperatures.

[13] Wetter, Pfister, Spring summer temperatures.

[14] Ault, Open-field farming, 28–34, Bennett, English manor, 110–111.

[15] Ault, Open-field farming, 60–63.

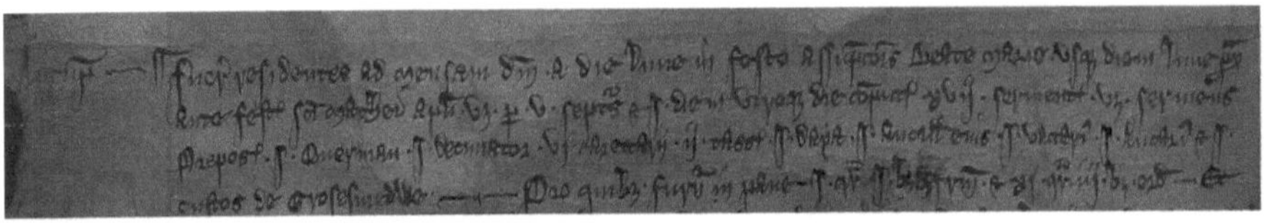

Fig. 3.1 Hindolveston NRO, DCN 60/18/23: *autumpnus*/harvest account for 1323. During this harvest 17 workers ate at the lord's table from the feast of the Assumption of the Blessed Virgin Mary (AM) [15th August] which fell on a Monday, to the Monday before the feast day of the Apostle Matthew [19 September] for five weeks and one day

with other holdings in the open fields, was under surveillance of the hayward or *messor*, a manorial officer.[16] The grain harvest was a communal activity.[17]

In the manorial records the period 1 August to 29 September, St Peter in Chains to Michaelmas, would usually be referred to as *autumpnus* (autumn), the harvest season. However, the real, variable date of the beginning of the grain harvest, which would relate to the cutting of the winter corn, wheat and rye, was recorded in some manorial accounts. The custom was widespread in northern East Anglia, as the information is to be found in the rolls of Norwich Cathedral Priory, the Abbey of St Benet's of Hulme and St Giles's Hospital of Norwich. It also appears in the *compoti* for the Norfolk manors of Lewes Abbey, Ramsey Abbey and the Le Strange family of Hunstanton, and at least occasionally in the rolls of Castle Acre Priory and also in the Suffolk accounts of the Abbey of Bury St Edmunds. This might reflect a regional preference, since neither the Winchester Pipe Rolls, nor the manorial accounts of the manor of Cuxham belonging to Merton College in Oxford, or the Bolton Priory *compoti* list the actual date of the grain harvest.[18]

Many of the East Anglian accounts, especially the Norfolk *compoti* of Norwich Cathedral Priory, that record the harvest date, do so in relation to the cost and expenses of the harvest. The information can be found at several places in the rolls: the *autumpnus* account, amongst other things, relates the cost of the lord's table, which was maintained for the permanent estate labourers, the *famuli*, the reeve, the hired labourers and the customary tenants performing their services during the harvest time.[19] The dates for opening and closing of the lord's table, and consequently for the harvest are given (Fig. 3.1). The account rolls of Norwich Cathedral Priory document the start of the lord's table until 1389–1390. After 1349–1350 the date of

[16] Bennett, English manor, 178–180.

[17] Evans, The farm and the village, 65.

[18] Harvey (ed.), Manorial records of Cuxham, 163–604, Kershaw, The Bolton Priory *compotus*, 35–570. The manorial accounts of the Bishopric of Winchester were checked for the years when the rolls were edited and published, in: Hall (ed.), The Pipe Roll of the Bishopric of Winchester, 1208–1209; Holt (ed.), The Pipe Roll of the Bishopric of Winchester, 1210–1211; Page, The Pipe Roll of the Bishopric of Winchester, 1301–1302 and idem, The Pipe Roll of the Bishopric of Winchester, 1409–1410. Stern, A Hertfordshire demesne, although considering a wide range of climate related agricultural activities, never refers to the grain harvest date.

[19] Dyer, Food consumption, 212. The lord's table formed a high point in the diet of the workers.

Fig. 3.2 Hindolveston NRO, DCN 60/18/23: dairy account for 1323. Cheese was produced until Sunday before the feast of the Assumption of the Blessed Virgin Mary (AM) [14 August]. First the scribe had noted that cheese was produced until the feast itself [15 August], but since on that day it was already destined for the harvesters, he scrupulously erased his words 'Monday in' and replaced them with 'Sunday before'. Normal cheese production was resumed with the last day of the harvest, Monday before the feast day of the Apostle Matthew [19 September]

the closing of the lord's table is normally omitted; it can be roughly determined by using the length of the harvest. From the mid-1350s onwards, though, the references on the duration are usually rounded up or down to the whole week; a trend that started after the Black Death. This also applies to other indications of the harvest length throughout the account roll. When the account lists the work units performed and the quantity of food consumed by the workers, it sometimes appears as if a high number of workers cut the harvest within one day; this is an accountancy device facilitating the counting of works performed and the food, grain, dairy produce, fish, meat and ale, consumed during the harvest.[20]

For the food at the lord's table, produce of the manor was used as much as possible. Hence cheese, butter and milk produced during the harvest time would go into the provision of the harvest workers. The cheese account lists the date after which the dairy products were not destined for the market or merely the landlord's household any more, but for the enlarged lord's table, as well as the date when commercial cheese and butter making was resumed (Fig. 3.2). This information is in accordance with the period of the lord's table recorded in the *autumpnus* account. Usually the normal dairy production would stop on the day of the beginning of the harvest or one day before and would resume at the end of the harvest, or one day later. At the manors of Norwich Cathedral Priory a great move towards the farming out of the dairy sector occurred in 1327–1328, thereafter the harvest information ceases to be given in the cheese accounts. It is the data supplied in the cheese accounts that were used by Hallam in his comparison of timing and quantity of the grain harvest.[21]

With the accounting reform in 1354–1355, works accounts become common in the rolls of Norwich Cathedral Priory. Under *opera autumpnalia* the harvest works are detailed. Start and end of the harvest, duration and the number of days actually worked and *opera* performed, as well as *opera* performed by and numbers of mowers and reapers are given (Fig. 3.3). The direct reference to the end of the harvest was increasingly omitted from 1363–1364 onwards and had dropped out totally by the early 1370s, but indirectly the information is supplied in the rolls since the duration of the harvest is always noted down. As long as parallel data on the start of the

[20] Stern, A Hertfordshire demesne, 28–29.

[21] Hallam, The climate of eastern England 1250–1350, 125.

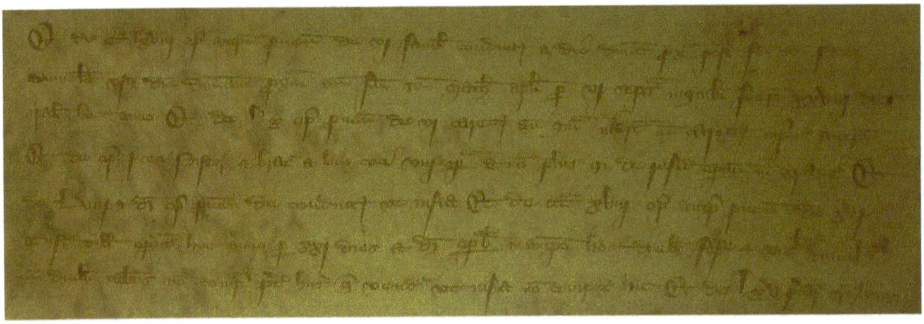

Fig. 3.3 Sedgeford NRO, LEST/IB 24: first part of *opera autumpnalia* for 1357. Works performed by the various groups of harvesters are given. For the six *famuli* the harvest time is specified to have lasted from Sunday after St Peter in Chains (SPC) [6 August] until Sunday before the feast of the Apostle Matthew [17 September], for six weeks, which included 28 actual working days. The information is identical with the harvest data given in the *autumpnus* section of the same account roll

grain harvest are provided in the *autumpnus* account and the *opera autumpnalia* account, this information is identical. From 1390–1391 onwards the works accounts alone state the harvest date; the information in this section is available until the end of demesne farming at Norwich Cathedral Priory.

Indirect references to harvest date or direct references on the harvest duration are often to be found in the section on wages and liveries: *corrodium, vadium* and *liberatio famulorum*. The former two refer to food allowances and other payments to the sergeant or other officials, such as the keeper of the grange. If employment of the officials was for the whole year, allowances would stop during the tenure of the lord's table. The duration of the suspension of corrody and *vadium* is coherent with the harvest duration specified in the *autumpnus* account. Early corrody entries sometimes detailed the date of beginning and end of food allowances. *Corrodium* paragraphs appeared the last time in the manorial accounts of Norwich Cathedral Priory in 1353–1354. *Vadium* entries emerged earlier for some estates, then gained ground in the early 1350s and replaced the corrody with the accounting reform 1354–1355. The *liberatio famulorum* paragraph lists food allowances to the permanent staff of the manor; the information is phrased in the same way as under the *corrodium* and *vadium* entries. Occasionally the livery paragraph is headed by another name such as *multura*, or is subsumed under the barley entry in the grange account, but it was made from the thirteenth to the fifteenth century.

The harvest data in the *compoti* of the supplementary series, the manorial accounts of St Benet's Abbey of Hulme, St Giles's Hospital, Bury St Edmunds Abbey and further single manors, are organized along similar lines; only dairy accounts are generally missing. In the rolls of St Benet's the start, end and duration of harvest, the expenses for the manorial staff and partly their time in the harvest, and the actual work days within the harvest is given under *autumpnus*. The duration of the harvest is repeated under the headings *vadium* and *liberatio*, it is usually rounded up or down to the full week. Only the Shotesham accounts 1352–1353 and

1353–1354 possess a paragraph on *opera autumpnalia*, which details harvest dates.[22] The earlier run of hospital accounts used in this study was produced in the 1330s and 1340s. The harvest date is to be found in the *autumpnus* section, together with the end date and duration. References to harvest duration also appear under *vadium* and *liberatio*. The later series in the 1390s and 1400s still features a thorough *autumpnus* account, and occasionally also specify start and duration of the grain harvest under *opera autumpnalia*. The *compoti* of Hinderclay around 1300 give the harvest date under in both of these paragraphs, in later accounts it is only the harvest duration under *autumpnus, liberatio* or *vadium*. For the manor of Hunstanton the harvest start and end date and duration are recorded in the *autumpnus* section. The manorial staff that were employed are listed as well as the number of labourers hired for the harvest and the day when they began working.

The Heacham accounts cluster around 1300. They are very detailed and the exact time on the fields for the individual workers and groups – the reeve, overmen, the c.40 hired workers, the carters, the thatchers and finally the shepherd, who brought the sheep to feed on the stubble of the cleared fields – can be distinguished under *autumpnus* (Fig. 3.4).

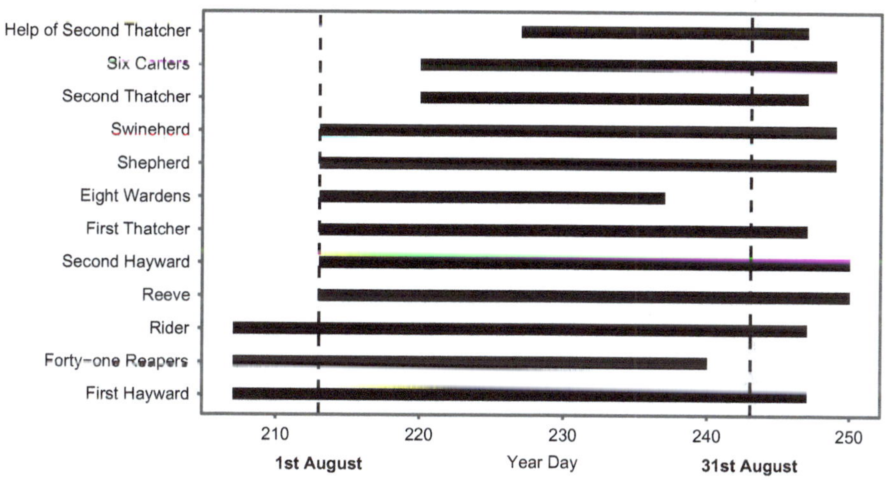

Fig. 3.4 Heacham: harvest 1296–1297. Plotted is the time various labourers spent harvesting. The harvest started 26 July with one hayward/*messor* and 41 harvesters. They were joined by the reeve, a second hayward/*messor* and a thatcher on 1 August, the official start of the harvest season in medieval England and in 1297 probably the start of the cutting of the spring corn, which was overseen by the second hayward/*messor*. Later carters came in as well as more thatchers. Most of the cutting must have been finished by the 28 August, when the 41 reapers finished work

[22] NRO, DN/EST 11/05 for 1352–1353 and NRO, DN/EST 01/10 for 1353–1354. For other manors no such paragraph exists, or as for Flegg it does not list these items.

3.3 Data Density and Security

Of a total of about 1240 manorial accounts checked between 1256–1431, 645 rendered harvest dates: 413 until 1349 and 232 after 1350. Of the 645 dates, 561 come from estates belonging to Norwich Cathedral Priory (Fig. 3.5).[23] Data density is, of course, strongly linked to the document survival rate and to the form of management of the demesne land. The number of harvest dates from all sources is comparatively low until 1290, most of these data come from Norwich Cathedral Priory, a few from the manor of Fincham. The survival rate of Norwich Cathedral Priory accounts is high for the period c.1290[24] to 1330 when many harvest dates are available, although some gap years remain. The manor of Hinderclay adds a series of harvest dates spanning two decades around the turn of the fourteenth century, Kempstone gives information in the 1320s. The supply of Norwich Cathedral Priory harvest dates during the 1330s and 1340s is low, but it is reinforced by dates from

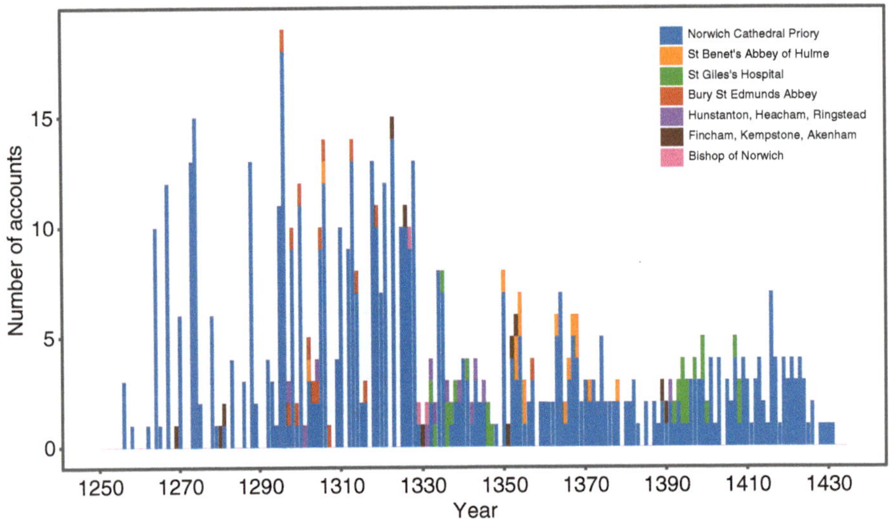

Fig. 3.5 East Anglian harvest dates 1256–1431: number of harvest dates per year. Plotted are harvest dates used in this study and the various sources of data

[23] Including the few dates from the stray episcopal accounts in NRO, DCN 95.

[24] For 1291–1292 the account of Thornham is included in the Sedgeford roll, NRO, DCN 60/33/09, as the Hindolveston account is incorporated in the Hindringham account, NRO, DCN 60/20/08. The following year the Hindolveston information again is set in the Hindringham account, NRO, DCN 60/20/09. For those accounts it is assumed that the harvest date merely refers to the manor the accounts were primarily made for, Sedgeford and Hindringham, and no information or account has thereby been registered for Thornham and Hindolveston in this study. The administrative personal responsible for the agricultural affairs in the Hindringham/Hindolveston rolls 1291–1292 and 1292–1293 is partly met again in the Hindringham roll for 1294–1295, NRO, DCN 60/20/10, but not in the contemporary Hindolveston account, NRO, DCN 60/18/11.

Hunstanton and the manors of St Giles's Hospital. After 1350 the number of gap years in the cathedral data diminishes and a low but steady number of harvest dates per year comes from the cathedral priory until the early 1430s. These data are supported by information from the manors of St Benet's of Hulme in the 1350s and 1360s and from estates of St Giles's Hospital in the 1390s and early 1400s, the manor of Akenham in the early 1350s and around 1390; the 1380s are not well covered.

Data security for harvest information gained from manorial accounts is high. The repeated listing of identical climate proxy data throughout various sub-sections of the rolls, allows for the cross-checking of this information in the account itself. In years when a number of *compotus* rolls survives, further comparison of harvest date and length information is possible. The use of accounts from a variety of sources raises the reliability of the climate proxy information further, because manors of different landlords were subjected to differing administrative structures and management decisions.[25] Hence similar dates and trends in accounts of different provenance demonstrate the independence of the grain harvest date from human decision and underline its tight relationship with the phenological phase of the grain development. The supplementary series also fill in some gap years, for which no information is given by the accounts of Norwich Cathedral Priory (Fig. 3.5).

3.4 Potential Non-climatic Influences on the Harvest Date

While the harvest date was dependent on the phenological state of the grain and hence on the mean temperature during the growing season, radically and rapidly altered socio-economic conditions or agricultural practices could disrupt this close relationship. Harvesting was labour-intensive and to avoid a negative influence of a shortage of labour on the harvesting process, a whole set of rules was in place to ensure a secure labour supply at harvest time for the lord. It was difficult for the village population to leave their home during that period. The lord could fall back on customary labour services, the day works and boon works; the harvest boon works were among the last services to be commuted to money payments and ecclesiastical landlords were especially conservative by avoiding commuting services much longer than their lay counterparts. Additionally the priority for hiring local labour lay with the lord.[26] Manors close to towns and cities also benefited from hired labour of the town people. With the accounting reform at Norwich Cathedral Priory in 1354–1355 itinerant harvest workers, so-called cockers,[27] are traceable for the first time. Their appearance in the post-1350 period is linked to the disruption and labour shortage resulting from the demographic crisis. On Norwich Cathedral

[25] On the importance of choosing manors that were not all owned by the same landlord, Stern, A Hertfordshire demesne, 31–32.

[26] Ault, Open-field farming, 33–34.

[27] Ernle, English farming, 12.

Priory manors cockers were present primarily in the second half of the fourteenth century. Between 1390 and 1400 the number of works performed by them dwindled.[28] For trends in the harvest work force and on the manor of Gnatingdon in northwest Norfolk 1256–1431, see Appendix 3. In extreme and rare cases plague could aggravate or cause a severe labour shortage, so that fields might be harvested too late or not at all, as was to be observed in England in 1349.[29] However, this was a short-term influence, on the long run the negative demographic trend after 1350 had no impact upon the harvest date. Možný et al. have recently shown, that the relationship of harvest date and temperature during the growing season is merely weakened during periods of acute political and demographic stress, e.g. in times of war and their aftermath, due to their direct impact upon agriculture and the rural population.[30] The plague, though causing a population loss of about 30% or more in 1348–1349, did not eliminate agricultural knowledge in the English countryside. Large scale political disruption in Norfolk during the study period was rare. The Second Barons' War 1264–1267 left its traces in Norwich,[31] but seemingly did not disrupt harvesting on the manors of Norwich Cathedral Priory. The events of the attack of the citizens of Norwich on the cathedral priory did result in the loss of some archive material of the monks, but did not affect the countryside. During the Peasants' Revolt 1381 there was widespread upheaval in western and eastern Norfolk during early summer; only two harvest dates survive for this year, they come from manors removed from the uprising's epicentre.[32] A second rising in Norfolk in the following year, 1382, took place around Michaelmas, long after the start of the grain harvest, and was quickly suppressed.[33]

The Norfolk manorial economy was characterized by a good degree of agricultural specialisation: some manors grew no or very little winter rye, a few others no winter wheat. Before 1350 rye was common, but was then increasingly marginalized in the decades following the Black Death, and the acreages sown with wheat on the other hand remained stable c.1250–1449.[34] The relative proportions of winter crops to spring crops also vary from estate to estate as well as over time, and the influence of this varying composition of the annual harvest on the harvest date is more difficult to discern. Generally in a grain harvest constituted by the cutting of

---

[28] For example the manors Sedgeford and Gnatingdon used cockers from the mid-1350s to the early 1390s respectively the middle of the first decade of the fifteenth century. The numbers of works performed by them, varied strongly from none to 174, the highest numbers were reached in the 1380s (especially the late 1380s), after which they dwindled quickly. As in Gnatingdon, cockers helped in the harvests of Hindolveston until c.1406. On the other hand Martham did not turn to cockers after the 1360s. Often sharp alterations in the number of works done by cockers in the harvest were associated with changes of the manorial management personnel.

[29] Knighton, Chronicon, vol. 2, 100–101.

[30] Možný et al., Cereal harvest dates, 814–815.

[31] Blomefield, History of Norfolk, vol. 3, 52–53.

[32] In the northwestern Norfolk: Sedgeford, NRO, LEST/IB 37, and close to the marginal Breckland Great Cressingham, NRO, MC 212/10.

[33] Walsingham, Historia Anglicana, vol. 2, 70.

[34] Campbell, Overton, Norfolk Farming c.1250-c.1850, 54.

winter and spring crops, the winter crops would be ripe and harvested first and consequently their cutting would mark the start of the harvest. Of the winter corn, wheat would be ripe slightly earlier than rye.[35] In Norfolk wheat and rye must have reached maturity within a short space of time because in the eastern part of the county *maslin*, a wheat-rye mixture, was successfully grown.[36]

In the classical three-course-rotation – winter corn, spring corn, fallow – the crops are annually moved around the fields. This process would alter the microclimatic and soil condition for the growing grain,[37] but exerts no significant influence on the grain harvest date in a region as flat as East Anglia. Field names are difficult to trace to the Middle Ages.[38]

Harvesting methods were subject to change during the Late Middle Ages. Until the mid-fourteenth century all corn crops were reaped with a sickle in England (see front cover). From c.1300 onwards mowing instead of reaping was practised in the Low Countries and appears after the Black Death also more often in East Anglia.[39] Mowing was a more specialized harvest method and required training. It allowed for a quicker and less work-intensive harvest process, but increased waste and loss.[40] In the Norwich Cathedral Priory manorial accounts, one mowing work is expected to replace two and a half reaping works. The increased waste related to mowing ensured that it would be restricted to the cheaper grains, in Norfolk particularly to barley and oats.[41] In the eighteenth and the early nineteenth century wheat was still usually reaped, whereas barley and oats were mown.[42] Wheat was simply too valuable for risking a high loss during the mowing process. Sometimes rye, though cheaper, would also not be mown to preserve its long straw, which was useful in

[35] Ernle, English farming, 9. Concerning the predominant wheat varieties he also states that on light land red rivet or a lost white variety would be used, on heavy soils red or white pollard and on clay soils 'gray' wheat, ibid., 8.

[36] Campbell, Seigniorial agriculture, 221.

[37] Nordli, Reconstruction of nineteenth century summer temperatures in Norway, 206, states that a temperature reconstruction based on grain harvest dates, should ideally be using the dates of one crop from always the same field. However, agriculture can not operate according to those lines. In the Middle Ages the three course rotation was widespread (although the productive regions in eastern and northern Norfolk would be cropped in four years out of five (which could include up to three courses of barley), Campbell, Eastern Norfolk, 28–29, idem, Seigniorial agriculture, 267–271); and also in modern agriculture it is for various reasons advisable to change regularly the annual crops on the fields.

[38] The first *compoti* of Norwich Cathedral Priory which name the fields, where a crop was sown, appear after the reform of the accounts in 1354–1355; the naming becomes regular later.

[39] Stone, Medieval agriculture, 250. In Sect. 7.2 and Appendix 3 more details are given for the use of mowing in the grain harvest of Gnatingdon and Sedgeford.

[40] Rösener, Bauern im Mittelalter, 126–127 and Stone, Medieval agriculture, 250. According to Stone mowing was employed on the East Anglian manors of Hinderclay and Wisbech Barton in times of crisis or when grain prices were low.

[41] The accounts distinguish between *metere*, to reap (wheat), and *falcare*, to mow. The different methods are described in Ault, Open-field farming, 28.

[42] Mowing wheat was established in Norfolk between 1820 and 1837, Wade-Martins, Williamson, Roots of change, 116–117.

Fig. 3.6 Carting grain. Luttrell Psalter, Lincolnshire, circa 1325–1340 (British Library, Add. MS 42130, f. 173v)

thatching.[43] The introduction of mowing with the scythe alongside reaping with the sickle was not relevant for the harvest date, and the grain cut at the beginning of the harvest, the winter corn, was subjected to a stable harvesting process. After cutting, the grain was bound into sheaves, these were dried before they were either carted to the barn or stacked in the field (Figs. 3.6 and 9.2). Good, dry weather during the cutting and drying time was essential, as wet grain is prone to spoiling. Hence rainfall would prolong the harvest; during very wet harvests the sheaves had to be untied again for allowing the grain to dry.

None of the abovementioned short and long-term factors and developments had an influence strong enough to disrupt the relationship between the East Anglian harvest dates and growing season temperature.

3.5 Dating the Harvest: Calendar, Work Management and Communication

Life in the Middle Ages was highly regulated by custom. The official harvest season in medieval England, *autumpnus*/autumn, stretched from St Peter in Chains (SPC), 1 August, to Michaelmas, 29 September.[44] The importance of 1 August was reinforced by this day also being Lammas Day, 'Loaf Day', when bread made from the first ripe wheat would be blessed. Ideally the harvest would fit inside the months August and September, and so many manorial accounts, especially from areas out-

[43] Campbell, Seigniorial agriculture, 220.

[44] Titow, Le climat à travers les rôles de comptabilité, 312.

side East Anglia, fall automatically back on these dates to circumscribe the harvest and no further specifications are given. Consequently such accounts reflect custom and do not provide any temperature proxy.[45] Manorial accounts that do list the real harvest start and end dates mostly do not define the date by numbering the days of the month, but employ the ecclesiastical calendar. This kind of dating relies on unmovable saints' days and other festivals, the date thereby falling on one of these feast days or the respective weekdays before or after the feast day.[46] As today the medieval days were organized in the seven-day week, which then began on Sunday.

3.5.1 *The Ecclesiastical Calendar*

The important feast days around the beginning of harvest which could be used as reference points were St James (SJ) on 25 July, St Peter in Chains (SPC) or Lammas Day on 1 August, St Laurence (SL) on 10 August, and the Assumption of the Blessed Virgin Mary (AM) on 15 August. *In extremis* there were St Margaret (20 July, SM) as well as St Mary Magdalene (22 July, SMM) and the very late St Bartholomew (24 August). For detecting changes in the setting of the harvest dates, the data will be studied separately for the main archival collections and over four sub-periods 1256–1300, 1301–1350, 1351–1400 and 1401–1431.

On the manors of Norwich Cathedral Priory the years between 1256 and 1300 are marked by the predominance of harvests that commenced on a saints' day or other commemorative festival; these are 56.86% of all harvest dates. Among the aforementioned main feast days SJ is of some importance, representing 9% of the data. However, SPC attracts a strikingly high share: 41% (Fig. 3.7). SL and AM do not stand out, though this might be primarily due to the early harvests of this period.

Obviously custom weighed heavily upon the decision to set the start of the grain harvest on the cathedral priory estates. Convenience in accounting and/or dating might also have played a role, so that harvest dates were rounded to the feast days. SPC, being the official start of the *autumpnus* season, is clearly over-represented. 1256–1300 was a phase of early harvests and SPC appears to have fallen often within the range of days, when beginning to cut the grain was possible. Harvest onsets up to three or four days before and after SPC are very rare, so one can con-

[45] On the inclination of medieval people to allot to each month its proper, representative (agricultural) activity see Henisch, Medieval calendar year, 1–4, and especially on the European hay and grain harvest, ibid., 107–118. The standard *autumpnus* season is also employed in some of the manorial accounts of Norwich Cathedral Priory. The late North Elmham *compoti* 1391–1392 to 1410–1411, NRO, DCN 60/10/28 35, limit their information on the harvest date in the works account to the standard harvest season, although the duration of the harvest remains variable and hence reflects reality. On the other hand, in the works accounts of the Taverham rolls between 1362–1363 and 1373–1374, NRO, DCN 60/35/33-42, first the standard harvest season with the standard duration is named, but then the real start and duration are specified.

[46] Grotefend, Zeitrechnung des Deutschen Mittelalters und der Neuzeit, vol. 1, 81–83; Cheney, A handbook of dates, 15.

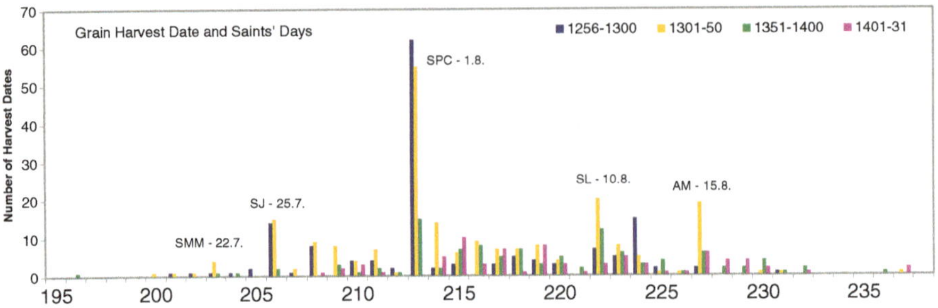

Fig. 3.7 Norwich Cathedral Priory manors 1256–1431: harvest date and saints' days. Plotted are the dates of the start of the harvest over the sub-periods 1256–1300, 1301–1350, 1351–1400, 1401–1431. Important ecclesiastical festivals are marked

clude that harvests that could have started on those days were postponed or advanced to SPC. To a lesser extent a similar structure emerges for SJ. The days before this feast day and the immediately following day are underrepresented in the data. Among the very early harvest dates neither SM nor SMM are prominent. The comparatively small number of harvests starting after SPC shows no tendency to fall on a feast day; this indicates that once within the official harvest season, no preference to special days was given.

In the first half of the fourteenth century 50.45% of the harvest dates coincide with a festival. SJ now holds 7%, SPC 25%, SL and AM each 9% of the data (Fig. 3.7). Although about half of the harvest dates are still feast days, the data are now more equally distributed. SPC exerts less pull upon the harvest dates than in the fourteenth century. The postponing of the harvest for up to three days for being able to start on SPC, was still frequent. The advancing of the harvests for matching the official onset of the *autumpnus* season, however, was no longer common practice, as is confirmed by the adequate representation of the days between the 2 August and 7 August. The data around SL display a similar pattern: villagers would wait one to two days for starting harvesting on SL, but usually would not advance harvests. The situation is somewhat different for SJ and AM. For SJ postponing the harvest for up to two days or advancing it one day, seems to have been possible. AM, lying at the end of the period when harvests could start, achieves its high share of harvest dates by the cutting of the grain being postponed as well as advanced for up to two days. Mid-August in the Julian Calendar was indeed very late to begin harvesting (Gregorian Calendar: 23 August) and apparently efforts would be made not to overstep this last important feast day. Apart from AM marking a kind of mental borderline, the fact that after this date weather conditions would rarely improve and contribute to a proper ripening of the grain, might have played a role in the tendency to avoid starting the harvest after AM.

Between 1351 and 1400 the percentage of harvests beginning on a feast day falls to 32.11%. Since this is a period of later harvests SJ represents merely 2% of the data. SPC takes a share of 14%, SL 10% and AM 6% of the data (Fig. 3.7). Obviously the saints' days and festivals were losing importance. SJ does not stand out; SPC attracts harvest dates that might otherwise have fallen up to two or three days before and one day later; for SL the harvests appear to have been sometimes postponed for one day. The same applies to AM, which does not mark the latest date for starting the harvest any more. The drop of data falling on feast days might not merely be due to a change in practice, but to a certain degree also to the surviving data being more evenly distributed over the half century. While the absolute amount of data is lower than before, there are fewer gap years and the amount of information available per year is quite steady.

This trend continues into the fifteenth century: although fewer data are available from Norwich Cathedral Priory, these data are spread evenly over the years. Merely 11.69% of the harvest dates now fall to a saints' day or festival. Harvest dates are generally late, none occur on SJ or the once so popular SPC (though there are some early harvests, starting in late July). SL represents 4% of the data, AM 8% (Fig. 3.7). It appears that for one or two days before AM, harvests might have been postponed, but there are not enough data for a conclusive analysis. For SL neither postponing nor advancing harvests was involved. Generally the data are now evenly distributed over the days and feast days hold no special importance any more.

The archival collections of the Abbey of St Benet's of Hulme and St Giles's Hospital contain too few data for allowing more than just the highlighting of trends. The manors of St Benet's supply data between 1350 and 1378; a higher percentage of harvests began on a feast day than would be expected from a random sample. The result is ambiguous to a certain degree, because some non-feast days are also over-represented. However, 47% of the surviving data are feast days (SPC and SL); consequently saints' days and festivals played an important role in setting the harvest date on the estates of St Benet's of Hulme. Of the two groups of harvest dates from St Giles's Hospital, the early group, 1332–1348, also displays the typical weighting towards feast days, which absorb 45.45% of the data. However, this is almost entirely due to the information from Hardley, on the other manors harvest dates coinciding with festivals are very rare. In the later group, 1392–1408, a certain degree of predominance of feast days is still visible: they take up a share of 28% of the harvest dates. The harvest dates of St Benet's and the manor of Hardley highlight the important role that feast days played in fixing harvest dates within the possible phenological time range.

As can be seen, the preference to start the grain harvest on saints' days and other festivals was widespread until the mid-fourteenth century; the custom existed on the manors of Norwich Cathedral Priory, on those of St Benet's' Abbey of Hulme and on some of the hospital's lands. Over the course of time, on the manors of Norwich Cathedral Priory after the mid-1330s, but latest after the Black Death, this preference was weakened until it finally almost disappeared around 1390. Consequently the error created by adjusting the harvest date to a feast day diminishes during the period studied.

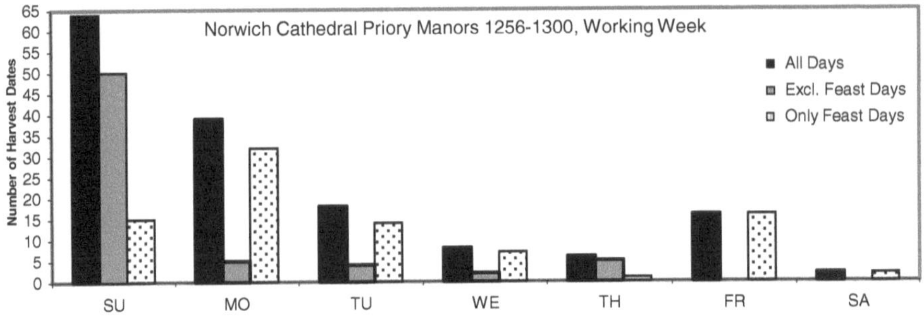

Fig. 3.8 Norwich Cathedral Priory manors 1256–1300: harvest dates and weekdays. Plotted are all harvest dates, harvest dates excluding the harvests that started on a feast day, and the harvest dates that fell on a feast day

3.5.2 The Working Week

In the second half of the thirteenth century the beginning of the week, Sunday and Monday, was preferred for the start of the grain harvest on the manors of Norwich Cathedral Priory (Fig. 3.8). The share of the individual weekdays drops steeply from Sunday with 43% to Thursday with 4%, only to constitute a local high again on Friday with 10%. Virtually no harvest ever started on a Saturday.

The notion that custom preferred Sundays as the start of reaping is underlined when only the harvest dates that did not fall on a feast day are considered. Of those, 75% are Sundays. Mondays to Thursdays represent between 3 and 8% of the data; Fridays and Saturdays do not occur at all. On the other hand the harvest dates that coincided with feast days were more equally spread over the week. 38% of them fell on Mondays, 18% on Fridays, 17% on Sundays, and 16% on Tuesdays.

Custom and practicality favoured harvests starting at the beginning of the week. The further the week progressed, the fewer harvests were started. If the time window for cutting the grain was too narrow for waiting for the new week to begin, Friday was chosen. Saturdays were avoided: so short before Sunday, one would simply wait one more day.

The four feast days, SJ, SPC, SL and AM, were also favoured harvest dates (56.86%). They come at intervals of seven to ten days and in most years did not coincide with Sundays. Consequently they provided convenient ‘stepping stones’ within the week for the start of such harvests, for which waiting for the next Sunday would have been too long and too risky. In this structure a harvest date would be pushed at maximum for three days, but usually less, to coincide either with a Sunday or a feast day. Most likely this would be handled by postponing the harvest, as the sudden rise of Sunday as harvest day and the consequent successive drop in percentages down to Saturday demonstrates.

Results very similar to the years 1256–1300 are obtained for the cathedral priory manors in the first half of the fourteenth century. However, two sub-periods can be distinguished. The change occurred in the mid-1330s. During 1301–1336 the

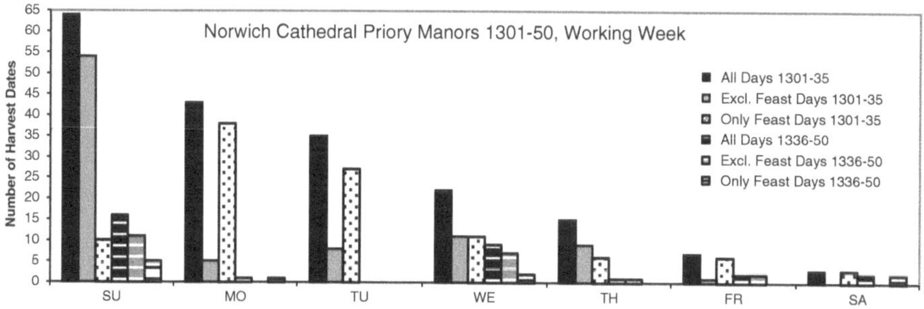

Fig. 3.9 Norwich Cathedral Priory manors 1301–1350: harvest dates and weekdays. As Fig. 3.8 but for the sub-periods 1301–1335 and 1336–1350

structure of the data strongly resembles that of 1256–1300. The overall drop in percentages from Sunday to Saturday is slower and smoother, but more continuous than in the preceding period. 33% of the harvests now began on a Sunday, 9% less than before. The local high on Friday has disappeared, probably because the percentage of the harvests begun between Monday and Thursday has increased (Fig. 3.9).

That the importance of Sundays was lessened to some degree is underlined by the non-feast day harvest dates. 61% of those fell on a Sunday, a drop of 14% compared to 1256–1300. Mondays were avoided, probably harvests that would have ideally begun on this day were started on the preceding Sunday. A small local high is presented by Tuesday to Thursday for the harvests that could not be adjusted to the beginning of the week. Very few harvests began on Fridays and Saturdays. Harvests that began on a feast day (50.45%) also tended to fall to the beginning of the week: Mondays predominate. From Wednesday onwards percentages are low and steadily fall to Saturday. Overall, Sundays still occupy the most dominant position, with a strengthening of the days Monday to Thursday.

This pattern is altered in the mid-1330s; unfortunately there are comparatively few data available for the period 1336–1350. The predominance of Sundays, especially among the non-feast days, is diminished further. Mondays and Tuesdays hardly occur as days for the beginning of the harvest. However, the mid-week high on Wednesdays, is clearly developed, due to the non-feast day dates (Fig. 3.9). On the whole the percentage of harvest dates coinciding with feast days has abruptly fallen to merely 32.26%.

On the estates of Norwich Cathedral Priory the main characteristic of the pattern – starting to cut the grain early in the week – is carried over to the next period, 1351–1400. A shift takes place at the end of the 1380s. From 1351 to 1389 the percentage of harvests starting on a feast day is already relatively low at 36.36%. The dominance of Sundays over other week days is strengthened, they account for 47% of the data. Mondays and Tuesdays are considerably less important than in the preceding period, but the mid-week high, on Wednesdays and Thursdays, remains. The importance of Sundays is accentuated in the non-feast day data; feast day har-

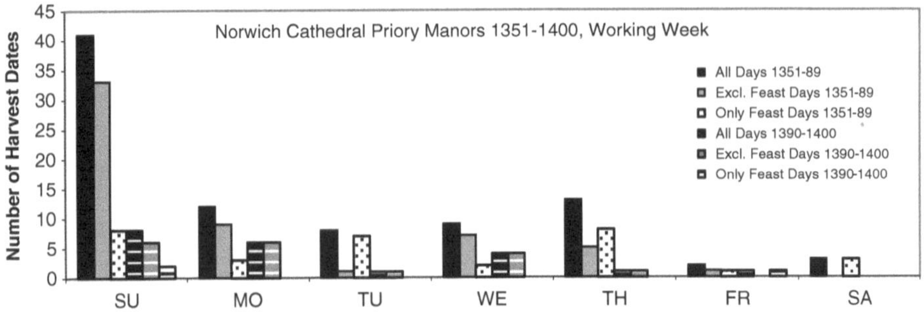

Fig. 3.10 Norwich Cathedral Priory manors 1351–1400: harvest dates and weekdays. As Fig. 3.8 but for the sub-periods 1351–1389 and 1390–1400

vest dates on the other side tended to coincide not only with Sundays, but also Tuesdays and Thursdays. Friday and Saturday remain underrepresented in both groups (Fig. 3.10).

In the final decade of the fourteenth century feast days merely represent 14.29% of the harvest dates. At such a low percentage they no longer exercise an influence on the distribution of the harvest dates over the week. Although there are relatively few data available, it is obvious that Sundays and Mondays are now of almost equal importance and that the Wednesday high is well developed. In the non-feast day harvest dates the overall situation is consequently closely mirrored: Sunday and Monday both achieve 33% and Wednesday 22% (Fig. 3.10).

The growing importance of Mondays in the last years of the fourteenth century led the way to the emerging predominance of this day in the fifteenth century. Whereas now 52% of the harvests began on a Monday on the priory manors, only 18% did so on a Sunday. Another 14% fell on a Wednesday, so the mid-week high persisted. Tuesdays and Saturdays are days unfavourable for the start of the harvest (Fig. 3.11).

Although harvest dates coinciding with feast days are very few in number (11.69%), they mirror the predominance of Monday and emphasize the importance of the beginning of the week, because none of them coincide with a Thursday, Friday or Saturday.

The data for St Benet's of Hulme 1350–1378 displays the strong inclination to start the harvest on a Sunday; 74% of the harvests that did not coincide with feast days fell on a Sunday. The rest of the dates occur towards the middle or the end of the week. The feast day harvest dates often corresponded with Thursdays. In this way a local high of mid- to end-week for the harvests that could not wait until the following Sunday was constructed. The harvests on the manors of St Giles's Hospital 1332–1347 also tended to start primarily on Sunday. However, Friday is also well represented. Both days figure strongly in the non-feast day data. On the other hand the feast day data could fall to any day of the week, Tuesdays stand out to some extent. During the later period, 1392–1408, the situation changed: feast days figure weakly in the data (28%) and of the non-feast day harvests 80% were Sundays and

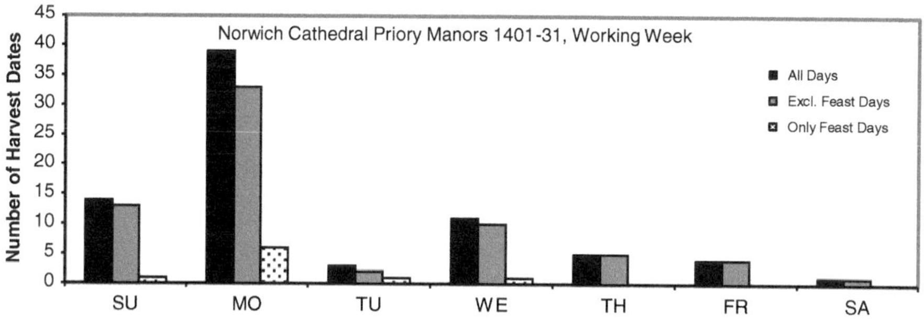

Fig. 3.11 Norwich Cathedral Priory manors 1401–1431: harvest dates and weekdays. As Fig. 3.8 but for the period 1401–1431

20% Wednesdays. Considering all harvest dates c. three quarters fell to a Sunday. The mid-week high is formed on Wednesday by feast days and non-feast days.

Thus the data from St Benet's of Hulme and St Giles's Hospital mirror the tendency of the Norwich Cathedral Priory harvest dates to start harvesting at the beginning of the week, especially on Sunday, as well as the existence of a smaller mid- or end-week high of harvest dates.

3.5.3 *The Harvest Date on Selected Manors of Norwich Cathedral Priory*

The individual manors do not diverge significantly from the overall Norwich Cathedral Priory harvest date-setting. Until c.1390 the percentage of feast days among the harvest dates is relatively high, ranging from 36% (North Elmham) to 68% (Monks' Grange). The average lies around 50%. After 1390 the share of feast days drops sharply to range from 0% (Taverham) to 25% (Plumstead) with an average of 12.2%.

In relation to the weekday distribution of the harvest date, most manors comply to the established picture until c.1390.[47] There was a strong preference for starting the harvest on Sunday, which was then followed by a decline from Monday onwards until Saturday, when almost no harvests began. Mid- or end week highs, developed to varying extents, interrupted this successive downward trend. Variation occurred in the steepness of the decline from Sunday onwards. The drop is very abrupt in North Elmham[48] and comparatively smooth in Eaton. Thornham is the only manor that preferred Mondays over Sundays. Another point of variability is the intensity of

[47] Eaton, Gnatingdon, Hindringham, Martham, North Elmham, Plumstead, Sedgeford and Taverham. These are almost all the places included in the analysis of individual manors.

[48] Between 1256 and 1390, 60% of all harvests began on a Sunday at North Elmham. This is due to the feast day data holding the comparatively low share of 36%. Since feast days often cover the normal week days Monday to Saturday, those are underrepresented here. A small mid-week high is situated on Thursday, formed as usually by non-feast day data.

the mid- or end-week high. In Martham the mid-week high was more pronounced than on other manors, whereas in Hindolveston there appears to have been almost no mid-week high. Depending on the individual manor mid- or end-week highs could focus on Tuesday and Wednesday, Wednesday alone, and also Thursday and Friday. In most cases the feast day data are distributed over the week days Monday to Saturday with a bias towards Monday and Tuesday. Non-feast day data are concentrated on Sunday; the mid-week high, too, was often constituted by non-feast days (Fig. 3.12).

The manor of Monks' Grange was the epitome of the rules for the harvest date-setting. From the most popular Sunday the data steadily drop down to a low on Wednesday and Thursday, to rise to an unusually well developed end-week high on Friday and Saturday (Fig. 3.13). The distribution of feast days and non-feast days is very rigid. On the one hand, virtually all the harvest dates coinciding with a feast day (68% of all the data) fell on a day between Monday and Saturday. On the other hand, all the harvests starting on a non-feast day, fell on a Sunday. Although most manors show similar tendencies (except for the mid-week high), Monks' Grange is the only place with such a clear-cut distinction between feast days and non-feast

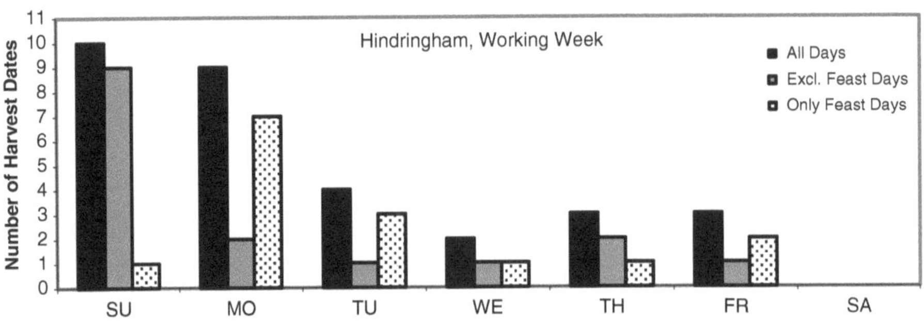

Fig. 3.12 Hindringham 1256–1390: harvest dates and weekdays. As Fig. 3.8 but for Hindringham. Hindringham has been chosen as a typical example

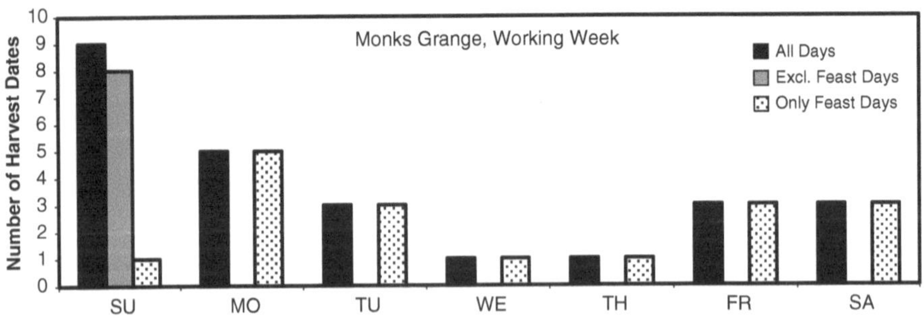

Fig. 3.13 Monks' Grange 1256–1335: harvest dates and weekdays. As Fig. 3.8 but for Monks' Grange. No data are available after 1335

days. The feast days were almost never Sundays, the non-feast days were always Sundays. It is also the only manor where beginning a harvest on Saturday, as long as it was a feast day, was not avoided. This distribution of harvest dates is probably connected to the fact that Monks' Grange was a manor without tenants. The harvest was performed by the *famuli* and hired labourers or tenants from other manors. Although hired labour was easily available so close to the city of Norwich, more organisation was needed for the mobilisation of this work force than for merely local hired labour and a clear and memorable date, like a Sunday or a feast day, would have to be communicated to ensure a timely supply of hands.

As was shown earlier in this chapter, the pattern in the week day distribution shifts after 1390. The trends of the following period are also clearly visible in the data of the individual manors. Now a small percentage of harvest dates, if any at all, coincided with a feast day. Usually Monday and to a considerably lesser extent also Sunday, were the most popular harvest days. Mid- or end week highs usually existed.

Unsurprisingly the individual manors of Norwich Cathedral Priory by and large reflect the general tendencies of all the priory's estates in respect to harvest date and ecclesiastical festivals or weekdays. Most likely the variations from manor to manor in choosing which day to start the grain harvest reflect differing local customs and work organisation, although Sunday, and after 1390 Monday, were almost always the most popular days for the beginning of the harvest. The mid-week high located between Tuesday and Thursday and constituted by non-feast days could be due to the distribution of the week-work or *opera* days. So harvests that could not wait to the following Sunday respectively Monday might be set to a feast day, if available, or to a normal weekday, when many *opera* were due. These days would have differed from manor to manor. This view is supported by the evidence from Monks' Grange, where neither a mid-week high consisting of non-feast days nor a resident customary tenancy to perform *opera* existed. The data of this place also seem to indicate that the more organisation was needed to engage harvest workers – in this case, because there were no customary tenants resident, in other cases, because larger manors also had to manage larger groups of customary tenants – the more prominent and clear dates, such as Sundays and feast days, would be used. The smaller the manor, the more flexible the harvest date-setting could be.

3.5.4 Harvest Date and Calendar

In late medieval northern East Anglia the setting of the grain harvest date within the short phenological time window of reap-ripe-state was dominated by the preference for starting the harvest early in the week, on Sunday or Monday, and by the preference for beginning to cut the grain on an important saints' day or ecclesiastical festival, such as SJ, SPC, SL and AM. Commencing cutting grain on the last days of the week, in particular on Saturday, was largely avoided on the manors of Norwich Cathedral Priory, St Benet's of Hulme and St Giles's Hospital. Mid-week highs show alternative days favoured for the start of the grain harvest. The trend of starting to harvest early in the week, though to some extent varying in its strength over the period, persists basically unbroken over the whole period 1256–1431. Tradition also favoured starting the harvest on an ecclesiastical feast day. The earlier in the study period, the more the harvest dates gravitate to these feast days. Probably already after the mid-1330s, but at the latest after 1350 their predominance is broken and after 1390 this pattern disappears altogether.

The preference for feast days and Sundays as days to start the grain harvest conflicts with the general nature of those days as holidays. Work was theoretically forbidden by canon law on such days, however, in reality adherence to that rule was lax. Minor holidays, contrary to Sundays and major feast days, were widely ignored on the English manors. The Norfolk data demonstrate that due to the importance of the grain harvest, permission was granted for harvesting even on Sundays and major holidays such as AM. For the lord of the manor this would prove quite convenient, since both tenants working on the lord's fields as customary labour as well as villagers working as hired labour were readily available at such a day, because they could not perform any major work for themselves. First the village community would assemble in church for rendering service to the Lord, then they would gather in the fields to do so for the lord of the manor.[49] The steady decrease over time of the percentage of harvests starting on a holiday – either a feast day or a Sunday – could be indicative of a rise of living standards and an improvement in working conditions of the common people in the Late Middle Ages. The watershed moment on the estate of Norwich Cathedral Priory, an ecclesiastical landlord, would not be the Great Pestilence in the mid-fourteenth century, but rather the Peasants' Revolt 1381 which marked the end of the time when conservative landlords could resist socio-economic change such as a reduction in customary labour dues. As workers were less and less willing to perform underpaid work, they might also have been less and less willing to work for the landlord on holidays.

The affinity of the harvest date to festivals and Sundays during most of the study period might be explained at least partially by the different use of customary and hired labour. The harvest date in the manorial accounts of Norwich Cathedral Priory primarily refers to the work of the *famuli* and the customary tenants. The days worked by hired labour are rarely detailed. Theoretically the performance of customary labour services was subject to mutual agreements between lord and tenants which were made a few days in advance. Such limitations were not in place for hired labour which was to some extent more flexible and could be called upon *ad hoc*. This is supported by the very few accounts of Norwich Cathedral Priory that state the day when hired labour began to harvest. In Eaton for 1304–1305 the harvest date for the *famuli* and customary tenants is 29 July 1305, but a small list is attached to the main account reporting that the hired labour came in on 26 July 1305.[50] The Bawburgh accounts for 1304–1305 and 1305–1306[51] give both SPC as

[49] Bennett, English manor, 115–118.

[50] NRO, DCN 60/08/11A.

[51] NRO, DCN 61/19 for 1304–1305; the account of 1305–1306 is in the account for Eaton, NRO, DCN 60/08/12.

the general start of the grain harvest, but then specify that the hired people were working a few days before. In the Hunstanton accounts the period of harvest work for customary and hired labour is often given. Here the hired labourers took on work either on the same day as the *famuli,* or a few days later, though hired labour still tended to start on a feast day.[52] The Bawburgh case in particular throws light on the many harvests until c.1300 that started on SPC – sometimes hired labour may have been employed a few days before. Customary labour for harvesting was perhaps easier to enforce, when the harvest season had officially started with SPC. However, it remains unclear if the events at Eaton, Bawburgh and Hunstanton are a glimpse of a more widespread practice, or are merely limited to those places during certain years.[53]

The system of 'stepping stones' of feast days and days early in the week was somewhat less refined until the end of the thirteenth century, when dates were fixed according to rougher scales, as the strong predominance of Sunday and SPC indicates. Several reasons might be responsible for this: there are still many gap years in this period, but in years with surviving data, the number of harvest dates is often very high (Fig. 3.5). The likelihood of reporting identical harvest dates in the accounts of such a year is thus raised, and data accumulate on these days. Probably this does not explain the whole extent of the fixation on Sundays and SPC in this period. More significant appears to be the role of custom. SPC as the official start of the harvest season in the Middle Ages led to a pronounced effort to actually begin harvesting on that day. The data during 1256–1300 in general indicate relatively early harvests, of which a substantial part already started in July before the official start of the harvest season. For making reality comply with the ideal some other early harvests might have been declared to have started with SPC. In the 1290s the financial situation of English landlords deteriorated and accounting procedures improved, hence the detail of the information supplied by the reeve or bailiff in the accounting process increased.

The fine grid of feast days and the days at the beginning of the week would also lose some of its precision in years when SJ, SPC, SL and AM actually coincided with Sunday or Monday. In such years the attraction of the aforementioned feast days would be increased and the normal push of up to two or three days could have been raised by another one to two days. This mechanism aided the general uniformity of the harvest dates on the manors of Norwich Cathedral Priory in 1288, 1295 and 1305. A similar feature could also apply to 1294, 1316, 1339, 1372 and 1389, although data density in these years is too low for allowing far-reaching conclusions. The clear cases of an increased effort to start harvesting on SJ, SPV, or AM are found in the years until c.1300, when the attraction of feast days was most per-

52 In the first three accounts 1331–1333, NRO, LEST/BG 2, 4–5, both groups started on the same date, but in NRO, LEST/BG 6, 9, 11–13 the different kinds of work are listed separately and differ either by zero, two, three, six or 14 days, though the last appears to be excessive and is probably a simple writing error.

53 The year 1305 was one of the rare cases when SPC fell on a Sunday. This increased the attraction of SPC as a harvest date. The summer 1305 and the growing season 1306 were also very warm, so normal arrangements might have been overtaken by the need for an early harvest.

ceptible. In some years during the fourteenth century the evidence is inconclusive, and towards the end of the study period no cases emerge any more at all.

Climate exerts the overarching control on the harvest date by setting the phenological time window in which harvesting is possible. Traditionally 1 August, SPC, was seen as the beginning of the harvest season in medieval England. It can therefore be assumed that this day would have corresponded closely with the average development of the reap-ripe-state in the grain in wide areas of England during the High Middle Ages; a notion that is supported by the data from Norfolk in the second half of the thirteenth and in the early years of the fourteenth century. Custom and social influences and the adjustment to the ecclesiastical calendar actually position the harvest date within the adequate pheno-state of the grain. The ecclesiastical calendar in combination with the seven-day week allowed for a generally fine adjustment of the harvest date to the phenological state of the corn. The pushing of the harvest onset to a preferred day did generally not exceed two to three days and was often achieved by waiting. If the short time window, when medieval man could bring in the harvest without much loss, drew to a close, custom and convenience in date setting would be overcome, as is demonstrated by the representation of all week days and all dates between 19 July and 20 August in the data.

Chapter 4
Farming in Norfolk Around 1800

The grain harvest was little changed from the Late Middle Ages to the nineteenth century. This is demonstrated in this chapter through an analysis of information contained in a selection of eighteenth and nineteenth-century farming diaries from across Norfolk. The county was at the forefront of the agricultural revolution and one of the most productive regions of England around 1800. The detailed farming diaries chronicle a time when agricultural improvement and change formed the basis of the industrialisation in England and fed a rising urban population.

For the grain harvest the decisive change came in the second half of the nineteenth century. In 1851 the American harvesters McCormick and Hussey were shown for the first time in England at the Great Exhibition and these models afterwards entered the English market and became widespread.[1] This mechanisation changed the harvest process in several respects. Whereas harvesting with sickle or scythe in the narrow time slot of eight to ten days as well as the harvesting success greatly depended on a run of good dry days, now the harvest would take place in a much shorter time and when the grain was over-ripe, so that the kernels would fall from the ear into the harvester.[2] Often a number of farmers shared a harvester and had to make arrangements on the availability of the machine.[3] Prior to the mechanisation of the harvest, potential influences on the homogeneity of the harvest date are constituted by the choosing of the stage of ripeness of the corn for cutting and the ongoing breeding of grain varieties. Neither of these factors exercised a substantial influence on the harvest date. Breeding was usually aimed at raising yields as well as increasing the resistance of the varieties against adverse weather conditions like wind and not at altering the harvest date. It is the mechanisation of the harvesting process that blurs the clarity of the harvest date as a temperature proxy to some

[1] Jones, Seasons and prices, 125.

[2] Dickel, Beginn der Mähdruschernte, 75 and Jones, Seasons and prices, 126. The speed of the cutting the crops with a harvester greatly reduced the farmers' vulnerability to weather.

[3] Dickel, Beginn der Mähdruschernte, 76–77. This might be less of a problem in English agriculture, since it is dominated by big farms operating with separate enclosed fields. Therefore a farmer might be more likely to afford a harvester on his own.

K. Pribyl, *Farming, Famine and Plague*, DOI 10.1007/978-3-319-55953-7_4

degree. Nonetheless, lately Možný et al. have shown that although the influence of the mechanisation as well as breeding progress and other potential changes is difficult to quantify, these factors do not override the temperature during the growing season as the main determinant of the harvest date.[4]

In Chap. 5 the medieval grain harvest date is used to create a temperature reconstruction for the growing season, the months April to July. For this a comparison series of modern harvest dates from East Anglia is required. Due to the change in harvesting method in the second half of the nineteenth century, the comparison series has to come from a time which predates the introduction of the harvester and for which instrumental temperature measurements are also available. The Central England Temperature series (CET) begins in 1659,[5] hence the comparison series of harvest dates has to come from the 200 years between the mid-seventeenth and mid-nineteenth century. The modern harvest dates analysed in this chapter can serve as this comparison series. Four sets of grain harvest dates are available for Norfolk in the years around 1800. They come from farms at Langham in northern Norfolk, Snettisham in the northwestern part of the county, Fritton which is south of Norwich, and Wymondham which is southwest of the city (Fig. 5.1).[6] Of those only the series from Langham is long enough to be used in a reconstruction of medieval temperatures via the calibration-verification approach. The data from Fritton, Snettisham and Wymondham primarily help to underline the validity of the Langham series to serve as a Norfolk harvest date series around 1800.

4.1 Langham Farm

A farm at Langham in northern Norfolk provides a long and continuous series of grain harvest dates between 1768–1861/1867. The village is close to the north Norfolk coast, and lies on the gentle slopes of the western end of Cromer Ridge (see Fig. 5.1 in Chap. 5). The farm at Langham was in the hands of first the Frost and then the Rippingall family, both families were bound by marital ties. The information on the grain harvest was recorded in a set of annual farming diaries which covers for Stephen Frost the years from 1768 to 1816. In 1788 Thomas Rippingall married the sister of Stephen Frost, Mary, and when he took over the farm in 1817, he must have been well acquainted with the land. In 1816 the farming diary was already in

[4] Dickel, Beginn der Mähdruschernte,75–78 on the problems caused by the introduction of the mechanical harvesters. Možný et al., Cereal harvest dates, 816–817; harvest dates of the period 1501–2008 were analysed.

[5] Manley, The mean temperature of central England, 1698–1952, 242–261, and idem, Central England temperatures: monthly means 1659–1973.

[6] All these records are held by the NRO. Also checked for harvest dates were Cornelius Stovin, Journals of a Methodist farmer 1871–1875; Cozens-Hardy (ed.), Mary Hardy's Diary; and Griffiths (ed.), William Windham's Green Book, 1673–1688, but no harvest date information of sufficient length or frequency could be found.

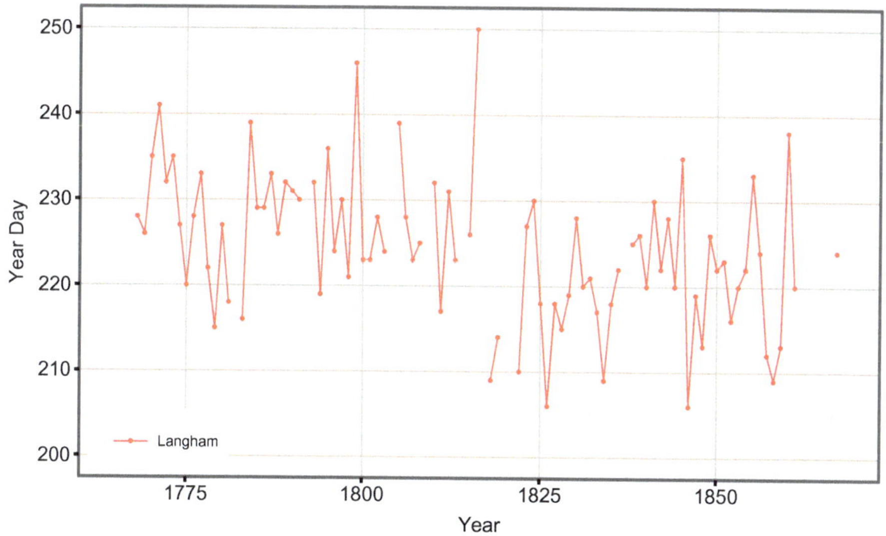

Fig. 4.1 Langham 1768–1867: harvest date. Wheat harvest date and the arrival of the harvest men usually coincide, but when wheat was cut before the arrival of the harvest men, the wheat cutting date has been used in this series

Rippingall's handwriting, even though the farm was officially still managed by Stephen Frost. The farming diaries of Thomas Rippingall run from 1817 to 1831; the 1832 diary was made by him together with his son Stephen Frost Rippingall, for whom information is available for the period 1833–1858. Three personal diaries containing the harvest dates 1859–1861 survive for William Rippingall. Additionally a harvest book was made 1847–1861/1867; in 1856, 1858 and 1867 it is the sole provider of the date of the grain harvest, and in the remaining years it gives identical information to the parallel diaries. Assembling the data from the various sources results in an almost complete harvest date series for the period 1768–1867 (Fig. 4.1).[7] For 86 of these 100 years harvest date information is available, gaps are rare and occur in 1782, 1792, 1804, 1809, 1814, 1817, 1820–1821, 1837 and 1862–1866.

The diaries of Stephen Frost and to a lesser extent those of Thomas Rippingall are very detailed and also contain weather references. Stephen Frost noted the rare cases when the start of the harvest took place either too early or too late in relation to the state of the crops. In 1795 he cut the wheat green, perhaps in line with other farmers for experimenting with early cutting.[8] In 1801 the harvest was begun too

[7] Farming diaries of Stephen Frost of Langham 1768–1816: NRO, MC 120/1-44, farming diaries of Thomas Rippingall 1817–1831: NRO, MC 120/45-57, farming diary of Thomas Rippingall and Rev. Stephen Frost Rippingall 1832: NRO, MC 120/58, farming diaries of Rev. Stephen Frost Rippingall 1833–1858: NRO, MC 120/61–85, personal diaries of William Rippingall 1859–1861: NRO, MC/120/121-123. The harvest book for the years 1847–1861/1867: NRO, MC 120/87.

[8] NRO, MC/120/26. Early cutting was mentioned and disapproved in Anonymous, The practical Norfolk farmer, 104–107.

late, because the turnips had not been hoed in time and the barn was not ready.[9] The diaries after c.1830, when Stephen Frost Rippingall took over the farm, record less information. Especially after the start of the meticulous harvest book in 1847, the farming diaries contain few references about the harvest beyond the start and end date. Until c.1835 the farm was mainly concerned with cattle and corn, afterwards the arable sector dominated.[10] In Norfolk around 1800 various four-to-six-course crop rotations were common.[11]

A Norfolk harvest around 1800 was organized in such a way that the grain crops would be cut in succession, and both stacking and carting would take place between the cutting. The cutting of white peas signalled the beginning and would be followed by the cutting of oats, then wheat and finally barley.[12] The cutting of the white peas and oats took place before the actual harvest. It was not included in the work of the harvest men and was partly performed by women. This concept was fully reflected in Stephen Frost's management of Langham farm 1768–1816. Frost's hired harvest men stayed throughout the whole wheat and barley harvest, also helping carting, stacking and sometimes cutting oats. Neither Stephen Frost nor his successors refer to sowing or harvesting rye in the farming diaries, but they grew wheat in all years. Hence the general harvest began with the cutting of wheat. Labour for harvesting was comparatively scarce in Norfolk due to the absence of greater population centres or industry from the county. Hence the harvest was performed somewhat 'slovenly'.[13] However, the carting of the corn was uncommonly quick and well organized in Norfolk.[14]

Itinerant harvest gangs were employed at Langham farm to help with the harvest of the grain, usually their arrival heralded the start of the harvest. The 'harvest men' normally 'come home' the night before the harvest began. Soon afterwards, often the next day, they were 'ordered at the Bell', presumably they took a celebratory drink in the public house in Langham which is still called 'The Blue Bell'. At the end of the harvest the workers were paid. The information in Stephen Frost's diaries confirms that he kept a close eye on the ripening-stage of his grain crops and often refers to it in the diary; occasionally there are also notes demonstrating the fine temporal adjustment of the commencement of the cereal harvest. Stephen Frost normally hired his harvest men in May, but shortly before the harvest he is sporadically 'warning' his 'harvest men home' for an evening in the immediate future.[15] Nonetheless it was occasionally necessary to begin cutting the wheat with the farm workers and local hired labour shortly before the arrival of the harvest gang. To ensure the capture of the signal of the harvest date in these years, the date of wheat

[9] NRO, MC/120/32.

[10] Afton, Investigating agricultural production, 238.

[11] Anonymous, The practical Norfolk farmer, 9–12.

[12] Anonymous, The practical Norfolk farmer, 99–117.

[13] Marshall, Review of the reports to the Board of Agriculture, 347.

[14] Anonymous, The practical Norfolk farmer, 115–117.

[15] For example 1777: NRO, MC 120/10, 1781: NRO, MC 120/14, 1783: NRO, MC 120/15.

cutting was used for the comparison series instead of the time of the arrival of the harvest gang. Red and white wheat were both sown at Langham.[16]

In the farming diaries the harvest duration is also recorded. The link between harvest length and weather, however, is difficult to determine. This is due to the scarcity of labour in eighteenth and nineteenth-century Norfolk, contrary to medieval times until the middle of the fourteenth century when labour had been abundant.[17] The labour shortage was particularly pronounced during the Napoleonic Wars and later in the 1830s.[18] The harvest process was adapted to the lower input of labour per harvested acre. Whereas in the eighteenth century wheat was generally reaped and the cheaper spring corn mown, between 1820 and 1837 the mowing of wheat took hold in Norfolk.[19] This shortened the harvest duration. Langham farm conformed to this new practice and turned to mowing wheat in the mid-1830s.[20] The spring corn had been mown already in earlier times. Whereas the mean value of harvest duration stands at 34 days for the period 1768–1832, after the introduction of mowing the wheat crop in the mid-1830s it averages at 24 days. The shift from reaping to mowing wheat does not cause a break in the Langham harvest date series. The measured Central England Temperature (CET) spring and summer mean temperatures 1817–1867 relate to the Langham grain harvest dates in the 1830s and 1840s as well as in the 1820s.

[16] Stephen Frost does usually not distinguish his wheat varieties in the harvest accounts, but occasionally mentions red wheat. It is likely that he sowed red and white wheat. Thomas Rippingall is more specific and refers regularly to red and white wheat. His successors hardly ever mention the variety of wheat in the harvest accounts. It can be assumed that red and white wheat were present at Langham farm between 1768 and 1867. According to a farmer cited in Lisle, Observations in husbandry, 132, red wheat that was sown at the same time as white wheat could be ripe a fortnight sooner. None of the Langham farmers distinguishes between red and white wheat with respect to cutting.

[17] Marshall, Review of the reports to the Board of Agriculture, 347.

[18] Collins, Harvest technology, 467.

[19] Wade-Martins, Williamson, Roots of change, 116–117. On the labour saving potential of mowing, see Collins, Harvest technology, 461.

[20] Stephen Frost referred to the wheat harvest with the words 'cut' or 'shear', all other grains were almost always 'cut'. The reaping of wheat and the mowing of barley or oats was so common that actively distinguishing between the two methods by refining the vocabulary was unnecessary. Thomas Rippingall was more precise in his wording, to the wheat harvest he referred using 'reap', 'shear' or 'cut', barley and oats were usually 'mown'. Only in the year 1818 part of the wheat crop was 'mown'. Stephen Frost Rippingall mowed the wheat in 1835, after 1837 all corn was indiscriminately 'cut'. Now 'cut' probably refers simply to mowing, since all grain crops were now harvested in this way in Norfolk.

4.1.1 The Working Week

In the medieval data the tendency to start the cereal harvests early in the week and during the thirteenth and early fourteenth century also on specific ecclesiastical feast days can be discerned (see Sect. 3.5). In the eighteenth century the ecclesiastical festivals had lost their social importance and could hold no influence over the setting of the harvest date any longer. For the weekday analysis the period 1768–1867 is subdivided into the years when the land was managed by Stephen Frost, 1768–1816, and the years when it was run by his successors, the Rippingalls, 1817–1867. Under Stephen Frost there was a noticeable tendency to start major works at the beginning of the week, preferably on Monday. This included activities such as the start of the harvest and the cutting of wheat, barley and oats. The start of the harvest, which meant the arrival of the harvest men, was most frequently a Monday (42%) and alternatively a Thursday (25.5%), but never fell to a Saturday or Sunday. The start of wheat cutting concentrated with Monday (29.5%) and Tuesday (24.5%) on the beginning of the week, but Friday (16%) forms an end-week high. For barley an almost continuous drop of importance of the weekdays from Monday (31.5%) to Saturday (5.5%) can be observed. Compared to these patterns the end of the harvest was more equally distributed over the week.

In the period after 1817 the importance of the beginning of the week as the time to start the cutting of grain was reduced. Although the harvest mostly still began between Monday and Wednesday, Wednesday (31%) was now more dominant than Monday (24%). After 1833 even Saturday was not avoided as an onset of the harvest (1817–1867: 12%). For wheat cutting, Monday (32.5%) was still strongly overrepresented as a start day, Tuesday and Wednesday (both 17.5%) were also popular, and the end week high fell on Saturday (15%). Wednesday in fact gained importance after 1833, perhaps in connection with the introduction of the mowing of wheat in the mid-1830s, which accelerated the wheat harvest considerably. The beginning of the cutting of barley is independent of the days of the working week. Again the end of the harvest is more or less equally distributed over all weekdays, but a stronger tendency to end the harvest at the end of the week (Friday: 29%), than under Stephen Frost, is visible. Unsurprisingly starting to cut the crops on Sundays is avoided throughout the whole period 1768–1867.

The preference of certain week days, usually at the beginning of the week, for the onset of the harvest and the cutting of individual crops found in much of the Langham data 1768–1867, existed already in the Middle Ages. The mid-week or end-week high of the Langham series were also known in medieval times. Obviously these features were by no means limited to the Middle Ages, but continued for reasons of work organisation at least into the middle of the nineteenth century. These characteristics are more strongly developed in the older Langham data recorded by Stephen Frost 1768–1816; in the latter part of the series 1818–67 these features are weakened.

Table 4.1 The modern harvest date series and the geographically nearest medieval series

Harvest date series	Mean	Min	Max	N
Langham 1768–1816	228.45	215	250	44
Langham 1818–1867	220.17	206	238	42
Hindringham-Hindolveston (Cromer Ridge) 1256–1423	228.93	214	246	56
Snettisham 1809–1827	224.67	207	247	12
Sedgeford-Gnatingdon-Thornham (Northwest) 1264–1431	226.55	204	247	106
Fritton 1803–1828	223.38	208	247	26
Wymondham 1795–1799	230.20	219	244	5
Manors around Norwich (Norwich) 1264–1426	220.66	208	235	58

The medieval series are organized in regional groups, medieval year days are adjusted to the Gregorian Calendar. For the regional pooling of the medieval series, see Sect. 5.1

4.1.2 The Break in the Langham Series

The transition of Langham farm from Stephen Frost to Thomas Rippingall in 1817 coincides with a break in the series of the grain harvest dates (Fig. 4.1). The mean of the series over the period 1818–1867 is about a week lower, than for the period 1768–1816 (Table 4.1). To test the significance of the difference in the means for this period, Student's *t* test was used; this test reveals the difference in means of the periods is significant at $p < 0.01$. A similar shift also occurs in contemporary data from Fritton and Snettisham, but is less pronounced there (Fig. 4.3).[21] Consequently the trend to earlier harvests, that was primarily due to higher growing season temperatures, must have been enhanced at Langham by other factors.

In 1815 Langham parish was enclosed.[22] Changes in the organisation of farming activities might have been the consequence, for example after 1815 the cattle at Langham farm was usually yard-fed.[23] Until 1813 oats were cut mostly shortly before or at the same time as the beginning of the grain harvest and the cutting of wheat.[24] After 1818[25] oats were usually harvested much later than the other grain crops including barley (Fig. 4.2). The difference between the general start of the grain harvest and the beginning of the cutting of oats in this period could amount to 15 or 20 days. Mowing oats at the end of the grain harvest might have saved some days labour before the cutting of the wheat and thereby helped to advance the wheat harvest. Another factor is the stage of ripeness of the grain. Thomas Rippingall might have decided to cut his grain generally at an earlier stage of maturity than

[21] Before 1817 the Langham harvest date was usually a few days after the harvest date of Fritton, after 1817 it usually coincides with or predates the Fritton harvest.

[22] Legislation.gov.uk, delivered by The National Archives, www.legislation.gov.uk/changes/chron-tables/private/25#f2

[23] Afton, Investigating agricultural production, 238.

[24] This is in accordance with the advice from Anonymous, The practical Norfolk farmer, 101–103.

[25] No information on the cutting of oats is available 1814–1817.

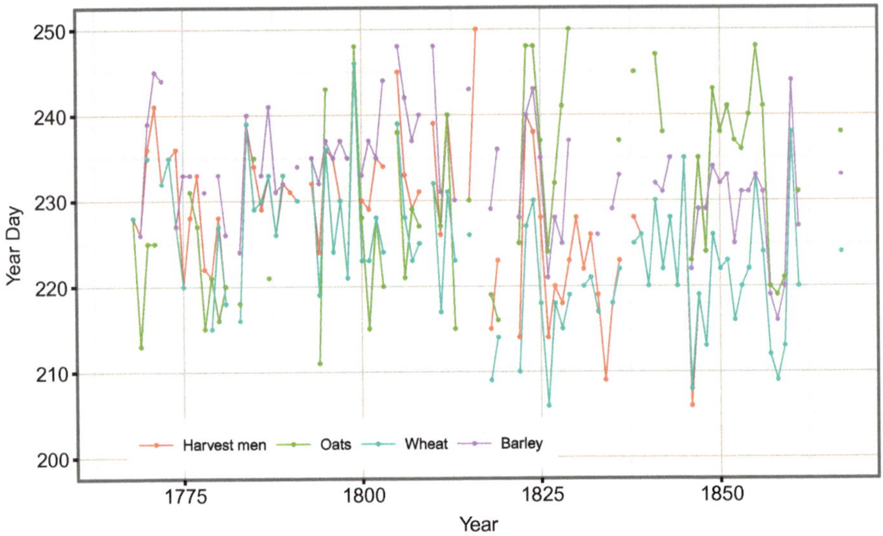

Fig. 4.2 Langham harvest 1768–1867: start of harvest and first day of cutting grain crops

Stephen Frost.[26] This practice gained popularity in the early nineteenth century[27] and would have advanced the harvest date somewhat.[28]

Since the Langham series 1768–1867 is divided by the shift of the mean value at 1817 into two sub-sets, it must be decided which part of the series can be used for

[26] According to Collins, Harvest technology, 460 this could add eight to ten days to the cutting season.

[27] Collins, Harvest technology, 456. He mentions that cutting at dead-ripe stage could occur in the eighteenth century.

[28] In some cases even cutting the corn green occurred. It is mentioned already in Young, General view of the agriculture of the county of Norfolk, 300 which was published 1804. Anonymous, The practical Norfolk farmer, 104–7 condemns the practice of cutting wheat green, which in the years leading up to 1808, the book's publication date, had been the custom of many farmers. Stephen Frost, too, notes in his diary in the week of Monday, 24 August 1795: 'Alway [sic] begin to Cut Wheat Green and you may all other Corn before it is too ripe, which prevent [sic] much spoiling, that often happening [sic] otherwise,' NRO, MC 120/26. However, 1795 was a cold and wet growing season, so that Stephen Frost might simply have had no other option, than to cut his wheat green because ripening was delayed. Cutting corn green is mentioned only a few times in Stephen Frost's diaries up to 1816: 1800 (part of the wheat at the beginning of September), 1802 (barley) and 1805 (oats). His successors reverted to it in 1818 (barley), 1823 (barley), 1836 (wheat), 1838 (oats), 1845 (wheat), 1848 (part of the barley at the end of August) and 1855 (wheat). Out of these years the harvests in 1795, 1805, 1823, 1845 and 1855 were late harvests, not starting before the later part of August. The part of the corn that was cut green was also cut late in 1800 and 1848. Therefore it rather appears that at Langham farm cutting green was practised not as a principle, but only in years of dull summers preventing the crops from reaching maturity even by mid- to late August. In 1845 Stephen Frost Rippingall describes such a situation: 23 August. 'Corn very green from cont[inuous] rain and want of sun.' He began the harvest that day, NRO, MC 120/72. It is likely that the chances for the corn ripening properly after the end of August were diminished due to the advanced season, so that the farmers were actually left with little choice.

calibrating the medieval harvest dates in Chap. 5. Comparing the mean value of the harvest date of the medieval East Anglian composite group (mean of year days 226.69) created by combining all the available harvest data for that period (see Table 5.2) to the mean values of the two Langham sub-series, a strong link between the grain growing conditions on the medieval manors and Langham 1768–1816 (mean of year days 228.45) is evident. Farming at Langham in the earlier part of the series is closer to the Middle Ages with regard to time, than the later part, and the general conditions of farming – as the preference for certain weekdays to begin working or the reaping of wheat – bear also a greater resemblance to the late medieval situation before the farm was handed over by Stephen Frost to Thomas Rippingall in 1817.

4.2 Fritton Estate

Fritton estate was held by Thomas Howes of Morningthorpe, it is situated about 15 km south of Norwich (Fig. 5.1). A continuous series of harvest dates for the years 1803 to 1828 is recorded in farming note-books of the estate (Fig. 4.3, Table 4.1).[29] In the Fritton farming diary a general start of the harvest is announced, which sometimes includes the pea and oat harvest. Where information is available, the wheat harvest often begins a few days later (about two days) than the general harvest. In other cases the two dates coincide. Until 1809 all, except one harvest, began on a Monday, during 1810–1813 all started on a Wednesday. Thereafter the weekdays for the commencement of the grain harvest at Fritton estate were highly variable.

4.3 Snettisham

A small and non-continuous collection of grain harvest dates comes from the lands belonging to Nicholas Styleman at Snettisham. Snettisham is a village in northwest Norfolk close to the coast of the Wash and about 5 km southwest of Sedgeford, one of the important manors of Norwich Cathedral Priory in the Middle Ages (Fig. 5.1). The Snettisham grain harvest dates are given in the journals of Nicholas Styleman between 1809 and 1827[30] and number twelve dates (Fig. 4.3, Table 4.1). The information comprises general references to the start of the harvest as well as references to the cutting of wheat, where both types of data overlap, they are identical.[31] No day of the week is over-represented amongst the grain harvest dates.

[29] Farming diaries for Fritton estate by Thomas Howes of Morningthorpe 1802–1827: NRO, MC 150/52/1-2. The entries were not made daily, but weekly.

[30] Nicholas Styleman, Journals 1809–1813, 1815–1827: NRO, LEST/LA 14 28, harvest dates in journals NRO, LEST/LA 15-16, 18–22, 24–28.

[31] The few times that oats are mentioned in the journal, they were usually harvested before the general harvest or the cutting of wheat.

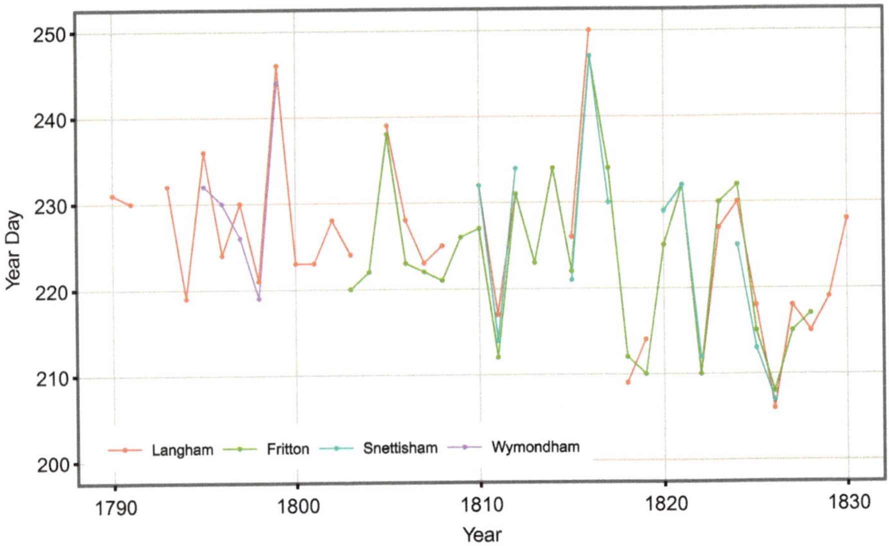

Fig. 4.3 Langham, Fritton, Snettisham and Wymondham 1790–1830: harvest date

4.4 Wymondham

The grain harvest dates found in the farming journal of Randall Burroughes constitute a series of just 5 years, 1795–1799[32] (Fig. 4.3, Table 4.1). Burroughes farmed in the parish of Wymondham, a market town 14 km southwest of Norwich (Fig. 5.1). He began his activities very close to Wymondham, in 1798 he acquired new land at Stannards Farm and focused his farming activities on this location.[33] The information for the commencement of the harvest is either given as a general remark on the start of the harvest or as a reference to the cutting of wheat. The grain harvest dates are distributed equally over the days of the week. The period 1795–1799 includes several cold years and late harvests, therefore the mean value of this short run of harvest dates is very high.

4.5 Medieval Versus Early Modern Grain Harvests

A comparison of the statistical properties of the grain harvest data from Norfolk around 1800 and the series of the medieval manors situated nearest to the modern farms in Table 4.1 shows the medieval and modern data to be in close agreement.

[32] Wade-Martins, Williamson, Randall Burroughes, 65–125. The journal runs from end of 1794 to the end of 1799, entries were not made daily.

[33] Wade-Martins, Williamson, Randall Burroughes, 3.

This applies to the early part of the Langham series 1768–1816 and Hindringham-Hindolveston (Cromer Ridge), to Snettisham and Sedgeford-Gnatingdon-Thornham (Northwest) as well as to Fritton and the medieval manors around Norwich (Norwich). Often the modern data have a slightly higher mean value, because they include the bad and cold years around 1800 and 1816, the ‘year without a summer’ after the volcanic eruption of Mount Tambora in 1815, whereas the medieval data often omit years marked by very cold growing seasons. As this comparison underlines, the medieval and eighteenth to nineteenth-century data are related and therefore comparable.

The modern harvest date series are highly correlated on the interannual basis (Fig. 4.3). Indeed they almost always display identical trends, and the spread of the dates within a year is reduced, when compared to the medieval data. There is no doubt about the modern harvest dates reacting to the same influence: the mean temperature during the growing season. This relationship forms the basis for the temperature reconstruction that is described in the following chapter.

Chapter 5
A Reconstruction of Medieval April–July Temperatures for East Anglia

A reconstruction of late medieval temperatures representative of the East Anglian region has been demonstrated by Pribyl et al. (2012). The medieval harvest date series used in the reconstruction has been discussed in Chaps. 2 and 3; those data were calibrated with the harvest dates 1768–1816 from Langham alongside the Central England Temperature series (CET).[1] In this chapter, a summary of the methods for the temperature reconstruction is provided. Since the initial reconstruction of the medieval temperature series, additional data have been found which result in a more complete series. Hence an updated temperature reconstruction is presented in this chapter.

5.1 Reconstruction Methodology

The temperature reconstruction follows the widely-used calibration-verification approach.[2] For this purpose the medieval data are expressed as year days from 1 January, and the values are adjusted from the Julian to the Gregorian calendar. Over the period 1256–1431 the number of harvest dates per year varies considerably, after c.1290 it is more stable (Fig. 3.5), and in total 645 harvest dates are available for late medieval East Anglia. These harvest dates come from 50 manors and each series of medieval dates contains a substantial number of missing values; some manors only yield very few data (Table 5.1). Hence the medieval harvest information has to be formed into a composite harvest date series for the whole of East Anglia by amalgamating the manors in regional groups that share broadly similar environmental conditions (climatic factors, soil conditions, altitude) (Fig. 5.1 and Table 5.1). For the twelve distinct regions the minimum value per year of each region was used (i.e. the earliest harvest date); substituting the minimum value with

[1] Manley, The mean temperature of central England, 1698–1952, 242–261, and idem, Central England temperatures: monthly means 1659–1973.

[2] Brázdil et al., European climate of the past 500 years. New challenges, 15–18.

K. Pribyl, *Farming, Famine and Plague*, DOI 10.1007/978-3-319-55953-7_5

Table 5.1 Regional groupings of the medieval manors

Region	Latitude	Longitude	Region	Latitude	Longitude
Northwest			**Taverham**		
Gnatingdon	52°53′	0°34′	Attlebridge	52°42′	1°08′
Heacham	52°54′	0°30′	Bawburgh	52°38′	1°10′
Hunstanton	52°56′	0°29′	Taverham	52°41′	1°12′
Ringstead	52°55′	0°32′			
Sedgeford	52°53′	0°33′	**Plumstead**		
Thornham	52°57′	0°34′	Plumstead	52°38′	1°24′
			South Walsham	52°39′	1°29′
Cromer Ridge					
Hindolveston	52°49′	1°00′	**Norwich**		
Hindringham	52°52′	0°56′	Arminghall	52°35′	1°18′
			Catton	52°38′	1°17′
Northeast			Costessy	52°39′	1°12′
Calthorpe	52°51′	1°13′	Cringleford	52°36′	1°14′
Hevingham	52°45′	1°15′	Eaton	52°37	1°15′
Little Hautbois	52°44′	1°19′	Heigham	52°38′	1°16′
North Walsham	52°49′	1°23′	Lakenham	52°37′	1°17′
Scottow	52°46′	1°22′	Monks' Grange	52°38′	1°19′
Witton	52°49′	1°28′	Trowse Newton	52°35′	1°19′
Worstead	52°46′	1°24′			
			Southwest		
North Elmham			Fincham	52°37′	0°29′
Gateley	52°46′	0°54′	Gr. Cressingham	52°34′	0°43′
Kempstone	52°42′	0°47′	Hardingham	52°36′	1°01′
North Elmham	52°44′	0°56′			
			Southeast		
Norfolk Broads			Aldeby	52°28′	1°36′
Hemsby	52°41′	1°41′	Hardley	52°33′	1°30′
Martham	52°42′	1°37′	Shotesham	52°31′	1°18′
Ormesby	52°40	1°41′			
Scratby	52°40′	1°42′	**Suffolk**		
			Akenham	52°06′	1°07′
Flegg			Denham	52°18′	1°13′
Ashby	52°41′	1°34′	Henley	52°07′	1°08′
Flegg	52°40′	1°36′	Hinderclay	52°21′	0°58′
Ludham	52°42′	1°31′	Redgrave	52°21′	1°00′

The allocation of manors to a regional group is based on geographical proximity and mean values in grain harvest date (where available). Adapted from Pribyl et al. (2012)

the mean value of the dates results only in minor differences. Linked to the different environmental conditions the average of the individual regional series varies across East Anglia, e.g. the group of manors around Norwich tended towards early harvests, the group of harvest dates from Cromer Ridge was normally the latest.

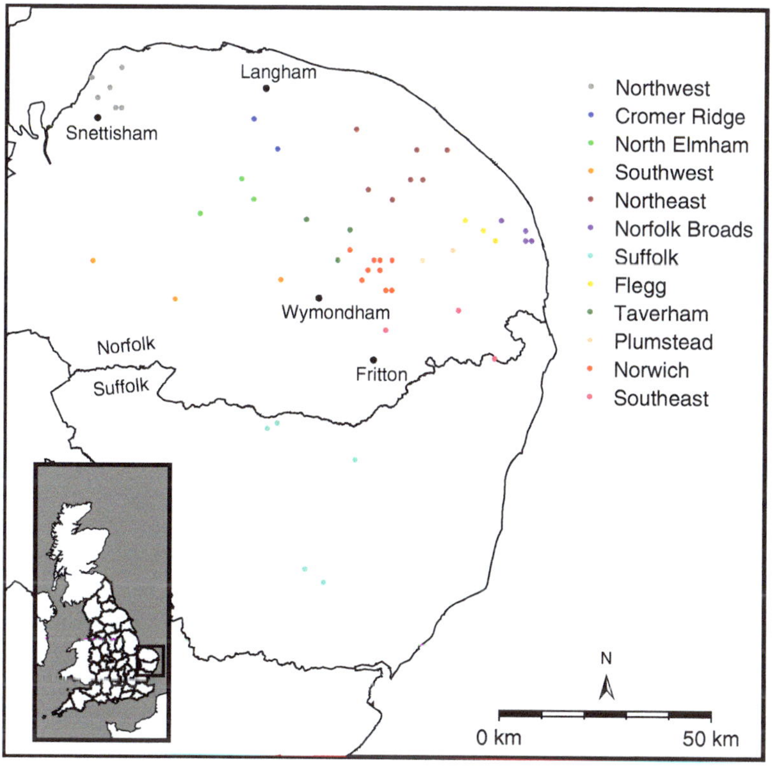

Fig. 5.1 Harvest date series: medieval regional groups and the modern comparison data from Langham. Also shown are the locations of the other farms (Fritton, Snettisham and Wymondham) providing eighteenth and early nineteenth-century harvest dates. For the composition of the medieval regional groups and the coordinates of the individual manors, see Table 5.1. (Adapted from Pribyl et al. 2012)

The most complete and longest series of harvest dates comes from the Northwest region which includes the manors of Sedgeford and Gnatingdon. To achieve a homogenous composite series for East Anglia, the other regional series were regressed to the level of the Northwest group, and the missing values in the Northwest series were filled in with regressed data of the other regional groups. The information was chosen in a hierarchical manner based on the strength of the relationships (r^2 value of each group) relative to the Northwest group. The resulting East Anglian composite series provides harvest dates for 147 years between 1256 and 1431. Forming the composite series by regressing the data to a specific regional group is preferable to a composite series consisting simply of the mean values of the regional data available per year, since the latter method would suppress the interannual variability. However, apart from extremes, it was the interannual variability that was most keenly felt by the medieval people and that impacted directly on agricultural success; hence this time series component is essential for the historian.

The medieval Northwest group and the home of the modern comparison series, Langham, are both close to the North Sea and are separated by a distance of about

Table 5.2 Medieval East Anglian composite series and the modern comparison series (Langham): statistical characteristics

	East Anglian composite series 1256–1431	Langham 1768–1816	Langham 1818–1867
Mean	226.82	228.45	220.17
SD	7.10	7.52	7.60
Min	204	215	206
Max	247	250	238
N	147	44	42

Adapted from Pribyl et al. (2012)

25 km. The medieval Cromer Ridge group is geographically closer to the modern comparison series, however, Cromer Ridge has merely half of the number of harvest dates of the Northwest group, making a regression to the level of Cromer Ridge statistically more risky. Langham also shares actually more geo-physical properties with the Northwest group than with Cromer Ridge. The Northwest region and the village of Langham are surrounded by similar, generally well draining soils, whereas Hindringham and Hindolveston of the Cromer Ridge group are situated on mainly loamy or clayey soils, which are subject to slight seasonal waterlogging.[3] The Cromer Ridge group is not only more vulnerable to high rainfall levels, but also lies on slightly higher ground than Langham and the Northwest group.[4] As a result the growing season is shorter for the Cromer Ridge group (<260 days), than for Langham and the Northwest group (>270 days).[5] Concerning mean values the early part of the Langham series 1768–1816 (mean value of year days 228.45) and the region Cromer Ridge (mean value of year days 228.93) are actually almost identical, whereas the value for the medieval Northwest region (mean value of year days 226.55) is c. two days lower than at Langham (Table 5.2). This tendency of the modern Langham harvest data towards slightly later harvests than in the medieval Northwest group can indeed be explained by the coinciding of the Langham series with a period known to have been cold, particularly around 1800, probably partly due to the Dalton Minimum, and also by the modern comparison series catching the disastrous impact of the eruption of Mount Tambora (1815) in the very cold summer

[3] Soils of England and Wales: Sheet 4 Eastern England, the fields of Sedgeford-Gnatingdon are dominated by shallow well drained calcareous sandy and coarse loamy soils over chalk or chalk rubble belonging to the soil association of Newmarket 2, and to a small extent include deep well drained coarse loamy, coarse loamy over clayey and sandy soils (Barrow). At Langham similar soils are involved (Newmarket 1 and 2, Barrow). The soils at Hindringham and Hindolveston are mainly deep loamy with slowly permeable sub-soils prone to slight seasonal waterlogging (Burlingham 1 and 3), or slowly permeable seasonally waterlogged fine loamy over clayey soils (Beccles 1 and 2).

[4] According to the Ordnance Survey: 132 North West Norfolk and 133 North East Norfolk, Langham and fields are at 30-40 m, the Northwest region at 30-50 m, Hindringham at 50-80 m and Hindolveston at 50-70 m.

[5] MAFF, Sheet 125 Fakenham, 4–5 and MAFF, Sheet 124 King's Lynn, 4.

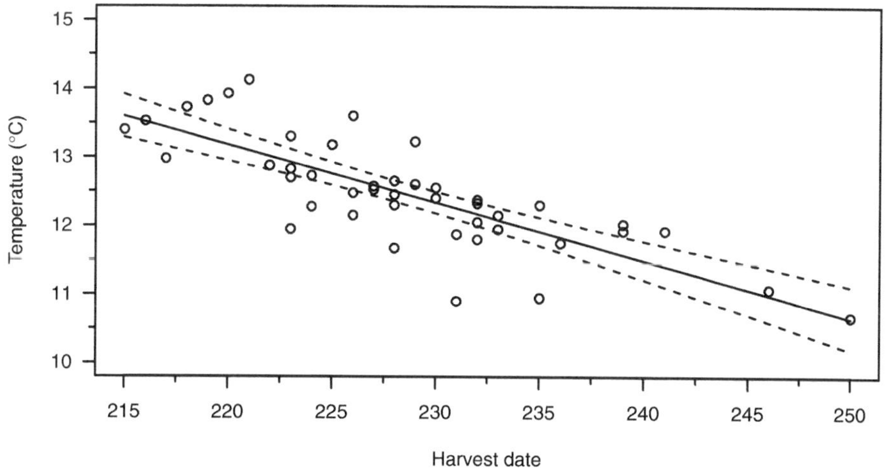

Fig. 5.2 Scatterplot of the Langham 1768–1816 harvest dates and CET temperatures. Plotted are the mean April–July temperature derived from the CET series, the linear regression line and the 95% confidence interval. (Taken from Pribyl et al. 2012)

half year 1816. For the aforesaid reasons it is preferable to regress to the level of the Northwest region.

As shown in Table 5.2 the statistical properties of the medieval East Anglian composite series regressed to the Northwest group are close to those of the Langham harvest date series 1768–1816. Consequently the correlation of this earlier part of the Langham harvest date series (n = 44) with the CET series can form the basis of the medieval temperature reconstruction.

The main factor determining the harvest date at Langham is the mean temperature during the period April to July, which is responsible for 62% of the variance in the harvest dates ($r = -0.79$, Fig. 5.2). However, correlations remain high, when March and/or August temperatures are also included in the analysis. March temperature is connected to the onset of the growing season. The cold northerly or easterly winds which are frequent in Norfolk, can delay plant growth in early spring.[6] It is indeed the temperature during the grain growing season that displays a strong linear relationship with the grain harvest date. A substantial percentage of the unexplained variance is probably connected to the sowing time and the rainfall during the growing season.

The reconstruction of April to July mean temperatures was achieved by using the harvest dates from the medieval composite series and the regression coefficients established over the period 1768–1816. As the break in the Langham series only affects the mean value, the data over the years 1818–1867 were used to verify the prediction of temperatures (Fig. 5.3). Over that period the correlation between recorded and reconstructed temperatures stood at $r = 0.86$ ($r^2 = 0.73$).

[6]This problem was already described in 1796 by Kent, General view of the agriculture of the county of Norfolk, 10. It is a Norfolk-wide phenomenon, see for example MAFF, Sheet 124 King's Lynn, 4 or MAFF, Sheet 126 Norwich, 5.

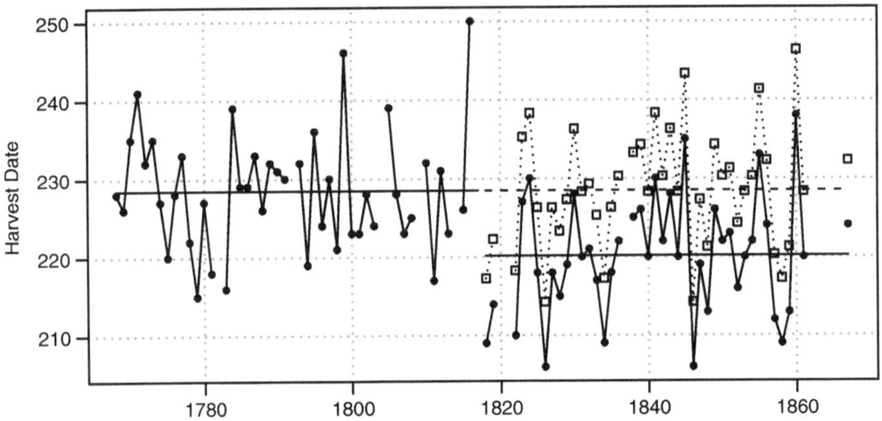

Fig. 5.3 Langham harvest dates with adjustment of the later part of the series. The *dotted line* represents the adjusted mean value of the later part of the series (1818–1867) so it equals that of the earlier part (1768–1816). The *horizontal lines* indicate the means of the early and late parts. (Taken from Pribyl et al. 2012)

5.2 Reconstructed Medieval April–July Mean Temperatures

The reconstructed April–July mean temperatures (Fig. 5.4) show a long term trend of decline over the period 1256–1431. Overall the initial version of the temperature reconstruction as presented in Pribyl et al. (2012) and the new version are very similar despite adding about 30 new harvest dates. Most of these harvest dates fall into years for which data already existed, they either confirm these data or belong to regional groups such as 'Suffolk' which due to its distance to the Northwest region and its greater geographical scope, can only supply the harvest date for the reconstruction, if no date is available from any other region. Five previous gap years could be filled in this revised reconstruction: 1269, 1280, 1307, 1330 and 1351.

Before 1335 temperatures above 13 °C were much more common than in the later decades of the reconstructed period (Fig. 5.4, Appendix 4). Until c.1315 springs and early summers were rarely colder than 12 °C. The hottest spring-summer in late medieval England, that of 1361 (14.5 ± 1.0 °C), was an outlier during the cooler decades 1350–1431. This value appears exceptional when compared to the more than 350 years of modern instrumental data (Fig. 5.5). However, it extends beyond the range for which the linear regression relationship was established in the 1768–1816 period. The lowest growing season temperature was reached in 1428. The coldest year of the reconstruction comes towards the end of the study period, and indeed the frequency of growing seasons colder than 12 °C also increased over time; they were common after c.1360. During the reconstruction period the average April–July temperatures decreased from about 13 °C at the beginning to 12.4 °C at the end. Within the context of this long-term trend, April–

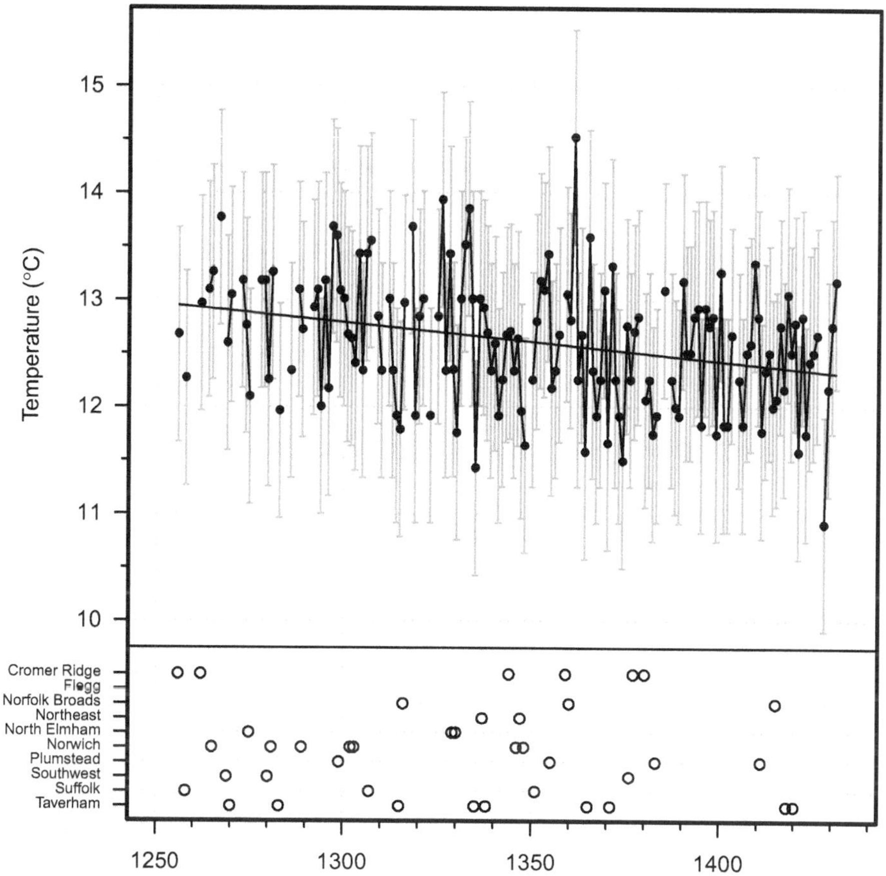

Fig. 5.4 The reconstructed April–July mean temperatures. The uncertainty range for each reconstructed temperature (the grey error bars) represent ±2S.E. The *circles* from the lower panel indicate values that have been filled with data from other regional groups regressed to the Northwest series. (Adapted from Pribyl et al. 2012)

July mean temperatures considerably higher than the preceding or following years occurred in 1267, 1297, 1298, 1307, 1318, 1326, 1332, 1333, 1361, 1365, 1371, 1385, 1390, 1400, 1409 and 1431. Grain growing seasons characterized by very low temperatures fell to the years 1275, 1283, 1294, 1314, 1315, 1319, 1323, 1330, 1335, 1348, 1364, 1370, 1374, 1421 and 1428. However, the decrease of the April–July mean temperatures is not steady over the study period. In the years c.1300–1310, 1326–1334, in the 1350s, the 1390s and the late 1410s local highs with respect to temperature are perceptible; whereas lows occurred in mid-1290s, c.1313–1323, in the late 1340s, from the mid-1360s to the mid-1370s, in the 1380s, the first decade of the fifteenth century and the early 1420s.

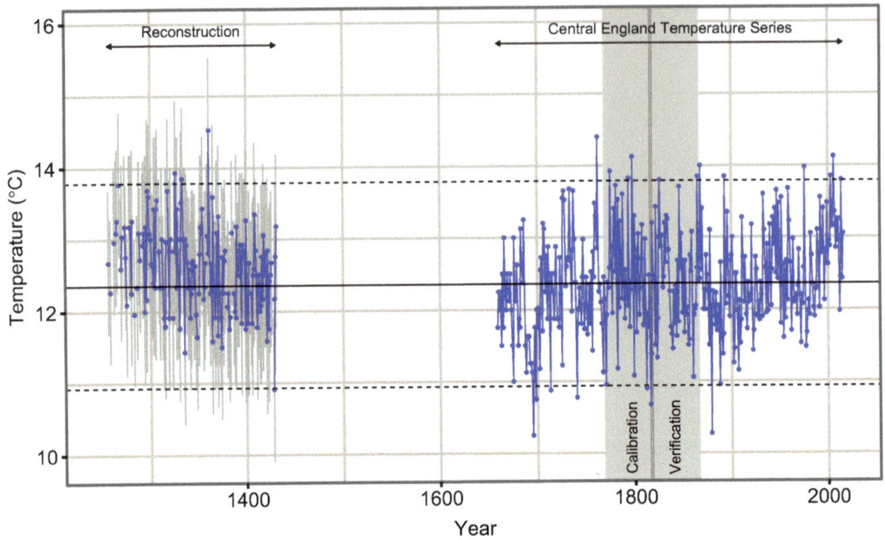

Fig. 5.5 The reconstructed temperatures in relation to the instrumental CET data. As with Fig. 5.4 the error bars represent ±2S.E. Displayed are the mean (*horizontal line*) and the ±2 standard deviations (*horizontal dashed lines*) of the instrumental data (1659–2016). Calibration (1768–1816) and verification (1817–1867) periods are also indicated. (Adapted from Pribyl et al. 2012)

The interannual variability was also subject to changes. Until 1290 it was low, the majority of the growing seasons were warm with temperatures between 13 °C and 13.5 °C. Due to this low data density few data come in consecutive years, a conclusive analysis of interannual variability before 1290 is therefore not possible. Periods of low interannual variability returned between the mid-1330s and the late 1340s, in the second half of the 1350s, from the mid-1370s to the early 1380s, throughout most of the 1390s, and in the 1410s. Shifts of up to 1 °C, in the medium range of the interannual variability, fell to c.1290–1315, c.1405–1411 and the early 1420s. Remarkably high levels of interannual variability mark then the years 1315–1335 and 1360–1375; the year 1428 interrupts the comparatively calm time of the second half of the 1420s and the early 1430s; in those times changes in growing season temperatures of 1.5 °C or more are possible. Due to the grain crops being annual plants, that means their growing speed was not influenced by preceding years, and the grain harvest date being closely linked to the phenological phase of the grain ripening, the interannual variability in the growing season temperatures is captured very well in the grain harvest data.

When comparing the reconstructed medieval April–July mean temperatures with the instrumental temperature observations of the CET series 1659–2016, the mild conditions of the Medieval Climate Anomaly, and the transition towards lower temperatures in the mid-fifteenth century in association with the onset of the Little Ice Age are clearly visible (Fig. 5.5). The rate of interannual variability in the Late Middle Ages was comparable to the conditions in the instrumental measurements.

5.3 Comparison with Other Documentary Reconstructions

The medieval reconstruction provides the earliest annually-resolved temperature series for the British Isles. In this section the reconstruction is compared against two temperature indices based on documentary information: the monthly indices by Ogilvie and Farmer for England 1200–1432[7] and the seasonal indices by van Engelen, Buisman and Ijnsen for the Low Countries 751–2000.[8] Neither series included the evidence about the East Anglian grain harvest dates and they therefore offer a useful independent comparison.

The Ogilvie and Farmer indices can be considered an extension of Lamb's indices.[9] The indices were made on a monthly level; unreliable information was removed and new weather information included. The indices cover the spectrum from −3 (for temperature: very cold, for precipitation: very dry) to 3 (very warm, respectively very wet). Unfortunately due to the nature of the medieval documentary data, many gaps remain. Medieval narrative sources, such as chronicles, and the direct references to weather they supply, focus on extreme events. Normal conditions were generally not recorded, so that it remains unclear if gaps in the available data represent normal conditions or if references to more extreme weather have simply not survived the centuries. The grain harvest date series does not suffer from this problem, although gaps also occur. With regard to the indices, especially those on a monthly level, the gaps lead to difficulties when the creation of supra-monthly or seasonal indices based on them is attempted. Hence, the available information in spring and summer temperature is too scarce in the Ogilvie and Farmer indices, so that a useful quantitative comparison with the reconstructed April–July mean temperatures is not possible.

The work of van Engelen et al. for the Low Countries provides a more complete set of temperature indices based on documentary sources for this period. They cover the summer (May–September) and winter season (November–March). Between 1256 and 1431 information is available for 158 summers. It ranges from 1 or I (extremely cool respectively 'obviously cool') to 9 or III (extremely warm respectively 'obviously warm'). The data from the Low Countries are dense, they are also geographically close enough to East Anglia to allow a meaningful comparison. Even though the time of the year represented in the summer index from the Low Countries is not identical with April to July, the Spearman rank correlation between the two datasets stands at rho = 0.47. Summers that are identified as very warm in both the reconstructed East Anglian growing season temperatures and the summer season index from the Low Countries are 1267, 1297, 1304, 1326, 1333, 1352, 1361, 1371, 1385, 1390 and 1400; summers identified as very cold are 1275, 1283, 1294, 1314, 1315, 1330, 1335, 1406 and 1428 (Fig. 5.6a).

[7] Ogilvie, Farmer, Documenting the medieval climate, 124–128. The indices set by Ogilvie and Farmer for temperature between 1256 and 1431 are sparse and mostly relate to winter.

[8] van Engelen et al., A millennium of weather, winds and water in the Low Countries.

[9] Lamb, Climate. Past, present and future, vol. 2.

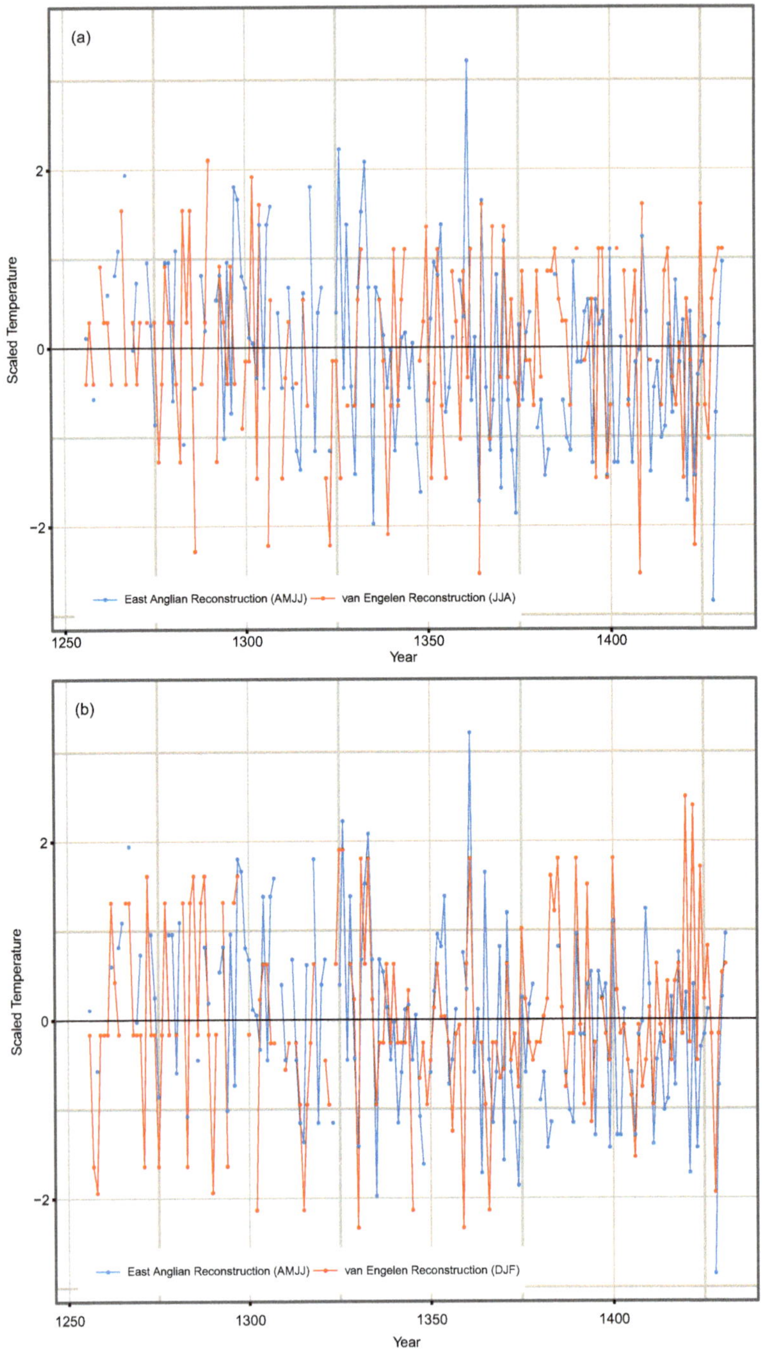

Fig. 5.6 A comparison of the van Engelen temperature series for the Low Countries and the reconstructed East Anglian temperature series for April–July. (**a**) Low Countries June–August and (**b**) Low Countries December–February. Data are presented in normalized units. The van Engelen indices covers May–September and November–March, the temperatures series, however, covers June–August and December–February. (Adapted from Pribyl et al. 2012)

In the light of the context provided by the van Engelen et al. index it becomes apparent that before c.1330 summers classified as very cold in the Low Countries index are poorly covered by the East Anglian data. This includes the years 1290 and 1322. These are categorized as very wet in the precipitation index for England by Ogilvie and Farmer, hence the cold weather most likely affected not only the continent but also the southern and eastern parts of the British Isles. The summer half years 1294 and 1330 are marked by very adverse weather in the Low Countries index, Ogilvie and Farmer note high rainfall levels again, and merely one East Anglian datum is available for each year confirming the very cold conditions. The time of the Great Famine 1315–1317 is also badly represented in the East Anglian series. Very few data are available for 1315 and 1316 and none for 1317. This lacuna in the East Anglian information during these periods of crisis is rooted in the low supply of manorial accounts available for years of bad weather and subsistence crises in East Anglia. Particularly the normally very reliable accounts for the prior's manors of Norwich Cathedral Priory are missing.

On a decadal level the datasets display broadly similar albeit offset trends before 1325 and between c.1360 and the mid-1370s (Fig. 5.6a). From 1325 to the late 1350s and also 1395–1420 the conditions in East Anglia and the Low Countries match closely. The trends diverge considerably in the 1380s and after 1420, but in the 1380s the East Anglian data density is too low to allow a meaningful comparison. Whereas the long-term trend in the data from East Anglia shows increasingly cool conditions over the study period, this development is not mirrored in the Low Countries.

In two periods the datasets diverge, even though data density is high. This is the case to a moderate degree between the late 1350s to c.1370, and to a higher degree after 1420. The datasets from East Anglia and the Low Countries do not represent the same period of the year, but overlap only between May and July, hence the cause for the break-up might lie outside these months. The months August and September are not included in the grain growing season, hence the focus here lies on April and potentially March, which are not covered by the summer index from the Low Countries. A winter season index, however, was also compiled; it is valid for November to March. During the study period 136 values are available in the winter index, and there are indeed occasions when the East Anglian data display a stronger connection to the winter instead of the summer conditions in the Low Countries (Fig. 5.6). The warm East Anglian growing seasons of 1278, 1316, 1365 and 1409 (Fig. 5.6b) are not mirrored in the summer index of the Low Countries, but did follow on from mild winters. For cold conditions in East Anglia a similar constellation involving preceding hard winters occurs in 1364, 1367, 1399 and 1423. An average East Anglian growing season is associated with a cold summer across the North Sea following on from a mild winter in 1302, and in 1420 and 1424 the situation was reversed. This connection between the winter index in the Low Countries and the East Anglian growing season mean temperature is due to the role of the length of winter. The onset of the growing season in March and April can be held up by cold conditions in early spring, from this interference follows a delayed harvest. Mild winters and early springs result in an early start of the growing season. Independent documentary evidence from England exists for the long winter 1363–1364 which was extremely hard and lasted well into March,[10] and for the winters 1422–1423 (Sect. 8.4) and 1423–1424.[11]

East Anglian data showing warmer conditions in the 1360s than those prevailing in the Low Countries are possibly related to a sequence of mild winters (except 1363–1364) that conceivably resulted from a predominance of westerly conditions across northwest Europe. At least during the later parts of the decade mild winters also prevailed in Central Europe.[12] Abundant precipitation in the summers[13] and partly also in the winters[14] add substance to this hypothesis. An early start of the growing season would have led to an earlier harvest, even though the summers were merely average. In addition, in 1369 the weather was marked at least by phases of warmer and drier weather previous to harvest time, but during the harvest it turned, and rain dominated the last part of the summer, influencing the summer index for the Low Countries but not the April–July temperature reconstruction (Sects. 6.3 and 8.4).

High spring-summer mean temperatures in East Anglia in connection with early onsets of the growing season and mild winters are distributed over the whole study period. Low temperatures in the East Anglian series without an adequate reflection in the Low Countries, but connected to severe winter conditions, however, only occur after c.1350 and cluster in the 1420s. In the Low Countries around 1420 summers were getting warmer and winters were cooling. This suggests that the long winters 1422–1423 and 1423–1424 were only extreme, but not atypical for the 1420s. Norfolk is particularly vulnerable to cold springs, because the northerly and easterly winds affect the county greatly, especially the land on the north coast and the manors of the Northwest group,[15] which supply the majority of the harvest dates for that decade. The high summer temperatures could not compensate for the delay of the growing season, thus the cool early springs noticeably cooled the reconstructed April–July mean temperatures. A shift in seasonality, that means a delay of the onset of the growing season as in the 1420s, was typical for the Little Ice Age and may be connected to the cooler conditions that have been associated with the Spörer Minimum.

The cause of the divergence of the data from East Anglia and the Low Countries in the 1380s cannot be determined; the growing season in East Anglia appears as much cooler than the summers in the Low Countries, particularly around 1380. In the Low Countries the winters were mild, although the central European winters are

[10] Ogilvie, Farmer, Documenting the medieval climate, 127.

[11] Titow, Le climat à travers les rôles de comptabilité, 338. Most references for the winter 1423–1424 are to rainfall of long continuation and subsequent flooding, so the winter was very wet. Overton in Hampshire, however, mentions tempests of snow and rain which continued for a long time, so the winter temperatures were probably below average.

[12] Pfister et al., Winter severity in Europe, 104.

[13] Ogilvie, Farmer, Documenting the medieval climate, 127, see Sects. 6.3, 8.4 and Fig. 7.5.

[14] Ogilvie, Farmer, Documenting the medieval climate, 127.

[15] MAFF, Sheet 124 King's Lynn, 4.

classified mostly as average or cold.[16] Documentary evidence from England for late springs survives only for 1388,[17] however, mostly information about long winters and delayed growing seasons comes from manorial accounts, and few of them are available for that period. Possibly in some severely hot and dry summers the vegetation development was hindered by drought, and drought was a feature of the summers in the mid-1380s in England (Sect. 8.3).[18]

5.4 Comparison with William Merle's Weather Diary 1337–1344

From 1337 to 1344 information on weather is available that is more continuous and detailed than the narrative or administrative sources: the 'Consideraciones temperiei pro 7 annis' by William Merle, one of Europe's oldest weather journals.[19] This long and detailed record of weather was most likely connected to Merle's work on predicting the weather.[20] Merle was the rector of Driby in northern Lincolnshire from 1331 onwards and probably a fellow at Merton College in Oxford after 1335; he died in 1347.[21] Merle's weather observations were partly made at Oxford and partly in Lindsey, northern Lincolnshire. He regularly recorded the weather, first weekly and after November 1339 almost daily. He noted all weather, not only extremes, as is generally an inherent problem of narrative sources. Towards the end of the journal Merle also tried to establish the spatial coverage of some weather events.

The weather diary offers parallel data to the East Anglian temperature series for the seven summer half years 1337–1343, a phase in which in the reconstruction as well as in the summer indices by van Engelen et al. for the Low Countries (2001) saw mostly average conditions and an extraordinary low interannual variability: the English reconstructed temperatures varied only on the scale of 1 °C (Fig. 5.7). Extreme events in the summer seasons of this period are also absent in Britton's weather compilation.[22] Nonetheless a year-to-year comparison between the Merle data and the reconstructed temperatures can prove instructive. A general

---

[16] Pfister et al., Winter severity in Europe, 104.

[17] Ogilvie, Farmer, Documenting the medieval climate, 127–128.

[18] Pfister, Variations in the spring-summer climate, 69.

[19] The oldest weather journal is also from England and covers the period of August 1269 to February 1270, Long, The oldest European weather diary?, 233–234. The references were written in the margin of an astronomical calendar for 1269 and might have been recorded by Roger Bacon. Another English weather diary, again in the form of notes taken on the margin of an astronomical calendar, survives for October to December 1439, Mortimer, William Merle's weather diary, 42.

[20] Meaden, Merle's weather diary and its motivation, 211, Mortimer, William Merle's weather diary, 42–43.

[21] Mortimer, William Merle's weather diary, 42–43. Driby is in the Lindsey district and within five miles of the northeast coast.

[22] Britton, Meteorological chronology, 138–140.

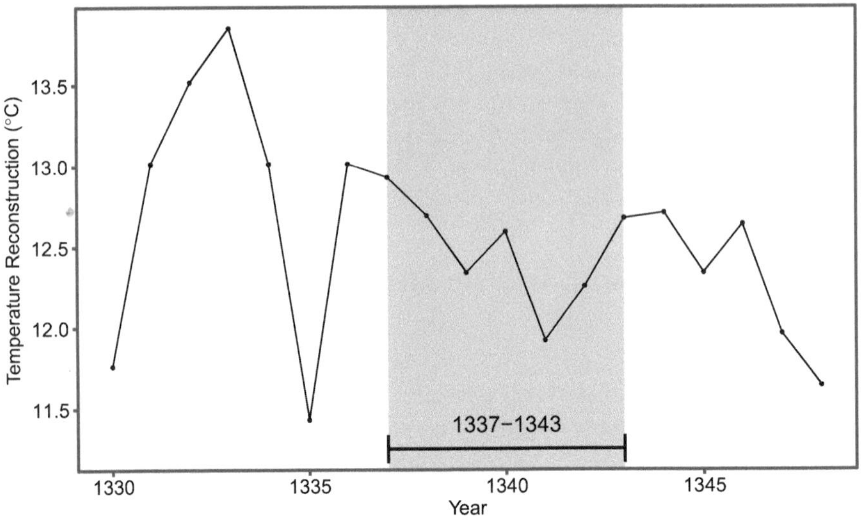

Fig. 5.7 Reconstructed April–July mean temperatures in East Anglia 1331–1348. The period within the *vertical lines* indicates the period covered by William Merle's weather diary 1337–1343

convergence between Merle's observations and data from manorial accounts from southern England and the London area has been demonstrated before.[23]

The two warmest reconstructed growing seasons overlapping with the Merle data are 1337 and 1338, though in the context of the whole series 1256–1431 these years are not extreme, but only somewhat above the long term average. In 1337 (see also Sect. 8.1) Merle describes April and May as moderately warm. Whereas April was wet, May was mostly dry and the dry weather lasted into July. June was hot, but July was not and this month also was rainy over long stretches.

During the summer 1338 heat was more dominant, it lasted together with drought throughout July and August. However, the spring had begun cold and hoar frost persisted throughout April. Rain came in the first week of May and in the second half of June. The dry parts of May and June were warm. The cold April slowed down vegetation growth and balanced out the warm end of the growing season, so that the mean temperature between April and July was on the whole average. The meteorological conditions during the growing season were excellent for grain cultivation, 1338 had one of the best harvests of the later Middle Ages.[24] Such a harvest success even found its way into the collective English memory. The most popular chronicles of the fourteenth and fifteenth centuries, the Brut, the Polychronicon and the London chronicles, all mention a very good farming year in the second half of the 1330s, some date it to 1337, but considering the grain price and Campbell's data

[23] Mortimer, William Merle's weather diary, 44–45.

[24] Campbell, Nature as historical protagonist, 299.

on harvest quantity[25] clearly 1338 is meant. Prices were also kept low by a scarcity of money and deflation. The Chronicle of London notes:

> Rex Edwardus Tertius, anno xj [...] Also in this yere was gret plente of vitaile, that a quarter of whete was sold at London for ij s; and a fat oxe for vj s. viij d.; and vj pegons for a peny: but natheles it was ful gret scarcste of money.[26]

Spring 1339 followed on the hard winter 1338–1339.[27] Some cold weather still marked early April and the second half of May and the first half of June were rainy, but at least during most of April and from mid-June onwards it was warm, respectively hot.[28] As a result, the reconstructed growing season mean temperature in 1339 was lower than in the preceding years, but still average. The bad harvest 1339, which was caused a substantial rise in the grain price, was not so much due to the weather during spring and summer, but rather to the weather at sowing time in the preceding autumn, which was extremely wet, and made the soil 'watery'. The cold winter – frost lasted for 12 weeks and started in early December – then changed the appearance of the saturated fields into that of sheets of ice, and by spring 1339 most winter corn had perished. According to Merle the summer season 1339 was marked by higher rainfall levels than 1337 and 1338, but April – after the long winter the time for sowing the spring corn barley and oats – was very dry, which led to reduced yields in these crops (Appendix 1).[29] Hence, the year following the harvest 1339 was dire for the common people, and the rise of the price of grain would have been more decisive if it would not have been the time of a severe shortage of bullion.[30] Although no clear line can be drawn from the increase in poverty and malnutrition, it is probably not a mere coincidence that in summer 1340, before the next harvest, when provisions were at their lowest ebb and most expensive, a widespread disease took hold in England. It caused people great pain and made them emit sounds resembling those of barking dogs.[31]

---

[25] Campbell, Nature as historical protagonist, 299.

[26] The Chronicle of London from 1089 [sic] to 1483, 56. A similar section is found in William Gregory's Chronicle of London, 80. The low prices of goods are also noted in the Brut, 292. A shorter version is contained in Higden, Polychronicon, vol. 8, 334 for spring 1339, but for the grain price with regard to the harvest 1338. The Polychronicon is the oldest text of the chronicles mentioned here.

[27] Merle, Consideraciones temperiei pro 7 annis, Titow, Evidence of weather, 397. This winter is also described as being long.

[28] Mortimer, William Merle's weather diary, 43. Further north in Staffordshire, the summer was wet, see Lynam (ed.), Croxden Chronicle, x. This description is obviously referring to the weather in late May and early June.

[29] Campbell, Great transition, 270–271.

[30] Murimuth, Continuatio Chronicarum, 88–89. Weather conditions are observed in detail in Merle, Consideraciones temperiei pro 7 annis, they entirely fit the description of Murimuth.

[31] Knighton, Chronicon, vol. 2, 36–37. Creighton, Epidemics in Britain, 59, considered the possibility of an outbreak of ergotism, but the harvest failure of winter grain in 1339 makes this unlikely. The editor of Knighton speculates about diphtheria, a throat infection, which could explain the barking voices, see Knighton, op. cit., 37.

The year 1340 (Sect. 8.1) followed the trend of increasing summer precipitation; the generally hot and dry weather was interspersed with a week of rain in May and two more wet weeks in July. However, the warm conditions were preceded by a cold and long winter. Merle reports frost well into April and occasional hoar frost throughout this month. As a result of the late start of the growing season the reconstructed East Anglian April–July temperature is average.

The coldest reconstructed growing season in the 1337–1343 period is 1341. Merle not only notes frost, snow and hoarfrost in the first half of April – his references to warm weather in the following months are few, only May had warm weather for more than a week – but Merle's references to rain are all the more frequent. This rather unpleasant growing season resulted in a late harvest, especially in northwestern Norfolk, where the cold April must have been felt keenly.[32] Although the cold and wet growing season did not impact severely on the harvest on a national level, the wheat harvest was poor on the Winchester manors in southern England.[33]

The trend of increased precipitation during the growing season continued in 1342. April and the first half of May are still characterized as mild or warm and dry by Merle. A few complaints about a dry spell in the Sussex accounts and the Pipe Roll of the Bishopric of Winchester refer to this period. Afterwards rainfall was frequent, especially in June and July.[34] In August precipitation levels remained high; southern England experienced some rainfall during the harvest.[35] Due to the prolonged dull and rainy period the reconstructed growing season temperature was fairly cool, but warmer than the previous year.

Rain was also a dominant feature in 1343. There was recurrent rainfall in April and May, the first three weeks of June witnessed light rainfall daily, the latter part of July was rainy and August also saw frequent rainfall, partly in the form of showers. According to Merle this rainfall at the end of July and later was a rather local phenomenon of northern Lincolnshire. Possibly this applies also to some of the precipitation in the earlier months, because some dry weather was then experienced in the Winchester area as well as in Sussex.[36] At the beginning of September Merle notes that in northern Lincolnshire he had not seen a single serene day in the preceding five weeks. The reconstructed growing season temperature was average, indicating that south of Lincolnshire was indeed not only wet and cool but also partly drier as warm air masses likely crossed over England.

---

32 Even the Pipe Roll of the Bishopric of Winchester, normally a very informative source for meteorological conditions, hardly records any dry weather in 1341. Only one manor refers to it, another mentions 'waters in summer', Titow, Evidence on weather, 398.

33 Campbell, Physical shocks, 25, Titow, Evidence on weather, 398.

34 Brandon, Late medieval weather in Sussex, 3, Titow, Evidence on weather, 398. Mortimer, William Merle's weather diary, 45 considers Merle's diary, which is reporting generally wet conditions, and the manorial accounts of the Winchester estates and from Sussex, which are recording some dry weather for the growing season, as incompatible for 1342. However, April and May were considered as part of summer by the medieval agriculturalist, especially if the weather was fine, hence the reported dry spells in the accounts can refer to a dry period in April and May.

35 Titow, Evidence on weather, 398.

36 Titow, Evidence on weather, 398–399, Brandon, Late medieval weather in Sussex, 3.

Repeated thunderstorm references by Merle underline the high propensity for convective cell development that summer, which is connected to higher temperatures.

Although summer and partly also spring 1342 and 1343 were gloomy and rainy, the rainfall in England was not excessive and no floods are reported. In central Europe, however, the situation was catastrophic: the continuous rain caused a major crisis. Southern and eastern Germany, Switzerland, Austria and the Czech Lands were hit by a sequence of partly extreme flood waves in 1342[37] and the Carpathian Basin in 1343,[38] including the worst central European inundations of the last millennium, the St Mary Magdalene's Flood in July 1342.

Thus the reconstructed temperatures in East Anglia and the weather evidence provided by William Merle correspond largely and complement each other. Although it is difficult to estimate rainfall amounts from the 'precipitation days' given by William Merle,[39] it is obvious that precipitation levels were comparatively low in 1337 and 1338 and above average in 1341, 1342 and 1343. The increased humidity in 1341 and 1342 coincides with cooler spring and summer reconstructions.

[37] Brázdil, Kotyza, History of weather and climate (1000–1500), 168. Rohr, Extreme Naturereignisse im Ostalpenraum, 226–228.

[38] Kiss, Floods and weather in 1342 and 1343 in the Carpathian Basin.

[39] Lawrence, The earliest known journal of the weather, 498–499, compared the average monthly frequencies of rain days 1337–1343 with those for 1901–1930 and concludes that for May to October the two periods are comparable, November and December 1337–1343 are average or slightly below the average in 1901–1930, January to April are also below average.

Chapter 6
Temperature Extremes 1256–1431: Independent Evidence and Context

6.1 Temperature Extremes and Agricultural Production

As described in the previous chapter, spring and early summer temperatures were falling over the course of the Late Middle Ages in England. Except in extremely cold and wet years this would not have created major problems for the agricultural production,[1] because in general in southern England the growing season is of sufficient length for the corn crops. Temperature determines the grain harvest date, but exerts only a minor influence on the grain yield.[2]

Extreme temperatures, however, indicate extreme meteorological conditions in general, precipitation included. Such extremes were of course dangerous for the agricultural sector, and could ruin the crops for an individual year or, worse, for a sequence of years. In England cool conditions during the growing season often go hand-in-hand with raised levels of precipitation, whereas warm spring-summer conditions are more likely associated with drier weather. The grain crops, wheat, rye, barley and oats, favour different meteorological conditions, and the severity of the impact of heat, cold, rain or drought is dependent on the timing. Nonetheless in England's maritime climate the primary risk factor for agriculture is cold and wet weather during the growing season. Sequences of cold years frequently correspond with times of high grain prices and even famine. These temperatures consequently indicate weather conditions detrimental to crop growth.

The phases of high interannual variability of the growing season temperature 1315–1335 and 1360–1375[3] overlap partly with periods containing very cold years and also correspond with high grain prices. The frequent shifts in weather patterns and the prevalence of cold and wet years raised the vulnerability of medieval

[1] Campbell, Four famines and a pestilence, 10.

[2] Weather patterns that correlate strongly with high or low grain yields are described by Titow, Evidence of weather, 363.

[3] The years are generally known to have witnessed weather conditions rendering farming difficult, Ogilvie, Farmer, Documenting the medieval climate, 127.

K. Pribyl, *Farming, Famine and Plague*, DOI 10.1007/978-3-319-55953-7_6

agriculture, because they prevented an adaptation of the agricultural production to the weather conditions.

Independent English documentary evidence for the extreme warm and cold reconstructed East Anglian spring-summers can not only provide more information on the meteorological conditions, but can also give an indication of the vulnerability of the agricultural sector in England during the medieval period and highlight wider societal impacts, particularly with regards to famine and susceptibility to epidemic disease. For the detection of poor harvests and price rises for grains the data by Phelps Brown and Hoskins, which have been reworked by Munro are used.[4] Apart from establishing the regional footprint of an extreme event, its spatial extension will be considered by comparing the reconstructed English temperature conditions to the temperature indices constructed by van Engelen, Buisman and Ijnsen in 'A millennium of weather, winds and water in the Low Countries'.

Independent contemporary evidence on English weather is provided in numerous narrative[5] as well as administrative sources. Britton's compilation 'A meteorological chronology to AD 1450' has been used as the starting point to seek out weather references in narrative sources. However, chronicles not listed in Britton have also been employed, as well as occasional weather references overlooked by Britton. All of Britton's information used in this work has been verified with the original source and the original text is used.

Evidence from administrative sources is mostly constituted by direct information on weather in manorial accounts. The largest surviving collection of these comes from the estates of the Bishopric of Winchester, which were mostly situated in south-central England.[6] The direct references to weather were collected by Titow in 'Evidence of weather in the account rolls of the bishopric of Winchester, 1209–1350' and 'Le climat à travers les rôles de comptabilité de l'évêché de Winchester, 1350–1450'. The first of Titow's works also includes material from Glastonbury Abbey. Information for Sussex was assembled in 'Late-medieval weather in Sussex and its agricultural significance' by Brandon.[7] The evidence for the Westminster Abbey manors in Hertfordshire, the most important one being Kinsbourne, comes from 'A Hertfordshire demesne of Westminster Abbey. Profits, productivity and weather' by Stern. Weather references from East Anglian accounts are also used throughout this chapter (Appendix 1), as well as the reconstructed East Anglian growing season temperature (Fig. 5.4, Appendix 4) and the precipitation index for July–September (Fig. 7.5, Appendix 5).

[4] Munro, Revisions of the Phelps Brown and Hoskins 'Basket of Consumables' commodity price series. The values are advanced by one year so that in this study they date to the harvest year.

[5] For an analysis of the sources available and the interconnection of some chronicles during the study period, see Gransden, Historical writing in England, vol. 1, 356–517 and idem, Historical writing in England c.1307 to the early sixteenth century, 1–248.

[6] Titow, Winchester yields, 38.

[7] When comparing the weather evidence from Sussex with other information, it appears that in Sussex, just as at Norwich Cathedral Priory (note 109 in Chap. 2), there was a dating problem in the early years of the reign of Henry IV, c.1400–1408. Hence the year of Brandon's weather evidence between 1400 and 1408 has been adjusted.

6.2 Warm Growing Seasons 1256–1431

6.2.1 Weather Conditions in 1267

For the warm growing season 1267 no additional English documentary evidence is available, but the Low Countries summer half year index agrees with the April-July mean temperature reconstruction and stands at maximum (III).

6.2.2 Weather Conditions in 1297 and 1298

The manorial accounts of the Bishopric of Winchester and the Westminster manors in Hertfordshire supply information on the dry conditions in 1297 and 1298. In both years the Winchester manors had to repair their ploughs which had broken during the summer ploughing because of the dry weather and consequently hard soil. Drought impacts on hay production and the productivity of pasture land were also common. The plough problems were repeated in the Westminster accounts for the growing season 1298 and in autumn a 'great drought' was noted.[8] The harvest in Norfolk and on a Winchester manor close to London was shorter than average in 1298 and hence also indicates dry weather in the harvest period. The summer half year 1297 scores III in the Low Countries and thereby confirms the reconstructed warm temperatures of spring and early summer in England, for 1298 no information is available. In 1297 the Norfolk barley harvest was poor; this is a typical drought impact.

6.2.3 Weather Conditions in 1304–1307

Another warm and dry phase were the years 1304–1307. In 1304 the evidence from Norfolk for a warm spring and early summer period is supported by other manorial accounts from the demesnes of Westminster Abbey. The manor of Kinsbourne reported a great drought for summer 1304, ploughs were broken. In the Low Countries the summer half year is indexed as 7. The harvest length in Norfolk 1304 was shorter than average as also in 1305 and 1306, thereby confirming dry conditions in late summer and early autumn in these years.

In 1305 another severe drought took hold in England. The manors of the Bishopric of Winchester and of Glastonbury Abbey report drought impacts. Ploughs broke frequently, the dairy production was low and pastures did not produce enough

[8]The 'great drought' either refers to autumn 1298 or spring/summer 1299, Stern, A Hertfordshire demesne, 93. In the light of the evidence listed above, it can be assumed that autumn 1298 is meant.

herbage. The Flores Historiarum also refer to heat and water shortage.[9] Whereas the heat is well reflected in the summer temperature index for the Low Countries which again stands at 7, the reconstructed temperature for East Anglia is below average. This underrepresentation of the summer temperature is connected to the winter 1304–1305 being long in Norfolk (Appendix 1) and consequently delaying the onset of the growing season. Norfolk's soils are also vulnerable to drought, which can result in a vegetation delay. The backbone of the East Anglian temperature series is the long harvest date series from the Northwest region, which is, however, situated on sandy soil close to the north coast and consequently not only more susceptible to drought than other Norfolk regions, but also suffers more than the inland manors from the cold north winds in spring time. In 1305 the combination of these factors resulted in northwest Norfolk producing the latest harvest date of all available harvest dates, which is unusual. Most Norfolk manors harvested around 1 August, whereas the Northwest region started harvesting 10 August. The temperature reconstruction, however, remains tuned to northwest Norfolk.

For 1306 the warm Norfolk growing season temperature went hand-in-hand with drought impacts upon pastures and dairy production on the Winchester manors, but conditions during the summer half year in the Low Countries were average.

In 1307 another warm growing season completes the run of dry and warm springs and early summers. In the Low Countries the summer half year was average, but the report of drought by a manor of the Bishopric of Winchester in Somerset and a probably dry late winter and early spring in Hertfordshire underline the dry and potentially warm conditions during spring in England. Conditions during the harvest, however, seem to have been very wet, the harvest was very long in East Anglia (Sect. 8.2).

6.2.4 Weather Conditions in 1318

Extremely warm and dry conditions returned to England in spring and summer 1318, just in the aftermath of the Great Famine. The year is part of the volatile weather conditions during the agricultural crisis 1314–1323 and is considered in detail in this context in Sect. 6.4.

6.2.5 Weather Conditions in the Mid-1320s

The April to July mean temperature 1326 is the second warmest in the reconstructed temperature series for East Anglia. The year is part of a severe drought phase in England and probably also in the Low Countries. Britton gives 1325 and 1326 as drought summers and the Norfolk harvests in those years were a rapid process.

[9] Flores Historiarum, vol. 3, 127–128.

Across the North Sea, the summer indices are both categorized as 9. The summer of 1324 was already good (7), and Britton states that several compilers also list 1324 as a drought year, but the only available medieval drought reference from England is non-contemporary, it is a copy of another chronicle which is actually describing 1326.[10] It is not supported by the administrative sources, the accounts of the Bishopric of Winchester are silent for spring and summer 1324 and in Hertfordshire there are merely indications for late winter or early spring, which was mild and dry. Unfortunately there is no harvest date for the reconstruction of the mean growing season temperature of 1324, but the East Anglian harvest was longer than average which is corroborated by a Winchester manor experiencing a rainy harvest.

The reconstructed East Anglian growing season temperature for 1325 is average, whereas it is very high for 1326. The Historia Anglicana, written decades later, describes the year 1325 as a time of extreme drought. The chronicle remarks on very low water levels in rivers and fountains, and the loss of domesticated and wild animals due to the lack of water. The heat and drought were so severe that the author thinks it fit to compare England to Africa.[11] Dry conditions were certainly present in East Anglia, which due to climate and soil is more vulnerable to drought than other regions of the British Isles. In East Anglia the growth not only of legumes, but also of oats, rye and grass suffered at least locally, ploughs broke in the hardened soil. Near Framlingham Castle in Suffolk holes were dug in the ground of a pasture to provide water for grazing animals (Appendix 1). The dry conditions put further pressure on the vulnerable dairy sector, and at Norwich Cathedral Priory the idea took hold to lease out the dairy production. The Winchester accounts list many drought impacts.

Nonetheless the drought reference in the Historia Anglicana could well be a summary of a drought description in the Annales Paulini, a London based chronicle, which actually details the conditions of 1326, and which was composed much closer to the events of the mid-1320s. According to the Annales Paulini a drought had taken hold in England, not only in summer, but also during other seasons, and had forced people to lead their animals to a water source for three or even four leagues. Waterbodies, rivers and wells, which before had withstood drought, now dried up. Low water levels caused a disruption of water transport and the dying off of fish. The water of the Thames was salty for almost the whole year, that means the water level was so low in London that sea water could push up the river channel through London Bridge. The fruit and vine harvests, however, were very good.

---

[10] Britton, Meteorological chronology, 134. Britton collected the drought reference from Burton, Chronica monasterii de Melsa, 349, but also suspects misdating. The Melsa Chronicle was written around 1400 and its interest in weather is normally very limited. The reference to a drought 1324 is also not independent, but very close to the wording of the various London chronicles, as the Chroniques de London, 50, which is cited below and which is an older text. The reference of the Melsa Chronicle is an abbreviated version and it excludes the paragraph on the salt water in the Thames, but includes the fires at Royston, Wandsworth and the Abbey of Croxden without giving the date for the conflagration.

[11] Walsingham, Historia Anglicana, vol. 1, 177. He is citing a classical text actually referring to Ethiopia, but in this context Ethiopia represents Africa.

Anno Domini MCCCXXVI, et anno regni regis Edwardi XIX [...] Eodem anno tanta siccitas per totam Angliam fuit, tam in aestate quam in aliis temporibus illius anni, quod homines duxerunt sua animalia ad aquandam, in aliquibus partibus regni, per iii. leucas vel iiii^{or}. Fontes et torrentes, putei et paludes, qui antea nunquam siccitatem sustinuerunt, omnino exsiccati fuerunt. Stagnum de Neuport in comitatu Essexiae, continens in circuitu unam leucam, siccatum fuit, ita quod pisces omnes perierunt. Eodem modo aqua de Haveringmere, aliquando portans magnas naves, in tantum fuit subtracta quod vix potuit parvum batellum portare. Et Thamesis fluvius fere per totum annum salsa fuit. Hoc anno maxima fuit copia frugum et vini.[12]

This description of 1326 in the Annales Paulini is echoed in the Chroniques de London, another chronicle from London. This text adds that as a consequence to the salty Thames water, the London citizens had to content themselves with salty ale. Several places fell victim to fire in June.

xix [Edward II]. En cele an fut graunt secheresse de rivers et de fountaigne, issint qe il avoit graunt defaute de ewe en plusours paiis. En cele temps, devant la feste seint Johan, ardoit la vile de Roiston et partie de Wandlesworth, l'abbeye de Croxtone pres de Leicestre, et autres arsouns furent adonke en Engletere. En cele temps, pur defaute de ewe douce, la mer surmonteit issint qe le ewe de Tamyse fut salé, dont mult de gentz se pleinoient de la servoyse fut salé.[13]

Manorial accounts give further detail. Not only is the list of drought references in the Pipe Roll of the Bishopric of Winchester much longer in 1326 than in 1325, but by 1326 the cumulative drought stress posed a threat even to trees in East Anglia. Norwich Cathedral Priory tried to sell a high quantity of desiccated ash-trees at its main wood, Hindolveston in northern Norfolk (Appendix 1). Lack of pasture limited the production of cheese in wide areas of England, as reported for southeast England, by the Winchester manors and by a manor of Christ Church Canterbury also situated in the Winchester region.[14]

The heat and drought of 1326 stretched to Ireland, because the London reports are echoed there:

Annus autem iste siccus fuit ultra modum Hybernie consuetum; sic quod in yeme quasi parum pluvie fuit, in vere estate et autumpno quasi nichil, tanta fuit siccitas et tantus calor, quod fontes et magni rivuli (ubi semper emanabant aque copiose) penitus siccabantur.[15]

Hence, on the British Isles, the dry and warm conditions in 1326 were considerably more pronounced than in 1325. Notwithstanding the unusual weather grain

[12] Annales Paulini, 312–313.

[13] Chroniques de London, 50. The chronicle was compiled in the mid-fourteenth century. For the genesis, sources and interconnection of the London chronicles, see McLaren, London chronicles of the fifteenth century, 15–48, and including also earlier specimens from the thirteenth century onwards: Gransden, Historical writing in England, vol. 1, 508–517. On the perception of weather in the London chronicles, a popular subject, see McLaren, London chronicles of the fifteenth century, 71–72.

[14] CRU, Bickersteth, Minister's account rolls of Christ Church Canterbury 1305-1386.

[15] Clyn, Annalium Hiberniae Chronicon ad annum MCCCXLIX, 18.

prices were low in both years, but in the comparatively dry climate of Norfolk with its often well draining soils, barley harvests were poor.

After a short break in the dry and warm conditions in 1327, these returned in 1328. The reconstructed temperature is high and the Low Countries temperature index for the summer season is above average (7). Again the warm weather stretched to Ireland, where non-contemporary information also hints at summer temperatures high enough for the formation of destructive thunderstorms.[16] Across England, in Norfolk and around Winchester, the harvest was short, so the late summer must have been dry. The grain prices rose sharply in the year following the harvest 1328, but it is unclear, if the (winter) grain suffered during the long winter 1327–1328, which in its later stages was very hard on the Winchester manors, or during a drought at a critical phase of crop development or from other conditions.

6.2.6 Weather Conditions in the Early 1330s

During the first half of the 1330s, 1331–1334, England experienced a sequence of dry and warm springs and summers. 1331 was a major drought year. According to Merle there was only light rain in spring, that did not penetrate the ground,[17] but winter had been very wet so that flooding was frequent on the Winchester manors. The Annales Paulini specify that before c.17 June no rain had fallen for 15 weeks. That would indicate that in the London region precipitation was very low since the end of February. Shortly before the 17 June rainfall set in.[18] After the wet and cold weather in 1330 and the wet winter, the drought 1331 took people by surprise and references to drought impacts are plentiful in the Pipe Roll of the Bishopric of Winchester: ploughs were broken, pasture and hay were scarce. Similar problems were recorded in the rolls of Christ Church Canterbury.[19] On the Westminster manor of Kinsbourne drought reduced the pea harvest. The van Engelen et al. summer index scores 8, the English April–July temperature was average. Most likely the drought held back the growth of vegetation. The combination of wet and probably also little seed corn from the harvest 1330, wet weather at the sowing time of winter wheat and rye at least on the Winchester manors in the west of England, a very wet winter and then the prolonged drought in 1331 were very problematic for grain cultivation. Prices for wheat and barley rose in the year following the harvest 1331, in Norfolk the barley harvest was very bad indeed.[20]

In 1332 the dry conditions were weaker, but it was a very warm growing season, and summer was very dry. All East Anglian harvests in the first part of the 1330s were shorter than average, but 1332 was the quickest amongst them. The speedy harvest

[16] Annals of Loch Cé, vol. 1, 608–609.

[17] Merle, Consideraciones temperiei pro 7 annis, under June 1340.

[18] Annales Paulini, 354.

[19] CRU, Bickersteth, Minister's account rolls of Christ Church Canterbury 1305–1386.

[20] Hallam, The climate of eastern England 1250–1350, 129.

was a widespread phenomenon, as it was also observed in Staffordshire.[21] References to dry weather in winter and during the growing season come from the Winchester and Westminster account rolls. The summer in the Low Countries stands at 6.

April–July mean temperature in 1333 was even higher than in 1332. In this year the summer warmth stretched at least from the Low Countries (index 8), to Ireland, from where a temperate summer and low grain prices are reported.[22] April was dry in England.[23] Several Winchester manors experienced difficulties with the ploughs and pastures because of dry weather.

The year 1334 was rather unremarkable. The reconstructed temperature is average,[24] the summer half year across the North Sea is indexed as 6. The East Anglian harvest was shorter than average, indicating low precipitation levels around harvest time, but this was not a severe drought.

6.2.7 Weather Conditions in 1354

Spring and early summer temperature 1354 was high, despite the preceding winter being hard and long and delaying the start of the vegetation growth, as information from accounts of Christ Church Canterbury[25] and of the Bishopric of Winchester point out. For the following seasons hardly any information is available for England in this year. Some rain hindered the hay harvest on the Winchester manors, and in the Low Countries the season was average. However, in later summer, the harvest time was dry in Sussex, the harvest length in East Anglia normal. Further west, Winchester manors report high levels of rainfall that set in during the harvest, causing a long and expensive harvesting process. Considering the dry harvest weather in Sussex and the average harvest length in Norfolk, it is likely that the rain arrived after the harvest in southern and eastern England was finished, i.e. after early September.

6.2.8 Weather Conditions in 1361

1361 was an exceptional year, the earliest harvest in the whole series 1256–1431 took place in 1361. In fact across large parts of Europe there is ample evidence for a very warm spring and warm summer, which led to an early vine harvest and in some regions damaged the grain harvest.[26] In the Low Countries the year was clas-

[21] Lynam (ed.), Croxden Chronicle, ix.

[22] Grace, Annales Hiberniae, 128.

[23] Merle, Consideraciones temperiei pro 7 annis, under December 1342.

[24] Most harvest dates for 1334 are on the 1 August. This is somewhat unusual and it appears that they were reduced to a common denominator for convenience purposes.

[25] CRU, Bickersteth, Minister's account rolls of Christ Church Canterbury 1305–1386.

[26] Alexandre, Le climat en Europe, 490–491 and Glaser, Klimageschichte, 67.

sified as 8. For England John of Reading describes a drought, that set in towards the end of May:

> [Anno gratiae MoCCCLXI] Postea, evolutis diebus sex, vio videlicet die Maii, in vigilia Ascensionis Dominicae, [...]. Sequebatur siccitas nociva, et ob defectum pluviae frugum, fructuum ac feni magna sterilitas.[27]

This is supplemented by the manorial accounts from Sussex and of the Bishopric of Winchester, which specify drought conditions, though not severe drought, and present here one of the very rare temperature references, noting also problems 'pro magno calore in estate'.[28] Portents recorded by John of Reading for continental Europe are likely to be connected to the presence of Sahara sand in the atmosphere and thereby indicate a southerly flow of air that contributed to the heat. In late May a blood rain fell in Burgundy, which could well have been rain mixed with Sahara sand. In Bologna a bloody cross was said to have stood in the sky for hours, this phenomenon might also have been linked to a raised amount of particles in the air.[29] According to John of Reading the heat continued in England into modern day autumn: after Michaelmas the roses flowered a second time and the birds bred again.[30] In this year a new fashion of 'indecent' – short and tight – clothing appeared, which made moralists fear for the worst,[31] and even though conservative chroniclers did complain about the introduction of new continental fashion at least since the marriage of Edward III to Philippa of Hainault in 1328, this sudden surge of the popularity of provocative garments in 1361 must have been linked to the unusually warm summer season.

The worst did indeed befall England: in this year the people of England and other European countries had to realize that the Great Pestilence had not been a singularity. The second national outbreak of plague affected England in 1361, it began in the south, then moved northwards, and caused a high death toll (Chap. 10).[32] Where the plague coincided with harvest time, the harvesting process was disturbed. This was a widespread phenomenon. In southern England, on the Winchester manors, harvests costs were raised '[...] propter pestilentiam subito supervenientem in autumpno.' A similar situation prevailed in southeastern England; on a manor of Christ Church Canterbury harvest costs were also high, because additional help had to be hired due to the epidemic.[33] The plague moved into Suffolk in late summer and autumn,[34] and probably affected Norfolk around the same time. In Gnatingdon, in

[27] Reading, Chronica, 148–149.

[28] A direct reference to the high temperatures is unusual and underlines the severity of the conditions. Normally early documentary sources focus on hydro meteorological extremes, because of their greater impact upon agriculture and the wider economy, Pfister et al., Documentary evidence, 2.

[29] Reading, Chronica, 149.

[30] Ibid., 149.

[31] Eulogium, vol. 3, 230–231.

[32] Continuation of Higden, Polychronicon, vol. 8, 360.

[33] CRU, Bickersteth, Minister's account rolls of Christ Church Canterbury 1305–1386.

[34] Bailey, Medieval Suffolk, 183.

northwestern Norfolk, the harvest took an extraordinary long time, although other, less isolated Norfolk manors were not affected in this manner and had harvests of average length. Grain prices rose over the agricultural year 1361–1362 and it is likely that the drought impacts and the interruption of harvest work and transport caused by the plague played a substantial role in this price rise.

6.2.9 *Weather Conditions in 1365*

For the events in 1365 see Sect. 8.4.

After 1365 the reconstructed temperatures drop. The warmest years of this sub-period were 1371, 1385, 1390, 1400, 1409 and 1431, although they do not reach the temperature levels of extremely warm growing seasons before 1365. These warm springs and summers mostly went unnoticed in the narrative sources in England.

6.2.10 *Weather Conditions in 1371*

During the warm April–July period 1371 the Low Countries summer half year scored 7 and in Sussex and Hertfordshire spring was dry. In Sussex a wet harvest followed, in East Anglia the harvest was longer than average. The rain may have affected East Anglia, too, but more likely this year's grain crop was a good and bulky one, since the grain prices fell from a very high level after the harvest.

6.2.11 *Weather Conditions in 1385*

In the summer 1385 it was very warm in East Anglia and in the Low Countries (8). The year is part of a sequence of warm and dry years that are described in Sect. 8.3.

6.2.12 *Weather Conditions in 1390*

Another warm summer followed in 1390. The Westminster Chronicle records intense heat between the start of June and late August.[35] It was also warm in the Low Countries (8). The warm weather was accompanied by dry conditions in spring and early summer as noted or implied by the *compoti* of the Bishopric of Winchester and the manor of Kinsbourne. The situation in Norfolk was similar: in Sedgeford oats were yielding extremely badly, in Gnatingdon straw was lacking; on the well-

[35] Chronicon Westmonasteriense 1381–1394, 438.

draining soils of northwest Norfolk these problems were most likely drought impacts (Appendix 1). In the summer, severe thunderstorms with partly extremely heavy rainfall crossed England,[36] and the curse of the Late Middle Ages, plague, appeared again. The Westminster Chronicle attributes the outbreak to the prevailing hot weather and the corrupted air. Since the plague wave affected extensive areas and resulted in an excessive death rate, it was counted as the Fifth Pestilence; primarily the young were infected.[37] It remained present amidst famine disease in 1391 and returned in 1393 to Essex (see Chap. 10).[38] The harvest 1390 was longer than would have been warranted by the generally warm and dry summer and the low yields; this must have been connected to the disruption of the labour market due to the plague, the damage caused by the thunderstorms to the corn crops and the drought impacts, such as the stunted growth of the grain in Gnatingdon. In 1389 the harvest had already been deficient and in 1390 it was so poor that pestilence was followed by dearth.[39] The mayor of London imported grain from across the sea.[40]

6.2.13 Weather Conditions in 1400

The reconstructed mean temperature for 1400 is high and is reflected in the van Engelen et al. summer index standing at 8. After a warm growing season the harvest was marked by rain in many parts of England, evidence comes from the Winchester manors, from Sussex and from northwest Norfolk itself (Appendix 1).[41] Walsingham notes an epidemic for the summer,[42] one Winchester manor confirms a shortfall of labour due to pestilence at harvest time (Chap. 10).

---

[36] Chronicon Westmonasteriense 1381–1394, 444.

[37] Chronicon Westmonasteriense 1381–1394, 438, Walsingham, Historia Anglicana, vol. 2, 197.

[38] Hatcher, Plague, 58.

[39] Walsingham, Historia Anglicana, vol. 2, 195, 198. Blomefield, History of Norfolk, vol. 3, 113 speaks of raised mortality, also in Norfolk, due to a dearth lasting from the harvest 1390 the following one. He attributes the dearth rather to monetary policy, than to harvest failure.

[40] Walsingham, Historia Anglicana, vol. 2, 203–204.

[41] The account roll of Gnatingdon, NRO, LEST/IC/29, also mentions that the fallow was ploughed only twice. This is a low number and could indicate a period of dry weather that hardened the soil around May and June.

[42] Walsingham, Historia Anglicana, vol. 2, 246 and Annales Ricardi Secundi et Henrici Quarti, 332 (probably also a work by Thomas Walsingham).

6.2.14 *Weather Conditions in 1409*

The very warm growing season 1409 was preceded by a mild winter (3) in the Low Countries. Widespread flooding is reported for February.[43] In England winter conditions were comparable, flooding also took place in Sussex and eastern Norfolk (Appendix 1). However, spring and summer were dry in the Low Countries.[44] Drought affected Flanders between March and August, and April was also very warm,[45] although in general the summer half year seems to have been cool, standing at 4. It is possible that the warmer conditions further north over Scandinavia[46] reached also over the northern parts of the British Isles. In England one manor of the Bishop of Winchester suffered from rain at hay making time, but regions further south and east mirror the dry conditions of the Low Countries. In the Portsmouth area there was a lack of pasture.[47] Across Norfolk vetches and peas perished 'for the lack of rain' and because of 'dry weather' (Appendix 1). In northwest Norfolk, the summer ploughing of the fallow was carried out merely twice, which is atypical and could be connected to dried-up soils.[48] It appears therefore that a mild winter was followed by an early onset of the growing season, and dry and probably warm weather during spring and early summer resulted in a generally warm growing season in East Anglia. As in Flanders the dry conditions continued throughout August, because the harvest was short. The prices for wheat and barley rose considerably in the year following the harvest 1409, and even though the exact reason remains unspecified, the price rise is probably connected to the prolonged dry conditions impacting on the grain crops in eastern England. In some regions of England the water quality was badly affected during the dry and hot weather, the water gushed forth from wells red as blood and dysentery broke out.[49] Newcastle-upon-Tyne was visited by plague (Chap. 10).

6.2.15 *Weather Conditions in the 1410s*

The summer half year 1409 was warm and dry. As such, it was the first year that displays features common to the dry phase of the second decade of the fifteenth century. It is also the first year for which drought impacts on the agricultural production in Norfolk are numerous in the *compoti* rolls (Appendix 1). Until 1420 almost in every year remarks about peas and vetches perishing in the field, about the weak

---

43 Alexandre, Le climat en Europe, 556–557.

44 Camenisch, Endless cold, 1062.

45 Alexandre, Le climat en Europe, 556.

46 Luterbacher et al., European summer temperatures.

47 Postles, Stubbington WCM 15389.

48 Gnatingdon NRO, LEST/IC 34.

49 Continuatio Eulogii, vol. 3, 414.

state of barley, rye and oats as well as about a lack of hay can be found in the East Anglian manorial accounts, partly even with the cause: drought. Only for 1411 no such entry survives. For a detailed description of the weather conditions during the 1410s, see Sect. 8.3.

6.2.16 Weather Conditions in 1431

The warm summer half year 1431 was also dry, as most years of the early 1430s (Sect. 8.3). In the Low Countries the index shows temperatures slightly warmer than average (6). Plague affected the estates of St Albans.[50]

6.3 Cold Growing Seasons 1256–1431

6.3.1 Weather Conditions in 1275

The first year in the April to July mean temperature reconstruction that was marked by a cold growing season is 1275.[51] As often is the case, no direct references to temperature are available in the contemporary narrative and administrative sources. Evidence about precipitation can be traced, however, and allows conclusions about the character of the growing season. The manorial accounts of the Bishopric of Winchester mention dry weather around April or May which impacted on the cheese production. Then the weather turned and rain hindered the harvesting process. Independently the Winchester Annals report continuous rainfall even earlier, for June, and then for winter:

> MCCLXXV: Eodem anno magna inundatio aquarum et pluviae continuatio mense Junii et pejus in hyeme.[52]

Walter of Guisborough notes, that in this year 'incepit communis scabies ouium in Anglia per loca diuersa'.[53] This outbreak of sheep scab might have been partly triggered by the wet conditions particularly in the winter months. The weather was foul not only in England. The East Anglian low growing season temperature is reflected in the summer half year in the Low Countries being categorized as I.

---

[50] Chronicon Rerum Gestarum in Monasterio Sancti Albani, vol. 1, 62.

[51] The growing season of 1258 which was a very bad year for farming in Europe due to a preceding volcanic eruption does not show as cold. This is in accordance with the general limited temperature response to this event, see Timmreck, Limited temperature response, L 21708. For the prolonged harvest in 1258, see Ch. 8.2.

[52] Annales Monasterii de Wintonia, 120.

[53] Guisborough, Chronicle, 215.

Although grain prices were high in the mid-1270s, no price rise can be connected directly to the harvest 1275.

6.3.2 *Weather Conditions in 1283*

For 1283 the sources are more informative. This time the cold and wet weather left a decisive mark on the agricultural and pastoral economy. In western England the harvest was ruined by the constant heavy rain in summer and at harvest time.

> Aestas tota, cum maxima autumni sequentis parte, vehementer continue pluviosa omnem fere spem satorum in viridi delusit in arido.[54]

Grain prices rose in the agricultural year 1283–1284 due to the disappointing harvest. Failure in the agricultural sector was joined by failure in the pastoral sector: the Dunstable Annals underline the detrimental influence of the continuous rainfall on the health of the lowland sheep and their consequent high mortality.

> Eodem anno ex maxima pluvia, quae quasi continue fuerat in aestate, fere omnes oves loca convallium pascentes, tantam humiditatem et pinguedinem interius conceperunt, quod generalis mortalitas ovium postea sequebatur: ita quod in partibus Dunstapliae vix habuimus ducentas oves in fine Martii subsequenti. Oves tamen nostre in Ciltria et in Pecco sanae et vegetes remanserunt.[55]

The Worcester Annals refer to much rain during summer and autumn 1284.[56] Britton ascribed this entry to 1283, which appears conclusive in the light of the other evidence.[57] In 1283 the reconstructed temperature for East Anglia is in accordance with the Low Countries summer index, which is equally cold, being I.

6.3.3 *Weather Conditions in 1294*

The cold and wet year of 1294 resulted in an agricultural crisis much more pronounced than 1275 or 1283. The summer half year was also cold in the Low Countries, it stands at a mere I. The mid-1290s still fall into the phase of high activity in historiographical writing in England, and especially the monastic chronicles provide plenty of information for the environmental conditions during the crisis.

The spring of 1294 was marked by severe weather. On 14 May London witnessed a snowstorm. It brought much snow and the strong wind damaged houses, trees and meadows.

---

[54] Continuatio chronici Florentii Wigorniensis, vol. 2, 231.

[55] Annales Prioratus de Dunstaplia, 305–306.

[56] Annales Prioratus de Wigornia, 489.

[57] Britton, Meteorological chronology, 120.

Rex Edwardus Primus, anno xxij. In this yere fel the grettest snowe that evere was seyn before this tyme; wherfore a vereyfyer made in metre thise vers:

C'stino tiburci s'c'or' Valariani
Nix cadit innanis vent' vehemens Borial'
Emulsit silvas ussit quas rep'it herbas
Edes dampnose detexit et impetuose
Quas clam p'stravit sic plurima dampna patravit.[58]

Growing season weather conditions were detrimental to crop growth and prices for wheat as well as barley were already high before the harvest 1294. Many chronicles comment on the dearth,[59] which must have been largely due to the harvest 1293[60] and partly also to the prospect of a late and bad harvest in 1294. Scarcity and high prices reigned in England in the months before the harvest 1294 which due to the rain in August and September procured only little or no new corn before Michaelmas.

Fames et inopia praegrandis per totam Angliam. Quarterium namque frumenti, quod vix et difficulter inveniebatur, ad xxiv. solid. per aliqua loca vendebatur; sed et mensis Augusti cum Septembri subsequente in tam imbribus continuis madidus fuit et pluviosus, quod usque ad festum S. Michaelis parum aut nihil de novo grano potuit reperiri. [61]

A very similar account is also in the Chronicle of Bury St Edmunds, which is situated in Suffolk, which was part of the East Anglian bread basket.[62] The late and long harvest, continuing in many regions of England at least till Michaelmas, is also reflected in the Norfolk precipitation index.

The Worcester Annals also ascribe the late cutting of the grain to the frequent rainfall. In western England the rain started in late June and continued until March 1295.

Anno Domini MCCXCIV. [...] Frequens pluvia maturitatem messis in tanto impedivit, ut quarterium frumenti pro xx. solidis Londoniis venderetur. [...] A festo Sancti Johannis

[58] Chronicle of London from 1089 [sic] to 1483, 35.

[59] Examples are: Guisborough, Chronicle, 232 and Annales Prioratus de Dunstaplia, 391.

[60] Since c.1290 the English weather appears very variable, the early 1290s show a tendency towards at least wet harvest times and partly also wet summers, see Britton, Meteorological chronology, 126–128. According to Campbell, Nature as historical protagonist, 289 all five harvests 1290–1294 were below average, 1293 was especially bad. Relating to 1293 Titow, Evidence on weather in the account rolls of the Bishopric of Winchester, 379–380, notes that the various grain crops ripened simultaneously which resulted in a shortage of labour on the Winchester manors. According to Ogilvie, Farmer, Documenting the medieval climate, 126 above average rainfall in May was followed by dry weather in June and again wet conditions in July and August. The April-July temperature reconstruction and the precipitation index for the period July–September show 1293 as a warm and rather dry summer season, but if winter and spring corn had to be harvested at the same time, this would have shortened the harvest and hence give the impression of a summer of below average rainfall.

[61] Continuatio chronici Florentii Wigorniensis, vol. 2, 273.

[62] Chronica Buriensis, 192.

> Baptistae usque ad festum beatae Mariae in Martio, in autumno metere vel in hyeme serere sicut oportuit, frequens pluvia non permisit.[63]

The Dunstable Annals connect the late grain and hay harvests to the 'intemperate air'.

> Item autumnus, propter aeris intemperiem, tardus erat, una cum tempore foenationum; et populus de anno subsequenti quamplurimum desperabat.[64]

This was not the end of the problems at the Priory of Dunstable. In summer 1294 the hay barn at Dunstable was destroyed by fire. In the context of the weather evidence for 1294 and the fact that the hay harvest itself was delayed by the weather it seems possible that the hay was stored too wet and then rotted. Decomposing hay can produce such heat as to ignite the hay and thus also the barn. The Dunstable annals also state that a new enclosure wall built in summer collapsed in winter 'propter aeris intemperiem in aestate', other enclosure walls shared its fate.[65]

The 'intemperate air' caused desperation among the people not only because it ruined the grain harvest and thus helped to maintain the high grain prices, but also because the unusual weather continued after the harvest throughout autumn and winter, hindering, as the Worcester Annals point out, agricultural activity. This lowered the chances of a good harvest in the following year.[66]

The rain in autumn and winter 1294 mentioned in the Worcester Annals, was responsible for inundations of the Thames at Bermondsey and Westminster in October.

> Anno Domini MCCXCIV., et anno regni regis Edwardi primi vicesimo secundo. Hoc anno gurges aquarum Thamisiae transcendit consuetos limites [xviii. die Octobris, et tunc accidit magna brecca apud Retherhithe], et planitiem de Bermundeseye et procinctum de Tothille superficialiter debriavit. Similiter et casas negotiatorum in nundinis Westmonasterii attingendo superius palos figere compellebat.[67]

The accounts of the Westminster Abbey manor of Kinsbourne in Hertfordshire also report a wet autumn and early winter.[68] Due to the partly heavy rain the wheat sowing rate had to be raised. With dearth conditions already present, this was surely not a cheap measure taken on the lord's demesne. If the peasant farmers still had the means to increase sowing rates likewise to ensure a better harvest in the following year, is open to doubt. The wheat and barley prices remain high for 1294–1295 as well as 1295–1296.

---

63 Annales Prioratus de Wigornia, 516, 519.

64 Annales Prioratus de Dunstaplia, 391.

65 Annales Prioratus de Dunstaplia, 388.

66 The success of the next harvest also depended on the quality of the seed corn used. After a wet harvest as 1294 the seed corn would not have been good, and might partly also have been consumed by desperate peasants. Even without rain at sowing time, these factors might lower the yields in 1295.

67 Annales Monasterii de Bermundeseia, 468.

68 The Winchester roll for 1293–1294 is missing, see Beveridge, The Winchester rolls and their dating, 96.

6.3.4 Weather Conditions in 1314–1323

The next exceptionally cold Norfolk growing seasons fall to 1314, 1315, 1319 and 1323. The first two are connected to the Great Famine 1315–1317 which was embedded in the agrarian crisis 1314–1323. Therefore these years will be considered under Sect. 6.4.

6.3.5 Weather Conditions in 1330

In the summer season 1330 cool and wet weather returned. It affected wide parts of western and central Europe, where the vines did not mature or only with great delay.[69] In the Low Countries the summer was extremely cold, an index of 1. From May till at least October England and Ireland suffered from rains.[70] Due to the weather the grain harvest was late and took a long time in the area between south-east England[71] and the northern Midlands. In Norfolk it also rained at harvest (Appendix 1), the gathering of the crops lasted till mid-October,[72] and in the area of the Winchester manors until early November. Further north, at Croxden Abbey in Staffordshire, it only began at Michaelmas, which marked the end of a normal harvest season, and lasted even until late November. The Croxden Chronicle also gives another poignant illustration of the prevailing extreme weather: in November fresh peas in the pod were offered to the monastery instead of apples or pears.[73] In Ireland the rain and storms gave summer and autumn 1330 an almost wintry character and as a consequence the harvest failed, corn stacks in the field were scattered by the autumn and winter storms, and famine ensued.[74] The wine of low quality and quantity which was produced on the continent this year was imported into England, where the incidence of taverns selling overpriced and unwholesome wine greatly increased, so that royal legislation tried to fix the price at a 'reasonable' level.[75]

---

[69] Alexandre, Le climat en Europe, 450–451.

[70] Clyn, Annalium Hiberniae Chronicon ad annum MCCCXLIX, 22, Lynam (ed.), Croxden Chronicle, ix.

[71] CRU, Bickersteth, Minister's account rolls of Christ Church Canterbury 1305–1386.

[72] The *compoti* of Norwich Cathedral Priory are missing for this year, information comes from Kempstone NRO, WIS 06.

[73] Lynam (ed.), Croxden Chronicle, viii–ix. The content of the Croxden Chronicle was later used in Holinshed's Chronicles of England, Scotland and Ireland, vol. 2, 598

[74] Clyn, Annalium Hiberniae Chronicon ad annum MCCCXLIX, 22.

[75] Cook (Bickerdyke), Ale and beer, 106.

6.3.6 Weather Conditions in 1335

After the warm and dry phase 1331–1334 (Sect. 6.2) the growing season of 1335 was exceptionally cold. No contemporary narrative sources supply information on the weather and its impacts, but the summer was also cool in the Low Countries (3). At the Westminster Abbey manor of Kinsbourne the harvest was wet and stormy in 1335. In the Low Countries, northern France and Germany storm and rain damaged the harvests and the sugar content of the vine harvest was low, the wine sour: it was a bad year for wine production.[76] In England a murrain decimated the cattle stock again.[77]

The English national grain price, although rising in the year following the harvest 1335, remained average, partly because by the mid-1330s deflation had struck. Also the Winchester yields did not diverge far from average,[78] but there is no information by Hallam about the barley harvest in Norfolk 1335.

Several London chronicles refer to great rains, high mortality amongst men and animals, and very high grain prices either in 1335, 1337 or 1338.[79] These London chronicles are mid-fifteenth century works closely linked to the English Brut.[80] In this case, however, no comparable reference is found in the English Brut or related texts, but almost exactly the same wording is used by the Polychronicon (in Latin) and its English translations for the Great Famine 1315–1317, raising the chances that the crisis described in the London Chronicles for a year in the mid-1330s is a transcription error,[81] and the problems of 1335 were not as severe as described by them.

6.3.7 Weather Conditions in 1348–1349

The arrival of the Great Pestilence in England in 1348 coincides with another cold mean temperature for April to July in Norfolk. The year was also marked by high rainfall levels, as was 1349. They will be analysed in detail under Sect. 6.5.

---

[76] Alexandre, Le climat en Europe, 459–461, Glaser, Klimageschichte, 65.

[77] Slavin, Cattle plague, 178.

[78] The usually verbose *compoti* of the Bishopric of Winchester, record no information for the spring and summer 1335. Hence it is possible that the meteorological conditions in south-central England were not as severe as in eastern England.

[79] For the bad weather etc. in 1335, see Chronicle of London from 1089 [sic] to 1483, 55, and William Gregory's Chronicle of London, 79, which has an almost identical text. The Chronicle of the Grey Friars of London, 4–6 and McLaren (ed.), London Chronicle, 165 put the crisis to 1337 respectively 1338, where it does not fit with any other contemporary evidence.

[80] Gransden, Historical writing in England c.1307 to the early sixteenth century, 221.

[81] Higden, Polychronicon, vol. 8, 308–309, including medieval English translations. The problems probably occurred due the similarity in the regnal years and kings' names of Higden's reference to the Great Famine and the year 1335. The references fall to c. Edward II 10 in the Polychronicon respectively to Edward III 9–10 in the majority of the London Chronicles.

6.3.8 Weather Conditions in 1364

It is clear that the low average temperature for the growing season 1364 is not so much due to the weather during late spring and summer, but to the exceptionally hard and long preceding winter. In the Low Countries the winter 1363–1364 was also extremely cold, being indexed as 9, the summer half year 1364 was average. The hard winter is noted by the Winchester and Christ Church Canterbury manors as well as in Sussex. From Kinsbourne in Hertfordshire there comes evidence of 'severe weather' in late winter or early spring.

That the winter was long and stretched into March is confirmed by many English chronicles including John of Reading.

> [Anno gratiae MCCCLXIV] Quam forte gelu incipiens circa festum Sancti Andreae Apostoli, quasi in fine Novembris anni proximo praeteriti, perseverans usque quartum decinum kalendas Aprilis hujus anni, quod opera campestria artesque manuales plurimum impedivit.[82]

The long continuation of the frost hindered the field work and of course delayed the onset of the growing season. The following summer was largely unexceptional; the Sussex and Winchester manors supply no weather reference to it. However, John of Reading writes also about heavy hailstorms that killed animals in many parts of England, even though the hail did not diminish the corn and vine.

> [Anno gratiae MCCCLXIV] Immediate, diversis in partibus Angliae cecidere grandines quae fortia animalia occiderunt; attamen blada et vina abundabant, fructus vero aborum et fenum modice.[83]

The formation of convective storms indicates warm weather, but the summer warmth could not compensate for the late start of the growing season, consequently there was a very late harvest in East Anglia.

6.3.9 Weather Conditions in 1368–1370

The years 1368 and 1369 were classified by Britton and in the Sussex accounts as very wet.[84] In the agricultural year 1369–1370 the grain price rose dramatically. Crop failures are also recorded in the manorial accounts of Norwich Cathedral Priory for 1368 and 1369. After the harvest 1370 grain prices decreased, even though the growing season had been very cold. The meteorological conditions and their role in the harvest success and failure between 1368 and 1370 need to be investigated.

---

[82] Reading, Chronica, 160.

[83] Reading, Chronica, 161.

[84] Britton, Meteorological chronology, 146 and Brandon, Late medieval weather in Sussex, 4.

The accounts of the Bishopric of Winchester report widespread continuous rain at harvest time 1368. The growing season, however, appears to have been rather unremarkable, one manor even refers to dry weather. The theme of drought recurs in the *compoti* of Kinsbourne in Hertfordshire, where ploughs broke, no surplus pasture for sale was available in springtime and the horses were fed oats longer than usual. From southeast England drought impacts are known.[85] The crop failures in East Anglia also relate closer to drought than rainfall impacts, although the cause of the problems is not mentioned (Appendix 1). On the freely draining soil of northwest Norfolk peas were lost when still green, the pea and bean harvests were disappointing, and the hay harvest was below average, so that *pulmentum* ('horsemeat') was used later as supplementary fodder. The number of fallow ploughings was reduced close to Norwich. Even though the area of the Winchester manors was subjected to long rainfall at harvest time, the harvest duration in East Anglia was average, indicating normal harvesting conditions. Therefore at least parts of spring and summer were dry in East Anglia; potentially the region also escaped the rain at harvest time. The growing season temperature was average to cool.

Rain was a much more dominant feature in 1369 than in the previous year. The Winchester accounts contain frequent flood references for the growing season. In Sussex flooding was severe 1368 and 1369. The Winchester manors and Walsingham describe the very wet conditions around harvest time; the rain damaged the crops and in the agricultural year 1369–1370 the price for wheat and barley rose steeply.[86] At a manor of Christ Church Canterbury the wheat was affected by mildew.[87] Nonetheless weather conditions were dry at least at one stage of the growing season in eastern England. At Kinsbourne in Hertfordshire dry weather is mentioned for spring, the pea seed dried up in the ground. During spring the draught horses were again fed oats longer than usual and in summer time straw had to be purchased. Similar impacts are recorded in accounts from northwest Norfolk for 1369 (Appendix 1). Peas partly failed, and probably in the winter 1369–1370 peas had to be given to the manorial livestock, because hay was scarce, indicating a small hay crop for 1369. Unsurprisingly the East Anglian growing season on the whole was warm. However, the wheat harvest was also bad in northwest Norfolk; this is not a classical drought

[85] Mate, Agrarian economy after the Black Death, 349.

[86] Walsingham, Historia Anglicana, vol.1, 309 and Chronicon Angliae, 65, the paragraphs are identical, both chronicles are connected to Walsingham. According to Gransden, Historical writing in England c.1307 to the early sixteenth century, 124, this text comes from the continuation of the Polychronicon to 1377, which might well have been another work of Walsingham, Continuatio Adami Murimuthensis, 205. In Britton the paragraph of the Chronicon Angliae is accidentally allocated to 1368, but this reference is in all three chronicles clearly dated to 1369. Also many fifteenth-century London chronicles report a 'dear year' around that time, obviously referring to 1369–1370, e.g. McLaren (ed.), London Chronicle, 171, Chronicle of London from 1089 [sic] to 1483, 68, William Gregory's Chronicle of London, 88, Chronicle of the Grey Friars of London, 4–6.

[87] CRU, Bickersteth, Minister's account rolls of Christ Church Canterbury 1305–1386. Regarding the general rise of fungoid diseases in the grain crops of eastern Kent in the second half of the fourteenth century, see, Mate, Agricultural technology in southeast England, 254–255. She concludes that a shift in climate as well as badly prepared fields before seeding contributed to the more frequent occurrence of fungoid diseases.

impact. Direct references to harvest conditions are not available for East Anglia and harvest duration was normal (Fig. 7.5). Consequently the region might have escaped the worst rainfall of 1369. The extent of the wet condition at harvest time 1368 and 1369 and during the growing season 1369 in southern England is difficult to assess. The summers in the Low Countries are average, and for 1369 even a good vine harvest in quantity and quality is reported for the southern Low Countries.[88]

It seems as if severely wet and dry phases were closely intertwined in 1369 and possibly also 1368, and it was this rapid sequence of extreme conditions that contributed to the harvest failure. The situation was aggravated by 1369 being the year of the Third Pestilence, which gained momentum around harvest time. Faced with a shortage of labour and plague-induced upheaval in the organisation of labour, a wet harvest was even more difficult to handle than under normal circumstances and part of the subsequent grain price rise must have had its root in the impact of the epidemic on the work force. These repercussions of the pestilence were not limited to the agricultural sector, but would have also affected the transport sector. Transport was under additional pressure from a zoonotic affecting 'larger animals', presumable cattle and horses.[89]

In 1370 spring and summer were cold in East Anglia and also cool in the Low Countries, where the summer index is 4. The preceding winter was somewhat colder than average in the Low Countries, being indexed at 6. Generally 1370 is an unremarkable year for the contemporaries, especially when compared to the inclement years before. Even the Winchester accounts are largely silent about conditions 1370. Late winter and spring were wet in Hertfordshire and in Sussex dry weather is said to have prevailed in summer, but autumn was rainy again. In East Anglia the harvest duration was nonetheless average (Appendix 6), and England-wide the harvest quantity was sufficient for substantially bringing down the grain price, even though it remained on a comparatively high level until the next harvest 1371.

6.3.10 Weather Conditions in 1374

The year 1374 was marked by a cool growing season and agriculture consequently suffered. In East Anglia the year had already opened badly. In 1373 a longer harvest duration hints at raised precipitation levels in the region, but the national grain price remained untouched by these problems.[90] The low growing season temperature in

---

[88] Alexandre, Le climat en Europe, 503.

[89] Walsingham, Historia Anglicana, vol.1, 309, Chronicon Angliae, 65, Continuatio Adami Murimuthensis, 205; the texts are as good as identical.

[90] Raftis, Ramsey Abbey, 258-259. Raftis indicates food shortages in East Anglia 1373-1374. He cites an unpublished chronicle manuscript (Fakynham MS held in British Museum). The text is in fact a chronicle of the Grey Friars in King's Lynn, and has since been published: Gransden, A fourteenth-century chronicle. Raftis listed the other meteorological events and the epidemics in the manuscript correctly, but overlooked that the food shortages did not affect East Anglia, but an English army in France.

England 1374 finds its equivalent on the other side of the North Sea, where the summer falls into the category 4. In general the grain crops and the vines in Europe were severely affected by the humid character of the year, particularly in southern Europe which was in the grip of a severe famine. Le Roy Ladurie suspects that the rains normally destined for England and northern France moved on a track further south in 1374.[91] Indeed in England not many weather references and no major complaints about rainfall are available for this year. The Winchester accounts contain no information on weather, even though the wheat harvest was far below average on the Winchester estates. In Sussex the summer appears normal. The Kinsbourne *compotus* implies some severe weather in late winter or early spring. The summer might even have seen some dry weeks in eastern England, in Kinsbourne pasture had to be purchased for the sheep. In Great Cressingham, on the edge of the sandy soils of the Breckland in western Norfolk, the wheat crop grew poorly, as on the Winchester manors. Indeed grain prices were up in 1374–1375. The cause of the problems in wheat growing is unclear, but Great Cressingham is vulnerable to dry weather. The Scandinavian summer 1374 was warm[92] and it seems possible that England lay at the periphery of the rains in continental Europe and the warm conditions further north. Latest in early winter the miserable weather caught up with England, for the fenland in western East Anglia, including the town of King's Lynn, but also for Sussex and Kent, storm and severe sea floods are reported; this was followed by hard frost until Christmas.[93] The mediocre harvest 1374 is the last before a sequence of bumper harvests which brought down the grain price in England during the second half of the 1370s.

Many cold growing seasons cluster in the following decades. In the context of the late fourteenth and early fifteenth century they are not extreme, but still remarkably cold. To them belong 1382 and 1383,[94] 1389, 1395, 1399,[95] 1401, 1402, 1406, 1411 and 1423.[96]

---

[91] Le Roy Ladurie, Histoire humaine et comparée du climat 79–88.

[92] Luterbacher et al., European summer temperatures since Roman times.

[93] For East Anglia see Gransden, A fourteenth-century chronicle, 278 and for Sussex and Kent see Mate, Agrarian economy after the Black Death, 349.

[94] The cold reconstructed East Anglian growing season temperature 1383 is confronted with a summer defined as 8 in the Low Countries; for England there is evidence of dry conditions during growing season and harvest time (Sect. 8.3).

[95] The summer half year 1399 was average in the Low Countries, but the growing season in eastern England was cold. The year was preceded by a cold winter in the Low Countries, harvest time was rainy (Sect. 8.4).

[96] 1423 was an average summer half year in the Low Countries. The year was preceded by a hard and probably long winter with the potential to postpone the onset of the growing season in England and the Low Countries (Sects. 5.3 and 8.4).

6.3.11 Weather Conditions in 1421

In 1421 the harvest around Norwich was late and in the northwestern part of Norfolk indeed very late. The cold weather conditions were at least at harvest time accompanied by heavy rainfall; at Sedgeford flooding was even reported (Appendix 1). The harvesting conditions, rainfall impacts, extra works necessitated by the weather in Norfolk, but also on the Winchester manors, are described in Sect. 8.4. Additionally there was flooding of pasture land in the district of Flegg, on the eastern Norfolk coast, during the agricultural year 1420–1421. Although the Winchester yields were more than 15% below average for all grains, no major rise of grain prices follows the sodden harvest.[97] The summer was average in the Low Countries.

6.3.12 Weather Conditions in 1428

Finally with the year 1428 the Norfolk manorial accounts supply data for another year when a major agricultural crisis caused by a cold and rainy growing season and harvest struck not only England, but also central Europe. The Low Countries summer half year is classified as merely 2.

In England rain started around the end of March and early April. Meadows and pastures submerged by endless rain during the growing season are reported by the manors of the Bishopric of Winchester. In Sussex, too, the summer was excessively wet.

The rain and the ensuing agricultural failure as well as murrains are alluded to by various London chronicles.

> Anno vj[to] [...] This same yere fro the begynnyng of April into Halwemasse was so gret abundance of reyn, where thorugh nought only hey was distroied, but moche corn, for it reyned almost every other day more or lasse.[98]

In William Gregory's London chronicle is stated:

> Anno vj – Henry VI. And that yere hyt was a wete for hyt raynyd for the moste party from oure Lady Day in Lentyn unto the feste of Mychelmas nexte folowynge. And that yere there was a grete morayne of bestys, and pryncypally of schyppe, for the more party of alle Inglondem for scheppe deyde ynne every contray of Ingelonde.[99]

[97] Across the Channel, in northern France food was scarce, particularly in the urban centres of Paris and Rouen, see Fagan, Little Ice Age, 83.

[98] Chronicle of London from 1089 [sic] to 1483, 116. Similar entry in another version of the London chronicle: McLaren (ed.), London Chronicle, 201–202.

[99] William Gregory's Chronicle of London, 162. The chronicle is organized in accounting years, which start at Michaelmas, 29 September. So the reference of the end of the rain at Michaelmas can simply indicate the end of the year and not necessarily the end of the rain. The type of sheep murrain is not specified in the text, but some animal diseases are triggered by rainy weather and long-term wet meadows, like liver fluke in sheep, Ollerenshaw, Climatic factors and liver fluke disease, 130–134.

According to the St Albans Chronicle the rain continued even until Martinmas.[100] The chronicle also describes how the first sheaves were brought into the barn on the 2 September (Gregorian Calendar: 11 September), and between the end of March (Gregorian Calendar: 8 April) and the 3 September (Gregorian Calendar: 12 September) there had not been three subsequent days and nights which had escaped the rain and wind.

> In crastino Sancti Egidii inducebantur garbae primo in horreum Sancti Petri, quia tempus pluviosum et incolis periculosum ab ultimo die mensis Maritii usque tertium diem mensis Septembris, ut non erant per idem tempus tres noctes et dies clare lucentes quin supervenerunt et venti cum porcellis, et imbres cum stillatione.[101]

In northwest Norfolk the harvest also started deplorably late on the 25 August (Gregorian Calendar: 3 September). This date is so late that it constitutes the highest positive deviation of the whole harvest date series. Extra harvest works had to be performed because of the rainy weather. Due to the rain and the shortness of the days, conditions were especially problematic at the end of the long harvest, which for the *famuli* of the manor of Sedgeford lasted until mid-October (Appendix 1).[102] In south-central England the Winchester manors also encountered harvesting difficulties caused by rainfall. It can be concluded that the rainfall continued at least to mid- or late October.

As a result of the ruined crop, prices for wheat and barley rose in late 1428 and in 1429. As it is put in William Gregory's London chronicle:

> Anno vij – Henry VI. And that yere hyt was a dyre yere of corne and pryncypally of whete and of alle maner of of vytayle, for a buschelle of whete was worth xx d.[103]

Times were hard.

6.4 Weather Conditions During the Agrarian Crisis 1314–1323

The very cold growing seasons of 1314, 1315, 1319 and 1323, as well as the warm year 1318 are related to the Great Famine and the prolonged agricultural crisis in its aftermath. The Great Famine 1315–1317 has been detailed in its various socio-economic aspects by Lucas, Kershaw, Jordan and Desai.[104] Whereas Kershaw and Desai concentrate on England, Lucas and Jordan also provide a comprehensive outlook on the continental conditions. Lately, Campbell and Slavin have turned their

---

[100] Chronicon Rerum Gestarum in Monasterio Sancti Albani, vol. 1, 26–27.

[101] Ibid., 27.

[102] NRO, LEST/IB/68. Most other harvesters left the fields by the end of September.

[103] William Gregory's Chronicle of London, 164.

[104] Lucas, The Great European Famine; Kershaw, The Great Famine and Agrarian crisis in England; Jordan, The Great Famine; Desai, Agrarian crisis in medieval England.

attention towards the same subject.[105] However, a more in depth analysis of the English weather conditions for the famine period and subsequent years is required. As Kershaw stated, a number of chroniclers are not very accurate in dating the consecutive phases of the Great Famine.[106] The same doubts are uttered by Britton with respect to weather,[107] and so for a chronology of the weather conditions it is best to rely on information from administrative sources, as these were produced regularly and close to the events, so doubling and misdating is avoided, and they report the conditions objectively.[108]

According to the manorial accounts already 1313 did not offer ideal conditions for the grain harvest. Although the reconstructed East Anglian growing season temperature was average, as was the Low Countries summer half year, several manors of the Bishop of Winchester experienced problems with harvesting because of rain. The year 1314 then marks the beginning of the agricultural crisis. The growing season was cool and the harvest time was beset by rain.[109] In East Anglia the harvest was longer than average. The Winchester manors as well as Monks' Grange near to Norwich complained about rain during harvest time; in northern Norfolk, on Cromer Ridge, rain and storm caused problems and the growth of oats was hindered by water (Appendix 1). Two Winchester manors state that the harvest was not finished before the end of September or later. When famine conditions arose in 1315, they came not without warning: the cumulative effect of the problematic harvest of 1314, the disastrous harvest 1315 and of the sheep murrain taking hold in England 1315–1317, has also to be taken into account.[110]

For the crisis years 1315–1317 only few accounts of Norwich Cathedral Priory survive, but those that remain clearly reflect the cold and wet conditions of the summer half year 1315: the grain harvest was very late as well as long. The index for the Low Countries stands only at 2. For 1315 plenty of independent documentary evidence is available. The Winchester accounts present a litany of rainfall and flooding references and impacts for the growing season and harvest time. They are joined by

[105] Campbell, Great transition, 191–198, Campbell, Nature as a historical protagonist, 287–293. Slavin studied the effect of the Great Famine in Norfolk in his thesis, Feeding the brethren, 128–135 and in general in idem, Communities of famine.

[106] Kershaw, The Great Famine and Agrarian crisis in England, 88.

[107] Britton, Meteorological chronology, 70.

[108] As Hallam, The climate of eastern England 1250–1350, 124 points out, in these times of upheaval in the English political scene, accounts of bad weather in chronicles or annals could be used as symbols of misgovernment or other political problems.

[109] Several chronicles claim that the rain set in May 1314 and lasted till Easter 1315, but as Kershaw, The Great Famine and Agrarian crisis in England, 88 explains, this is probably misdated and should refer to 1315–1316. He is very likely correct, because the manorial accounts for 1314 refer to a wet harvest, but not to a wet growing season, which makes continuous rainfall between May and July 1314 rather unlikely. In 1315 then the *compoti* are full of references to rain for this period.

[110] Kershaw, The Great Famine and Agrarian crisis in England, 88–9, 102–106. Sheep murrain is often directly triggered by weather, for example in the form of flooded pastures which can be linked to liver-fluke. Foot-rot is connected to warmth and/or humidity. Generally a lack of pastures and also hay can result in malnutrition, making the animals more susceptible to disease. On duration and effect of the sheep murrain, see Desai, Agrarian crisis in medieval England, 250–251.

information from Glastonbury Abbey and Kinsbourne in Hertfordshire which does not only indicate a very wet harvest, but also a wet and cold late winter or early spring. The failed harvest drove the grain price to unprecedented heights.

Surprisingly in 1316 the situation appears improved in the East Anglian data: the harvest was early in timing and shorter than average.[111] The data can be interpreted in the form of an average to warm growing season which was followed by a dry spell during harvest time. However, the Low Countries summer index stands only at 3, and in Saxony, Bavaria, Bohemia and Austria many inundations are recorded, those that are dated took place in the second half of June after abundant rainfalls. Timing of the inundation and geographical coverage are strong indications for a typical slow moving and rain intense Genoa Low (Vb track cyclone), which is associated with a high risk for large summer floods in central Europe.[112] A Genoa Low can not reach the British Isles and indeed the administrative sources from England lend support to the possibility that summer 1316 was unspectacular in meteorological terms.[113] The Pipe Roll of the Bishopric of Winchester, never lacking in words, does not complain about rainfall in spring and summer 1316. One manor's grievance was flooded pastures, which might also have been due to the preceding wet year 1315 and the wet winter 1315–1316,[114] and another manor referred without explanation to inadequate or inconsistent weather, but none reported actual rainfall; in any case the number of complaints is small.[115] For Kinsbourne, Stern found indi-

[111] The evidence used is from Scratby, NRO, DCN 60/30/05, Worstead, NRO, DCN 60/39/06, and Hinderclay, CUL, Bacon 446. Hallam, Rural England and Wales, 1042–1350, 1005 mentions a Sedgeford account for 1314–1315 and idem, The climate of eastern England 1250–1350, 125 cites a account from the same manor for the harvest 1316, which started on 2 August and ended on Michaelmas. Such a date would be early for Sedgeford, the harvest duration very long, but the start and end date could also indicate a standardisation of the harvest season. However, it was not possible to locate these account rolls. The NRO catalogues and the Manorial Documents Register of the National Archives list no Sedgeford account for 1314–1315 and only a tithe account for 1316, NRO, LEST/IB 79, which does not give any information on harvest date or length. In Hallam, The climate of eastern England 1250–1350, Tab. 2 on harvest date and size, the year 1316 is not given either.

[112] Messmer et al., Climatology of Vb-cyclones, 542.

[113] English narrative sources do refer to the period between c. May and September as very rainy, but it is not clear if their dating is correct, or if they do not double the events of 1315, Britton, Meteorological chronology, 70.

[114] At Waltham St Lawrence pastures were partly flooded and partly suffered from 'nimiam habundanciam aque', Titow, Evidence of weather, 387. However, Waltham St Lawrence was prone to flooding and experienced flooded pastures in many winters, see Titow, Le climat à travers les rôles de comptabilité. Titow therefore only used the references to severe flooding, when published in the weather references 1350–1450, ibid.

[115] In 1315 17 references for c. May to July are available from the manors of the Bishopric of Winchester, they come from 18 manors; every single reference mentions rainfall and/or flooding. In 1316 there are four references for the time c. May to end of June, none of them explicitly states rainfall; the information was produced by three manors. The different levels of severity of the weather conditions in 1315 and 1316 are obvious. The roll of 1315–1316 ends 29 June 1316, because bishop Henry Woodlock de Merewell died 28 or 29 June 1316. The next roll, 1316–1317, includes the quarter of the year between end of June 1316 and late September 1316, without bringing more weather references for this period, Beveridge, The Winchester rolls and their dating, 97, 112.

cations for wet conditions in late spring, but also for a dry summer. All these factors support the evidence from Norfolk and Suffolk, which points to 1316 itself being – after a wet spring – a year, that saw no exceptionally bad weather conditions during the grain growing season in England.

The data for 1316 from Norfolk, Suffolk, Hertfordshire and the Winchester manors indicate that the bad harvest in 1316 and the consequent high prices 1316–1317[116] were not solely due to the weather in 1316, but also to other factors,[117] such as the likely low quantity and poor quality of the seed corn[118] and the rainfall during winter sowing in 1315 and also at spring sowing 1316.[119] In 1316 Norwich Cathedral Priory was short of wheat, but not of barley, which confirms the problems at winter sowing time 1315 and in winter 1315–1316, whereas the spring conditions in East Anglia must not have been too contrary to barley growing.[120] A lack of seed corn might have led to less densely sown crops or a reduction in sown acreage, which was indeed taking place on many Norwich Cathedral Priory manors between 1313/1314 and 1318 (for Sedgeford and Gnatingdon, see Fig. 2.3).[121] On a national level the harvest size in 1315 and 1316 was pitifully small.[122]

However, with respect to East Anglia the reconstructed temperature and the harvest length for 1316 lend support to Hallam's theory that the Great Famine was a

[116] On harvest size again: Campbell, Nature as historical protagonist, 288. On the returns of the different grain crops on the Winchester estates 1315–1316, see the summary of Kershaw, The Great Famine and Agrarian crisis in England, 98–101.

[117] For a general assessment of the vulnerability of the various grain crops to weather, during the time between sowing to harvesting, but also during the preceding year producing the seed, see Hooker, Weather and crops, 120–121.

[118] For a detailed description of the impact a very wet summer has on the growing grain, see Pfister, Agrarkonjunktur und Witterungsverlauf, 117–118 who cites eye-witness accounts of the wet summer 1758 in the region of Bern, Switzerland. Although being temporally and geographically distant, the detailed impacts are probably comparable to those in England 1315 and perhaps already 1314. In the Bern region 1758 the corn partly sprouted in the fields, partly the kernels split open, and harvesting was extremely difficult. Fermentation set in; the grain could hardly be used to produce edible bread. It can not be assumed that such grain makes good seed corn.

[119] At least the winter sowing 1315 was very wet and pastures were flooded in winter 1315–1316, Titow, Evidence of weather, 386.

[120] Hallam, The climate of eastern England 1250–1350, 126.

[121] In Eaton, Gnatingdon, Hemsby, Hindringham, Monks' Grange, Sedgeford and Taverham. The average reduction was c.11%. The largest reductions occurred just outside Norwich, perhaps because of a greater lack of seed corn so close to the town. In Eaton the sown acreage was reduced by 15.5%, in Monks' Grange by 22.5%. Scratby is not part of this development, no downsizing took place there between 1302, the last available account before the Great Famine, and 1315. The reductions occurring on the Norwich Cathedral Priory manors are minor compared to what happened on the Christ Church Canterbury manor Appledore. The drop in Appledore between 1314 and 1316 stands at 30.3%, CRU, Bickersteth, Minister's account rolls of Christ Church Canterbury 1305–1386. There might also be a connection between soil type and reduced sown acreages; in the continuous rainfall heavy soils must have suffered worse than light and well draining soils.

[122] The wheat yields on the manors belonging to the Bishopric of Winchester were down by 36% in 1315 and 45% in 1316, Titow, Evidence of weather, 285–286; barley and oats often did not fare well either, idem, Winchester yields, App. C (49), D (59), E (69). On the national demesne harvest, see Campbell, Nature as historical protagonist, 288.

difficult, but not disastrous period.[123] Although documentation of the famine years is poor, it appears that, on a more local level, the Breckland also escaped the worst in those years.[124] Slavin used the few accounts rolls of Norwich Cathedral Priory surviving for the crisis years to reconstruct harvest sizes for Eaton 1315 and Sedgeford 1316.[125] Whereas the Eaton harvest 1315 was extremely poor for all crops and especially for wheat, the data used for Sedgeford 1316 show severely depressed yields only for wheat and beans, but rye, barley, oats and to a lesser degree also peas, rendered a normal harvest quantity.[126] These results lend credence to the assumption that the 1316 barley harvest was good enough to avoid disaster in Norfolk. All evidence from the English narrative as well as administrative documentary sources, the East Anglian growing season temperature as well as harvest length, and the Eaton harvest, show 1315 as a catastrophic year, when extremely bad weather ruined all farming efforts. The summer half year 1316 still saw phases of bad weather, but did not undermine farming entirely and allowed for good harvests of rain resistant crops, at least on manors on sandy soil. The only major grain crop that failed in Sedgeford 1316 was wheat, which was still sown during the deluge 1315 and which is a rain sensitive crop. Therefore it is likely that Norfolk's (including the Breckland) well draining soils and its reliance on the hardy barley reduced the vulnerability of this region to continuous rainfall and flooding, and avoided a general second harvest failure in 1316.

Even though Norfolk escaped the worst, socio-economic impacts of the Great Famine were severe. In the fourteenth century, during periods of raised grain prices, the level of crime fluctuated with the price of wheat. Hence in the famine period 1315–1317 crimes in Norfolk rose by 382% above the pre-famine level, these crimes included also an increased percentage of stolen foodstuffs.[127] At the same time the land market was very active, bearing witness to the difficulties of the peasants. At Hindolveston the surrender of property by tenants rose by 160% in 1316 compared to the preceding year and in 1317 the number of transactions was still 70% above the level of 1315. These surrenders were made for 'great hunger'.[128]

In 1317 the East Anglian data default; other manorial accounts hold no evidence on high rainfall levels that can be attributed to the growing season and harvest 1317.

---

[123] Hallam, The climate of eastern England 1250–1350, 126, 132, see ibid., 129 on sizes of the Norfolk barley harvests 1258–1349.

[124] Bailey, A marginal economy, 201–202.

[125] Sedgeford, tithe account, NRO, LEST/IB 79. Slavin, Feeding the brethren, 130, points out, that the computations based on the tithe roll are less reliable than those based on the normal manorial accounts.

[126] Slavin, Feeding the brethren, 129–131. Average harvest calculated on the basis of 'normal' harvests between 1311–1312 and 1318–1319. Sedgeford lies on the 'Good Sands' (note 3 in Chap. 5), and well draining soils are of advantage in years of dramatically increased precipitation levels. Eaton's soils tend to slight seasonal waterlogging, Soils of England and Wales: Sheet 4 Eastern England. The manorial accounts of 1315–1316 allow no assessment of the quality of the grain harvested.

[127] Hanawalt, Crime in East Anglia, 14.

[128] Hallam, The climate of eastern England 1250–1350, 126, Fagan, Little Ice Age, 39.

The summer half year in the Low Countries was average. Conditions appear to have approached normality; the harvest definitely was better than in the preceding years, bringing down the grain prices by 50%.[129]

In England the dearth found an end in the good harvest after the very warm and dry growing season 1318. In this year the harvest in East Anglia was generally very early and shorter than average, conditions were similar further west in Staffordshire and in the south on the Winchester manors.[130] In the Winchester Pipe Roll drought references are also frequent. Around Norwich the soil was so hard that it caused problems for fallow ploughing (Appendix 1). As a result of the dry and warm weather the barley harvest in Norfolk was poor,[131] but England-wide the grain harvest was good and prices fell to a very low level after the harvest. The English weather conditions were mirrored on the other side of the sea, where the summer half year index is 7.

Lucas considered 1318 as the end of the famine, whereas Kershaw extends the phase of agrarian crisis to include the years up to 1322 on the basis of the cattle plague 1319–1321 and renewed harvest failure in 1321.[132] In fact Jordan speaks of the Great Famine 1315–1322, which implies seven consecutive harvest failures, which in England was certainly not the case and for which – as we have seen – weather conditions as a basis for such a continuous failure were non-existent.

It is worth outlining the prevalent weather modes of the growing seasons in question. The year 1319 can be classified as rainy and cold again. The reconstructed East Anglian April to July temperature is low. The manorial accounts of the Bishopric of Winchester contain isolated references to rain in the growing season and a wealth of references to rain at harvest time, which is reflected by the longer than average harvest in East Anglia.[133] The barley harvest in Norfolk was, however, normal,[134] as was

[129] Kershaw, The Great Famine and Agrarian crisis in England, 95.

[130] Lynam (ed.), Croxden Chronicle, vii.

[131] Hallam, The climate of eastern England 1250–1350, 129. In the Breckland agricultural output was also reduced, Bailey, A marginal economy?, 201. Neither the poor Norfolk barley harvest nor the bad general grain harvest in the Breckland are surprising. The Breckland soils are very drought sensitive, and barley is vulnerable to prolonged dry conditions.

[132] Kershaw, The Great Famine and Agrarian crisis in England, 96.

[133] Hallam, The climate of eastern England 1250–1350, 126 and 128, claims, that narrative sources describe a summer drought, that preceded the wet autumn in 1319, without giving the sources. However, those references could not be found. Britton, Meteorological chronology, has nothing for 1319, Ogilvie gives no drought index in 1319, the Pipe Rolls of the Bishopric of Winchester mention no dry weather, Stern identifies nothing in the Kinsbourne (Hertfordshire) accounts hinting at drought, Bickersteth cites no drought evidence in the manorial accounts of Christ Church Canterbury. However, the reconstructed Norfolk growing season mean temperature is low, which makes prolonged dry conditions in the summer half year rather unlikely. No Dutch summer season index is available.

[134] Hallam, The climate of eastern England 1250–1350, 129. Considering Hooker's statement that barley likes a cool growing season between May and August and the temperatures in East Anglia are often too high for optimum conditions, the normal barley harvest 1319 does not come as a surprise, Hooker, Weather and crops, 121.

the English grain harvest in general.[135] As long as the narrative sources have not misdated the Great Famine, they remain silent for 1319.[136] The real calamity befalling England in this year was the cattle plague, arriving around Easter in Essex.[137] As Slavin points out, the frequent lack of pasture, hay and fodder in the preceding years will have raised the vulnerability of the cattle.[138] Due to the cattle plague the average English demesne cattle stock shrank by 63%.[139] While Desai sets the period for restocking somewhat optimistically to roughly 10 years,[140] Slavin found that oxen numbers reached c.80% of the pre-plague levels by 1331 and dairy cattle c.90% by 1341.[141] Whereas the effects of famines were relatively quickly overcome for the survivors, the rebuilding of cattle herds needed decades. The epizootic impacted on the agricultural sector by the loss of draught animals and manure, and on human diet directly via a severe drop in dairy and meat consumption.[142]

According to the Norwich Cathedral Priory harvest data the weather situation improved in 1320, the reconstructed April to July temperature was average, but the harvest duration was above average. The latter is easily explained: once more it was a rainy harvest and the Winchester manors experienced problems, too. The manorial accounts remain silent in respect to the months between early spring and harvest, which implies average conditions. The East Anglian harvest date complies with this. Neither for 1319 nor for 1320 are enough data available in the Low Countries to permit indexing the summer half year. Whereas Kershaw saw indications that the harvest 1320 was merely mediocre and therefore the grain price 1320–1321 was higher than in the previous year, Campbell's comparatively recently collected data from the demesne farming sector shows average yields.[143]

Then in 1321 harvest failure returned and England was part of a zone of poor harvests stretching from the British Isles, probably over northern France, and then from central Europe to northern Italy. On the island the grain prices again approached the levels of 1315–1317.[144] The chronicles in England do not refer to adverse weather conditions during the growing season, making Kershaw suspect a drought instead of the more striking rainfall as cause of the English harvest failure.[145] During

[135] Campbell, Nature as historical protagonist, 288.

[136] Kershaw, The Great Famine and Agrarian crisis in England, 88.

[137] Slavin, Cattle plague, 166.

[138] Ibid., 167.

[139] Ibid., 168.

[140] Desai, Agrarian crisis in medieval England, 257.

[141] Slavin, Cattle plague, 177.

[142] Ibid., 169–171.

[143] Kershaw, The Great Famine and Agrarian crisis in England, 97, Campbell, Nature as historical protagonist, 288.

[144] Alexandre, Le climat en Europe, 442–443, Campbell, Nature as historical protagonist, 288, Kershaw, The Great Famine and Agrarian crisis in England, 97.

[145] Ibid., 97. The Brut, 223, mentions snowballs in connection to the execution of Thomas of Lancaster on the 22 March at Pontefract Castle under the 1321. This would indicate a long hard winter 1320–1321, however, the Brut misdates the events, Thomas of Lancaster died in 1322.

the political upheaval in England in that year, a silent drought might not have commanded the chroniclers' attention. Indeed, apart from very few references to dry conditions sometime between May and June the Winchester estates also do not supply any weather information for 1321, but the given references are not sufficient for suspecting much more than a dry spell. The Kinsbourne accounts indicate wet weather at winter sowing time in 1320 and then again flooding in spring 1321, both would impede upon sowing and lower the chances of harvest success. In East Anglia the growing season temperature was slightly above average and would allow for a period of warmth and possibly dry conditions in spring and early summer, but the summer index for the Low Countries is slightly below average in 1321, at 4. The East Anglian harvest length was average in 1321. None of these references and proxies indicate a really severe drought that could have destroyed the grain crops. Looking across the Channel reveals a wet summer period in northern France, where in the Paris region processions took place in August to stop the rains, which were damaging the crops. The year counted as wet in central Germany. Hail and rain played a role in the harvest failures in Germany, in Italy they were due to rainfall setting in during late July. The rain affecting northern France in August might have reached into England, but no references to it survive. Tree-ring data from England point indeed to rainfall levels above average, but they were by no means excessive.[146] The evidence assembled points to the harvest failure 1321 being due to a combination of factors: wet seed corn from the harvest 1320, probably raised rainfall levels at winter sowing time 1320, increased wetness levels in spring 1321 followed by drier conditions and possibly rainy weather at harvest. This resembles strongly one of the weather patterns that forebode badly for the grain harvest according to Titow,[147] but still cannot fully explain the exploding grain price.

The harvest was average in 1322[148] and after the harvest the grain price began to fall, but corn stayed expensive until the mid-1320s. Not much contemporary evidence is available from England, also no manorial accounts of the Norwich Cathedral Priory estates survive for 1321–1322.[149] In Hertfordshire late winter and early spring were very cold; in late March the ground appears to have been still covered by snow at Pontefract Castle in northern England.[150] Conditions in northern France were similar, the frost lasted until mid-March, the snowcover even longer.[151] In addition to the extremely long harvest in East Anglia, information from Hertfordshire also indicates a wet late summer and autumn period. Bad weather also

[146] Alexandre, Le climat en Europe, 442–443. For the tree rings see Cooper et al., Hydroclimate variability.

[147] A wet autumn followed by an unremarkable and probably average winter and a dry growing season, Titow, Evidence of weather 363.

[148] Campbell, Nature as historical protagonist, 288.

[149] The accounts of the Bishopric of Winchester also default, the see was void, see Beveridge, The Winchester rolls and their dating, 98.

[150] When on the way to his execution snowballs where thrown at Thomas of Lancaster on 22 March 1322 at Pontefract Castle, Brut, 223 (in the Brut erroneously under 1321).

[151] Alexandre, Le climat en Europe, 443–444.

characterizes the summer half year in the Low Countries which was cool (3), in some areas inundations due to snowmelt lasted until May and new floods occurred in late June.[152]

The meteorological conditions of 1323 are obscure. The winter 1322–1323 was extremely cold. In Hertfordshire parts of spring seem to have been wet. For the summer half year the documentary sources in the Low Countries supply insufficient data to set an index. However, indications are that spring and summer were not ideal for grain cultivation. The winter 1322–1323 was probably not only very cold, but also long, thereby delaying the onset of the growing season, which in East Anglia started out wet and could never make up for the late start; April–July mean temperatures were low. In late June a tornado and a water spout did significant damage in the surroundings of Leeds.[153] Such phenomena arise, when advancing cold fronts meet warmer and wet air masses over the British Isles in summer. The Norfolk barley harvest was good,[154] as was the grain harvest on a national level,[155] but the grain price, even though sinking, remained high after the harvest 1323. The 1323 growing season was the last marked by a low mean temperature and probably at least spells of wet weather. With this year ends the cycle of recurrent cold and wet years between 1314 and 1323.

The analysis of the years 1320–1323 reveals a period when the rising or high grain prices seem to be strangely detached from the weather events (only 1322 saw clearly a wet growing season) and also from the harvest success of the demesne farming sector, which normally was a strong predictor for the grain price.[156] The Great Famine 1315–1317 and its effects generally stand in the centre of attention of studies dealing with the agricultural and pastoral upheaval 1314–1323, hence it is advisable to focus here on the cause and the socio-economic consequences of the price spike in the early 1320s. Campbell's data on the agricultural and pastoral sectors of demesne farming and Slavin's data on the cattle plague inform the following analysis. In 1320, 1322 and 1323 harvests were average or good, but the grain price first was rising and later not returning quickly to normal after the spike caused by the 1321 harvest failure. Part of this discrepancy can be explained with the harvests being wet in 1320, 1322 and possibly in 1321, and wet harvests are voluminous and heavy, while the quality of the grain is actually lower than that of grain gathered in under dry conditions.

However, the rapid succession of the cattle plague and the period of high grain prices suggests that the massive mortality amongst the cattle impacted on the grain

152 Ibid., 444.

153 Flores Historiarum, vol. 3, 216–217.

154 Hallam, The climate of eastern England 1250–1350, 129.

155 Campbell, Nature as historical protagonist, 288. In Cuxham, Oxfordshire, the barley harvest was normal whereas the wheat harvest was very poor, Hallam, The climate of eastern England 1250–1350, 131, based on the data from Harvey (ed.), Manorial records of Cuxham. The poor wheat harvest might be due to a wet growing season, but also to rainy weather at sowing time as indicated by, Stern, A Hertfordshire demesne, 94, who states that the wheat seed was damp when sown in autumn 1322. Wet seed corn is detrimental to the success of wheat cultivation, Hooker, Weather and crops, 120.

156 Campbell, Nature as historical protagonist, 297.

Fig. 6.1 Ploughing with oxen. Luttrell Psalter, Lincolnshire, circa 1325–1340 (British Library, Add. MS 42130, f. 170)

production and price. In its immediate aftermath the catastrophe in the pastoral sector raised the vulnerability of the agricultural sector to the vagaries of the weather. The loss of draught animals posed a problem to the time-intensive ploughing (Fig. 6.1). A reduction of manure reduced the available amount of fertiliser. The transport costs for grain must have increased, which in turn would be reflected in the grain price.[157] Regions where horses were already well in use as draught animals, as in Norfolk, were suffering less from the cattle plague, as were the peasantry who were quicker than the demesne sector at adopting the horse as draught animal. However, possibly due to stocking densities and transport links, the mortality of cattle was higher further west and south, where the use of oxen was still predominant.[158]

The spread of the cattle plague shows the sequence in which regions dropped into a state of acute shortage of draught power. After its arrival in Essex at Easter 1319 the murrain joined the supply trains for the war with Scotland and by August 1319 had travelled to northern England, in late summer it had spread across East Anglia, but its spread westwards was slower and Wales was reached around September 1320.[159] Therefore winter ploughing 1319 was primarily affected in the east, where horses were the dominant draught animal, spring ploughing 1320 was under threat everywhere. The cultivated acreage contracted for the harvest 1320, which was average on the acres that could be sown, thus explaining the different evaluation of that harvest by Kershaw and Campbell. For the harvest 1321 part of winter ploughing may still have been pulled through at some places in the west, but by spring ploughing time, the murrain had run its course. In this context, it is of interest that the grains failing worst in 1321 were the spring grains barley and oats, not the winter grains wheat and rye.[160] However, barley and oats are also more

[157] Ibid., 289.

[158] On the geographical distribution and the reasons for the varying mortality levels, see Slavin, Cattle plague, 168–169. For the use of oxen and horse as draught animals across England, see Langdon, Was England a technological backwater, 282–283, idem, Economics of horses and oxen, 40.

[159] Slavin, Cattle plague, 166.

[160] Campbell, Nature as historical protagonist, 288.

drought sensitive than the winter corn. When analysing the reasons for harvest failure and dearth Merle in 'De pronosticacione aeris' – even though the text focuses mainly on meteorological parameters – names the lack of ploughing or the wrongly timed ploughing as the first cause of harvest failure; without him specifying his motif, it appears as if the short and long-term effects of the cattle plague were still at the forefront of his mind 20 years after the cattle plague.[161]

Keeping the shortage of draught animals in mind, that what normally would have been simply bad weather with associated problems at the ploughing times in the early 1320s, now became an existential threat. Winter and spring ploughing were repeatedly disrupted by rainy weather or cold and long winters. It was raining in autumn 1320 and in spring 1321 there was flooding, this would have slowed down the ploughing work further. The winters 1321–1322 and 1322–1323 were both hard and long, scoring 7 and 8 in the Low Countries, and in 1322 spring ploughing was delayed by weeks. When speed was crucial there was a dramatic lack of draught power to compensate for the late start of the ploughing season, and tilling the land with spades is a slow, work intensive process, which can never make up for a lack of plough-animals. Harvests on the ploughed land were still acceptable, but the cultivated acreage declined sharply. Even on the estates of the Bishopric of Winchester arable acres were halved between 1319 and 1321.[162] No data are available for the decrease of cultivated acres in the peasant sector of farming, but for the peasants, so shortly after the last famine and faced with renewed high grain prices, resources for maintaining their cultivated acres must have been even scarcer than those of the lords. And they kept shrinking.

Already in 1320 the Bishopric of Winchester doubled the numbers of horses on its estates.[163] The nadir of the demesne oxen numbers was reached in that year, but from 1321 onwards they increased rapidly for about a decade, while the number of other cattle remained low for about the same time.[164] This recovery of oxen numbers was clearly not achieved by reproduction, but by acquisition. Such a restocking strategy of draught animals and the necessary purchases at a time of high grain and high cattle prices could only be pursued by wealthy lords, and the only source for this replenishment of demesne ox teams was the cattle of lesser landowners and peasants. Since the cattle plague had ravaged all herds, the replenishment of the great estates was a reallocation of the scarce resource of draught power. In those times of political upheaval not all transactions were by free will or entirely legal, as the events at Croxden Abbey in Staffordshire illustrate. The monasteries neighbour at Alton (Alveton), decided to compensate his own losses by taking not only one of the abbey's wagons and 160 of its sheep, but 20 oxen and 32 horses, when they were just under plough. The abbey and the lord of Alton made peace within a few months, but as a result of the robbery, the abbey could not plough the lands at the grange and

[161] Merle, De pronosticacione aeris, BLO, MS Digby 147, fols. 136r–136v.

[162] Slavin, Cattle plague, 169.

[163] Ibid., 176.

[164] Campbell, Physical shocks, 25, Slavin, Cattle plague, 178.

another manor, whereas the lord of Alton could probably do his ploughing.[165] In general at a time when there were too few draught animals, while pasture was plentiful, it would have been reasonable to literally cling on to oxen and to profit from their ploughing and carting services, and only the spectre of hunger that was raising its head with the harvest failure 1321 could have convinced peasants and small landowners to do otherwise. As the consumption of seed corn and the sale of land, peasants putting their oxen on the market equalled the decision to sell the future productivity of their land for the survival in the present. None of these actions were a sustainable coping measure, when faced with persistently high grain prices. By 1323, in the third year of raised prices, the poorer sections of the English people had met with their limits: land sales and the crime rate shot up. The sale of land and the recourse to crime were coping mechanism that had already been employed during the Great Famine.[166] Under those circumstances it is unlikely that the peasant agriculture could compete with the performance of seigniorial agriculture during the early 1320s. The situation improved in the mid-1320s, when weather better suited for grain cultivation succeeded the volatile phase 1314–1323 and the reproduction of the bovine population yielded the first generation of adult trained oxen born after the cattle plague; grain and cattle prices fell. The catastrophe in the pastoral sector therefore raised the vulnerability in the agricultural production and during the immediate post-epizootic period the cattle murrain enhanced the stress caused by inclement weather during agricultural activities. The murrain was itself an important factor driving up the grain price in the early 1320s.

The data on harvest date and length from East Anglia add new information about the severity and phases of the climatic conditions during the Great Famine, the time of the cattle plague and the subsequent difficult years in the early 1320s. As Kershaw stated for this period in England, it 'was not a single entity'. Nor were the climatic conditions present. They rather represent a jigsaw, containing extremely wet and cool years, as well as warm and dry years. Excessive rainfall set off a vicious sequence, causing first a wet harvest 1314 and then severe harvest failure in 1315. Since no major adverse weather is recorded in the administrative sources or mirrored in the East Anglian data for growing season and harvest time 1316, the continued harvest failure at least in this region was largely due to wet sowing times in 1315–1316 and a wet spring 1316 and to some extent also to the agricultural and socio-economic consequences of the preceding poor and wet harvests. The latter also applies to the harvest 1317 which was still 10% below average.[167] In other words it took medieval English agriculture two years to offset the cumulative impact of the sodden harvest 1314, the rain induced disaster of 1315 and the subsequent failure 1316. In England the grain prices fell to normal levels, helped by good grain growing conditions, in 1318, and high grain prices were not to return until the harvest 1321, though the harvests 1319 and 1320 were wet and in 1319 the growing

[165] Lynam (ed.), Croxden Chronicle, vii–viii; on the general lawlessness and plundering of the times, see Fryde, Edward II, 69–86, 149–152.

[166] Campbell, Nature as historical protagonist, 291–292, Hanawalt, Crime in East Anglia, 14–15.

[167] Campbell, Nature as historical protagonist, 288.

season also cool. After the fall of the grain prices to normal levels in 1318 it was less than a year left before the arrival of the cattle plague in spring 1319, which constitutes the second major factor in the agrarian crisis 1314–1323. The high mortality was linked to the long phases of malnutrition the bovine population must have suffered in the preceding years and also in the winter 1318–1319 due to flooding and the destroyed hay crop.[168] The harvest failure 1321 appears to be the result of a variety of factors such as the use of wet seed corn, wet ploughing times, a lack of draught power, and for the growing season neither a severe drought, nor heavy rainfall in summer can be ruled out. For this harvest and in the immediately following years, of which 1322 was once more wet and cold, the lack of cattle contributed to the high grain prices.

To extend the Great Famine in England from its core years 1315–1317 to cover the following cattle plague and the harvest failure 1321 or even longer[169] is confounding the term with respect to cause and outcome. There was no continuous climatic crisis in the form of those dangerous cold and extremely wet growing seasons and there was also no consecutive harvest failure for seven years. A major crisis in the agricultural sector was followed by one in the pastoral sector, both with mutual repercussions. Although the cattle plague had in the long-run more profound and severe consequences for agricultural production than the Great Famine, it did not result in the immediate death of c.10% of the population,[170] so that it could be adequately labelled 'Great Famine'. Although it created problems for agriculture, the difficult two decades for landowners after c.1325 were mainly caused by low corn prices[171] that means by good harvests.[172]

6.5 Weather Conditions During the Great Pestilence 1348–1349 and the Agricultural Crisis 1348–1352

The arrival of the Black Death in Britain and the agricultural crisis in the late 1340s followed a sequence of years with suboptimal conditions for farming. For 1345 Adam Murimuth describes a very wet harvest. It was raining throughout August and in some regions corn perished in the wet weather and could not be stored dry.[173] This continuous rainfall affected the west of England, where the manors of the Bishopric of Winchester and of Glastonbury Abbey complained about rainfall, but further east conditions were probably less severe as in Sussex the season counts as unremarkable and in East Anglia the harvest was longer than average, but not very long.

[168] Titow, Evidence of weather, 387.
[169] Slavin, Cattle plague, 179 extends it to 'at least until the 1330s and beyond.'
[170] Kershaw, The Great Famine and Agrarian crisis in England, 93.
[171] Desai, Agrarian crisis in medieval England, 253.
[172] Campbell, Physical shocks, 28.
[173] Murimuth, Continuatio Chronicarum, 173.

However, as is indicated by Murimuth[174] the rainfall persisted well into autumn and until the sowing time of the winter corn, as evidence from Hertfordshire and the Winchester and Glastonbury manors also shows. Flooding took place in winter. High levels of rainfall during sowing time are detrimental to crop growth, as is poor quality seed corn from a wet harvest, and in consequence of these conditions wheat and rye suffered. The harvest 1346 was not sufficient to avoid a price rise and dearth took hold in England about a month before the harvest 1347.[175]

With 1347 the temperature of the summer season began to drop towards the extremely cold summer of 1348. The growing season opened with dry weather as the manorial accounts from Hertfordshire, Sussex and Surrey attest. The conditions were so severe as to result in suffering from heat, a lack of fresh water and dysentery for the English besieging Calais.[176] By harvest time the character of the season had changed and rainfall had set in on the Winchester manors and in Sussex. Reflecting the cool-wet second half of summer the Low Countries index stands at 4. In East Anglia the harvest was short, showing drier conditions in this region but also demonstrating the resolution to finish harvesting before the end of the official harvesting season in late September. Nonetheless the grain crops, including the sturdy barley, responded badly to such weather, and the grain price did not sink in the year 1347–1348.

While the Black Death was spreading north from the European Mediterranean coast across the continent, the poor in England must have been facing difficulties in securing their daily bread, and indeed the weather in 1348, the year the epidemic crossed the Channel, would increase the pressure on grain cultivation. Apart from being more likely to rot, grain grown and stored in wet conditions is nutritionally inferior and more liable to infection with fungi. In short, many English must have encountered the newly arrived epidemic in a malnourished and immunodepressed state.

In 1348 the April to July average temperature was low (Fig. 5.4), the summer in the Low Countries was average. The year is also known to have been wet, at least from Midsummer onwards, in English chronicles references to continuous rainfall from about the Nativity of St John the Baptist (24 June) to Christmas are plentiful.[177] John of Reading gives one of the accounts written closest to the actual event.

> [Anno gratiae MCCCXLVIII] Eodem anno, circa partes meridianas occidentalesque inundaverunt pluviae a Nativitate Sancti Johannis Baptistae usque ad Nativitatem Domini, vix diebus vel noctibus cessantes quin plueret aliquantulum.[178]

---

[174] Ibid.

[175] Knighton, Chronicon, vol. 2, 88–89.

[176] Ibid., 86–87.

[177] For a list see, Britton, Meteorological chronology, 141. To his list can be added various London chronicles, which tend to misdate the event by one or two years.

[178] Reading, Chronica, 106. In a note the editor adds, that the reference bears a close similarity to a passage in Higden, Polychronicon, vol. 8, 347. The remark about the rain falling in the west and south, though, is unique to John of Reading and found nowhere else.

The high levels of precipitation are explicitly localized in southern and western England, and rainfall impacts were indeed recorded for Sussex and the Winchester manors, whereas in East Anglia the harvest duration was shorter than average and the Hertfordshire manors of Westminster Abbey did not report any rainfall impacts in 1348, as they do then in the following year 1349.

In 1349 when the Black Death raged in England, the weather remained very wet, in fact the rainfall might have been more severe than in 1348. The harvest was wet in Hertfordshire and on the manors of Glastonbury Abbey; the Winchester accounts for winter, late summer and harvest mention many problems caused by floods and rain. The harvest in East Anglia was long, though it is not clear if this was due to the rain or to the lack of labour during the epidemic.[179] In Taverham oats perished on the fields,[180] this might have been the effect of an unmentioned dry period in summer, or – since oats are not likely to suffer much from rainfall – it might also been a result of a lack of harvesters as the villagers fell victim to the plague or survivors demanded higher wages. In England unharvested fields were common in 1349.[181] Nonetheless the Norfolk barley harvests defied the weather conditions as well as the lack of labour and were normal or good in 1348 and 1349,[182] Henry Knighton also describes the corn as abundant.[183] However, the national English grain harvest fared considerably less well.[184] Major price rises were only avoided because of the great mortality which led to a sudden breakdown on the demand side.[185]

Another very bad harvest followed immediately in 1350, indeed in the seigniorial sector of English agriculture seed-yield ratios remained far below the long term average until 1357 and were even 40% below average for the years 1349–1352.[186] The difficulties in maintaining agricultural output are observed in the Polychronicon which complains about the land and sea in post-plague years being more barren than before.[187] Traditionally these harvest failures were attributed to the disruption of agricultural activities and the demographic downturn. Campbell widens this view and emphasizes the role of unfavourable environmental conditions which was also indicated in the Polychronicon.[188] As indicators he uses tree growth data, which slumped in the late 1340s, and information about temperature over western

[179] The harvest duration 1349 and 1350 on most Norfolk manors was actually exorbitantly long; which is a clear sign for a greatly diminished labour supply for (demesne) farming. Such data were excluded from the statistical analysis.

[180] NRO, DCN 60/35/30 for 1349–1350, but referring to 1349. For the difficult farming year 1349 almost no Norwich Cathedral Priory accounts survive.

[181] Knighton, Chronicon, vol. 2, 100–101.

[182] Hallam, The climate of eastern England 1250–1350, 129.

[183] Knighton, Chronicon, vol. 2, 100–101.

[184] Campbell, Nature as historical protagonist, 301; Dodds, Estimating arable output, 270.

[185] Hatcher, Plague, 21–25. Concerning the disruption of the grain market and the price system during plague years, see Campbell, Nature as historical protagonist, 304.

[186] Campbell, Nature as historical protagonist, 301.

[187] Higden, Polychronicon, vol. 8, 347.

[188] Campbell, Physical shocks, 20–24, idem, Nature as historical protagonist, 300–305.

Greenland, which shows a period of intense cold 1349–1353.[189] However, since the excursions of cold air over Greenland extended until 1353, this implies that bad weather played also an important role in the harvest failures in 1350, 1351 and 1352.

Yet evidence of prolonged rainy and cold conditions is scarce for these years. In Sussex the year 1350 appears normal, in the Winchester area a few pastures or meadows were flooded during winter or summer and in Hertfordshire there is evidence not only for a wet spring, but also for a dry summer, followed by fine weather in autumn. In the Low Countries the summer was warm and fine, rain affected farming in August.[190] The East Anglian growing season and the summer half year in the Low Countries were average to cool,[191] the latter is indexed 4. The bad harvest 1350 and the subsequent price rise for grains was partly still due to the social upheaval in the wake of the Great Pestilence[192] and partly to the seed corn stemming from the two wet and bad harvests of the preceding years, the wet spring and possibly a wet harvest time; but rainfall levels were not comparable to those in 1348 and 1349.

The grain growing season of 1351 is more difficult to assess. Again no references to weather events interfering with agriculture could be found in Sussex, and whereas the Westminster Abbey manors in Hertfordshire experienced difficulties in winter and then a wet spring, no references to summer exist, and October and November saw fine weather. The spring might have been wet in south-central England too, there the Winchester manors repeatedly reported flooding in winter and summer. The summer flooding seems to centre around hay harvest time, since hay making was interrupted, the hay was partly ruined or even carried away by the water. For southeastern England very dry growing season conditions are assumed.[193] At least from May onwards it was warm in the Low Countries, the vegetation was advanced, the grain harvest is said to have suffered from the heat. After August it was rainy.[194] The East Anglian April–July mean temperature and harvest duration were average. In fact in the years immediately following the Great Pestilence an average harvest length is, considering the labour reorganisation, comparatively short and could well indicate dry conditions (Sect. 7.1). The Low Countries summer stands at 6 and more continental Europe, France and Germany, witnessed a hot and dry summer *par excellence*, so hot and dry indeed, that here too the grain was damaged.[195] The meteorological character of the growing seasons in 1350 and 1351 is therefore very dif-

[189] Dawson et al., Greenland (GISP2) ice core, 431.

[190] Alexandre, Le climat en Europe, 478.

[191] Most harvests in East Anglia were excessively long 1350, possibly that was not only due to the social upheaval, but also to rainfall in August 1350. However, there is no English direct evidence for the rainfall of the Low Countries stretching indeed into England in August 1350.

[192] For example on Breckland manors in 1349 boon services were not performed and labour had to be hired, disrupting demesne farming, Bailey, A marginal economy?, 224. 770 out of 890 works owed by the tenants to the lord at Fornham were still not rendered in 1352–1353.

[193] Classic drought impacts are cited by Mate, Agrarian economy after the Black Death, 342–343.

[194] Alexandre, Le climat en Europe, 479.

[195] Le Roy Ladurie, Histoire humaine et comparée du climat, 68–70.

ferent from the preceding two years. The weather was less likely to have caused massive harvest failure by itself in England, although in 1351 an unpleasant spring and then drought in parts of summer probably played a role. In any case prices continued to rise further.

In 1352 then conditions are totally reversed compared to the years 1348–1349. The reconstructed growing season temperature in East Anglia is average to warm, and in the Low Countries the summer was similar (7). It was a very dry spring and early summer. The account rolls of the manors of the Bishopric of Winchester abound with references to drought and even report heat. Some of the manors attributed their problems with the growing of spring corn, which is drought sensitive, to the lack of rain. Barley and oats harvests were actually only at about half their normal yield. Drought information also comes from the manors of Christ Church Canterbury.[196] After the dry weather and the damage to the spring corn, rain fell in summer in Sussex, in East Anglia the harvest duration was average in the context of the whole series, so it might again indicate dry conditions in these chaotic post-plague times. In the Low Countries the summer was hot and dry, the drought was severe and greatly reduced the quantity of the hay, fruit, legume and oat harvests; however, under the warm conditions the wheat harvest was good and the wine of good quality.[197] These weather conditions extended further into continental Europe.[198] Some narrative English sources report a major drought in 1352, others in 1353. Henry Knighton describes under 1352, how in the extreme drought the cattle perished from a lack of water on the pastures and the marshes dried up, so that new pathways appeared.[199] John of Reading refers to the damage done by the dry weather from March to July 1353 to the corn and hay and mentions even relief efforts in the form of grain shipments from Zeeland to help the Londoners.

> [Anno gratiae MCCC quinquagesimo iij] Provenit et hoc anno tanta siccitas a mense Martii usque ad Julii mensem, quod non cecidit pluvia super terram, unde fructifera, seminata et herbae pro majori parte perierunt. Ob quorum defectum sequebatur magna miseria hominum et / jumentorum caristiaque victualium, adeo quod Anglia semper fertilis ab insulanis indiguit quaerere victui necessaria. Quorum misertus egestatis dux Willelmus de Selond plures naves onustas siligine Londonias direxit.[200]

Although in 1353 spring was very dry in Sussex, Hertfordshire, and probably also in Norfolk (Appendix 1), and these conditions reached into the Low Countries, the abovementioned drought descriptions in English chronicles refer rather to 1352, when the drought was most severe in the lands across the North Sea and also in England according to the administrative accounts.[201] Although the oat harvest was lost in some

---

[196] CRU, Bickersteth, Minister's account rolls of Christ Church Canterbury 1305–1386, Mate, Agrarian economy after the Black Death, 343.

[197] Alexandre, Le climat en Europe, 480.

[198] Le Roy Ladurie, Histoire humaine et comparée du climat, 72.

[199] Knighton, Chronicon, vol. 2, 118–119.

[200] Reading, Chronica, 117–118.

[201] For a list of the sources describing the drought, and on the subject of misdating this event, see Reading, Chronica, 117–118. The editor of the Chronica assumes that Reading's drought para-

places in Sussex, the general grain harvest in 1353 can not have failed, since the grain price was low over the following months. The temperature indicators point to an unspectacular year: the growing season temperature in East Anglia was average to warm, in the Low Countries the summer was average. The climatic factor causing the bad harvests in 1352 and potentially also 1353 was drought, and the socio-economic impact of the post-plague labour shortage as well as a reduction in the sown acreage (for Sedgeford-Gnatingdon see Fig. 2.3) must still have played a role.

Even though the agricultural crisis around the time of the arrival of the Black Death came to an end in the early 1350s, severe climatic conditions returned in the mid-1350s. The harsh and strangely contrary weather of 1356 is worth describing: spring and early summer were marked by severe drought, and from the end of June this was followed by a very rainy summer. In Sussex the year was dry. Robert of Avesbury and the manors in the hand of the Bishop of Winchester give an impressive list of drought impacts, before the latter embark on the problems of a rainy and long harvest. According to Avesbury the drought started in March and was ended by rainfall beginning around midsummer. The rain invigorated the withered crops.

> [...] a medietate mensis Marcii usque festum Nativitatis sancti Johannis baptistae, [...] valde modicum pluit, sed tanta fuit siccitas quod ordea, avenae, vescae, fabae, et alia semina quadragesimalia in multis locis Angliae modicum vel nihil crescebant. Post dictum vero festum sancti Johannis baptistae pluit in magna abundantia; et dicta semina, ordea, avenae, vescae, et fabae incipiebant crescere et fuerunt satis alta et spissa, et tamen edgrowe.[202]

One Winchester manor harvested until All Saints Day, 1 November (Gregorian Calendar: 9 November). This extraordinary duration and also the longer than average harvest in East Anglia can partly be attributed to the renewed crop growth after the end of June. Therefore the climatic conditions 1348–1353 in England can be classified as highly variable. In 1348, 1349, 1352 and possibly 1353 the weather, though first in the form of relentless rainfall and cold, and then in the form of excessive drought and heat, impacted greatly on the harvest success. In 1351 conditions were very variable, spring appears as very wet and summer as dry. Indications are that the crop failure 1350 was not so much the result of extreme weather during the growing season (August 1350 may have been wet in England as it was in the Low Countries), but was rather caused by the dearth and rain impacts in the preceding years (lack of seed corn, wet seed corn), and by the societal disruption in the wake of the Black Death. Quantifying the climatic and the demographic influence is impossible.

Considering the variable climatic conditions during spring and summer over England and the Low Countries for 1348–1353 the connection of the English harvest success to the low temperatures over western Greenland is unclear. The use of cold air excursions over Greenland and the associated North Atlantic sea-surface temperature proxy record as predictors for summer rainfall in Europe[203] is problem-

graph is referring to 1352 and his description of the withering fruit, seeds and grass indeed echoes the drought impacts in the Low Countries in that year, where fruit, hay and grass also suffered.

[202] Avesbury, De Gestis Mirabilibus, 468. 'Edgrowe' is the 'aftermath' according to the editor.

[203] Dawson et al., Greenland (GISP2) ice core, 430–433.

atic, the signal of annual or winter meteorological conditions in Europe contained within the North Atlantic sea-surface temperatures is not a good indicator of the European summer rainfall pattern.[204]

The wet and cool summer seasons of the late 1340s and the consequent harvest failures, could well have, as Campbell has stated, led to another Great Famine, if it were not for the Great Pestilence wiping out about one third of the population of England.[205] As things stood, the bad and wet harvests raised the vulnerability of the people to disease. Considering the demographic catastrophe and the climatic extremes, it is clear that in 1348–1353 all certainties and securities were obliterated, even the weather was entirely unpredictable, the experience and resources to cope with it were reduced, and the general character of these years appears as merciless.

6.6 Summary of Extremely Warm and Cold Growing Seasons

Complementing extreme reconstructed mean temperatures April–July with independent documentary evidence permits a fuller picture to be formed of the extreme years. Cold and wet growing seasons attracted much attention by the chroniclers as

---

[204] Shorter episodes of cold air incursions over Greenland given by Dawson et al., Greenland (GISP2) ice core, 430–433 for the early 1330s, the second half of the 1330s and the early 1360s did not go hand in hand with rainy and cold summer half years in England. This becomes clear from the annually resolved April–July temperature reconstruction and July to September precipitation index, as well as from evidence from other written records. The early 1330s were characterized by warm and dry growing seasons, 1331 was marked by a severe drought in spring and early summer (for the weather conditions of the summer half years in England see Sect. 6.2); dry conditions with average temperatures prevailed in the springs and summers in the second half of the 1330s. Campbell, Nature as historical protagonist, 299, notes that from 1332 to 1338 all grain harvests were favourable, 1333 and 1338 belong to the best in the period 1270–1429. In the early 1360s the spring and summer 1361 were extremely warm and dry in England. The period of high sea-surface temperatures 1323–1326 and the associated cold air incursion over Greenland 1323–1327 centres around the extreme dry summers 1325–1326, 1326 was also very warm. Interestingly, however, these dry and warm summer half years in northwest Europe which coincided with the warm events in the sea-surface temperature over the western North Atlantic and cold air incursions over Greenland often either started with or were predated by a wet and cold summer season. In the light of the indication by Jones et al., High-resolution palaeoclimatic records, 458 that dating in ice-core series is dependent on layer accumulation and layer counting, which becomes increasingly difficult with depth, i.e. increasing age of the ice core, it is important to outline this pattern. During the agricultural crisis causing the Great Famine 1315–1318, which coincided with a warm event in the North Atlantic, the extreme wet and cool summer conditions in England and northwestern Europe were concentrated in the earlier part of the crisis, 1314–1316. By 1317 weather normalized and 1318 saw a dry and warm summer season. The warm event in the North Atlantic 1323–1326 was preceded by the cold summer half year 1322, the high temperature event in the North Atlantic then started with the potentially cool and wet 1323, the event 1331–1333 was preceded by the appalling summer of 1330, the cold air excursion over Greenland 1336–1338 came after a wet and cold summer in 1335 and the warm event in the North Atlantic sea-surface temperatures 1350–1351 followed the extremely rainy summers 1348–1349, 1348 was also a cool summer half year.

[205] Campbell, Physical shocks, 29.

well as by the reeves and bailiffs who were eager to explain the severe impacts on agriculture and their unprofitable management of the manors. Most of the reconstructed very cold growing seasons are connected with bad harvests and partly even famine conditions; the incidence of murrains amongst livestock was also raised under very cold and wet spring-summer weather.

In warm and dry summer seasons the risks for harvest success were much lower and hence such years frequently escaped the notice of chroniclers and annalists. Additionally, wheat, the bread grain of the upper strata of society and also of the chroniclers, is drought resistant and much more likely to suffer from rain than dry weather. Barley can tolerate higher levels of humidity and is vulnerable to drought. In Norfolk it was the grain of the common people and exactly for that reason the state of the barley harvest was not of high interest to the literate classes.[206] Drought mostly only caught the attention of chronicles when it was extreme and hindered crop growth or impacted on human health, e.g. by affecting the reliability or salubrity of the water supply. It is noticeable that in England a number of plague waves occurred in years that saw high spring-summer temperatures. Contrary to the narrative sources, manorial accounts report the problems caused by dry weather for agriculture and pastoral farming. A lack of rain reduced the productivity of the hay meadows, pastures an d leguminous crops, and thus endangered not only the dairy sector but due to the lack of winter fodder the profitability of the pastoral sector as a whole.

6.7 Climate and Viticulture in Medieval England

The economically profitable cultivation of vines and the production of quality wine has become possible in England in recent decades; a period of global climate change and, in Europe, rising temperatures. Hence the popular mind takes the existence of viticulture in medieval England as an indication first of the temperature levels during the Medieval Climate Anomaly equalling contemporary climate, and second of modern day warming being neither due to anthropogenic influences, nor being outstanding in the historical context since the Medieval Climate Anomaly occurred in the pre-industrial period. The decay of English viticulture in the fourteenth century has been linked to the climatic deterioration at the onset of the Little Ice Age.[207]

The production of good-quality wine, however, was not the primary purpose of medieval English viticulture. Sugar-content and must-density are linked to temperature during the vine growing season, especially at its end in summer, so England was and remains at the northern limits of viticulture. Medieval wine-growing in climatologically disadvantaged regions similar to England, such as in northern or eastern Germany, was expected to give mostly sour wine, except in very warm and

[206] Hallam, The climate of eastern England 1250–1350, 124.

[207] Hyams, Viticulture in England, 33–35.

sunny years; this sour wine was consumed sweetened and spiced.[208] Thirteenth-century summers were warm (Fig. 5.4), and would at least partly allow for a high sugar-content in the grapes in England, but success or failure of the English vine harvest was by no means critical, as the country did not actually depend on home grown wine for drinking. Wine was brought to England from many French regions, Spain and the Rhineland; above all strong trade links with Gascony, where the English crown held extensive lands, favoured the mass import of quality wine.[209] The vines at Ely (Norfolk), named the 'l'isle des vignes' by the Normans,[210] illustrate the low importance of the quality of English wines. The grapes in the vineyard of Ely often did not ripen properly and only verjuice or sour wine could be made, nonetheless the vineyard was maintained, it was one of the longest-lived wine producers in England and is documented until 1469.[211] Given the climatological caveats it is not surprising that English wines did not impress foreigners, even though during the High Middle Ages they were occasionally fervently defended by the English themselves.[212] Nonetheless the products of the English vineyards found a ready market. Wine, which was not of the same quality as good drinking wine, found a use in cooking and in the celebration of mass. Grapes and verjuice, which was made from green, unripe grapes, were an essential ingredient in medieval cooking. Both, (sour) wine and verjuice were indispensable in medicine and certain special diets, for disinfection and (ritual) cleansing.[213]

In the rise and fall of English viticulture economic factors such as the price of wine, the price and availability of labour and commercial relations with the continent played an important role. With the Norman Conquest (1066) an invigorated interest in viticulture had taken hold in England and while in the 150 years after the conquest no preferred English 'wine region' can be detected, by the thirteenth century the southeast of the country had developed into the centre of English viticulture. However, wine-growing could be found as far north as Yorkshire,[214] and eastern England was dotted with vineyards. The Benedictine abbey at Bury St Edmunds (Suffolk) and the monastery at Thorney (Cambridgeshire) as well as the cathedral priories at Ely, Peterborough (Cambridgeshire) and probably Lincoln (Lincolnshire) possessed vineyards. More were to be found in Huntingdonshire, in Colchester

---

208 Rösener, Bauern im Mittelalter, 112–113.

209 Before the Hundred Years War, between 75,000 and 100,000 tons of wine were exported annually from the Gascon ports. England was the principal market for Gascon wine. Due to the war, wine exports fell drastically (70% from 1335–1336 to 1336–1337) and then fluctuated strongly, but never recovered to pre-war figures, James, Medieval wine trade, 9, 32–33, Unwin, Wine and the vine, 202–203. On the origin of the wines, see Henisch, Fast and feast, 119 and on the practice of the wine trade and the merchants, see Kermode, Medieval merchants, 207.

210 Henisch, Fast and feast, 118.

211 Jäschke, Englands Weinwirtschaft 374, Lamb, Climate. Past, present and future, vol. 2, 460.

212 Henisch, Fast and feast, 118–119.

213 Jäschke, Englands Weinwirtschaft, 286–289, Woolgar, Great household, 131, Henisch, Fast and feast, 120–121.

214 Jäschke, Englands Weinwirtschaft, 361, Henisch, Fast and feast, 118.

(Essex) and at the manor of Forncett St Mary south of Norwich.[215] Norwich Cathedral Priory itself, as many other ecclesiastical institutions, was engaged in viticulture. It owned vineyards in Sedgeford and Plumstead, and possibly vines were cultivated in the cloisters of the cathedral priory itself.[216] None of these vineyards were extensive – the first vines in Sedgeford were planted within one day. The vines for Sedgeford were sent from Ely in 1263–1264, and more plants were brought to Sedgeford in 1273–1274. Viticulture in Plumstead is first mentioned in the account for 1312–1313. The lifetime of these vineyards appears to have been short: the last reference to the Sedgeford vines stems from 1327–1328 and no records survives for the vineyard in Plumstead after 1334–1335. Sedgeford and Plumstead were not untypical, discontinuity was a specific characteristic of medieval English vineyards.[217]

The establishment of the vineyard at Sedgeford in the middle of the thirteenth century falls within a period that was economically favourable for viticulture in England. The price of wine doubled during the thirteenth century. Whereas in the twelfth century commercial vine growing was largely unknown in England, Ely cathedral started now to sell its wine on the market.[218] The population in England was growing and stood between 4.75 and 7 million,[219] labour was abundant and cheap. Most ecclesiastical vineyards were in the possession of monasteries, and many vineyards were established in extensive manors which held a high number of *famuli*, permanent estate labourers of low social status. This work force was indispensable for the labour-intensive wine-growing.[220] All these features apply to Sedgeford and Plumstead: Sedgeford was the biggest demesne of the Norwich Cathedral Priory and Plumstead was also one of the bigger demesnes under plough.

The first reference to the vineyard in Plumstead in the account of 1312–1313 is relatively late. Most likely it was a thirteenth-century foundation, but had left no trace in earlier manorial documents. An installation as late as c.1310 would be an outright argument against any causal connection between viticulture and climate in England, because the years c.1290–1315 were marked by a raised interannual variability in the growing season temperature of up to 1 °C, and April to July were generally not as warm any more as between 1256 and 1290 (Fig. 5.4).

If climate would have been the determining factor for English viticulture in general and the vineyards of Sedgeford and Plumstead specifically, the crucial point

---

[215] Jäschke, Englands Weinwirtschaft, 297, 329–332, 351–353, Davenport, A Norfolk manor, 26.

[216] Saunders, Obedientiary and manor rolls, 112.

[217] Very few vineyards existed for a hundred years, see Jäschke, Englands Weinwirtschaft, 360–361. He supposes that the interruptions might be partly due to a lack of research or a lack of sources. On the vineyard in Sedgeford: Yaxley, The prior's manor-houses, 22; on Plumstead: ibid., 6. These small vineyards are not included in the study of Jäschke, Englands Weinwirtschaft.

[218] Jäschke, Englands Weinwirtschaft, 279, 374. An exception were the estates of the archbishop of Canterbury which included vineyards that aimed at market production, ibid., 377–378.

[219] The different estimations for the English population around 1300 are summed up by Britnell, Economic development, 11–12.

[220] Jäschke, Englands Weinwirtschaft, 324–328.

would have been the disastrous climate-induced agricultural crisis of the Great Famine and the following difficult years until 1323, which included not only a further sequence of cold and wet summers (Fig. 5.4), but also of very harsh winters (Sect. 6.4). These cool and wet summers must have resulted in wine of very poor quality, and the extremely cold winters had the potential to destroy the vines. Since the Sedgeford vineyard is still accounted for in 1327–1328 and the Plumstead vineyard in 1334–1335 they survived this period of climatic turmoil and persisted into a time that frequently saw warm and dry growing seasons, although the interannual variability of the growing season temperature was high (Fig. 5.4). Therefore it appears that these two vineyards rather succumbed to a reassessment at Norwich Cathedral Priory initially of specific sectors and then of demesne farming as a whole. The period of 'high farming' had come to an end, the prices for agricultural products had fallen in the mid-1320s, and the price-wage ratio made labour intensive cultivation less profitable for the landlords.[221] Most likely some time between 1330 and the mid-1350s Norwich Cathedral Priory's wine production fell victim to this development. Maybe the cold growing seasons 1330, 1335, 1347–1348 and the high precipitation levels of the 1342–1343, the mid- and late 1340s contributed to the abandonment of viticulture, but they can not have been decisive since 1315–1316, 1319 and 1322–1323 had not been so. The increasing scarceness of cheap labour and dropping profitability of demesne farming after 1350 would have prevented a reactivation of the vineyards; Sedgeford's and Plumstead's demesne under plough had fallen by a third due to the Great Pestilence (for Sedgeford see Fig. 2.3). Agriculture in Norfolk after the Great Pestilence was less labour intensive than before[222] and for a task as laborious as viticulture resources were scarce. Since these conditions were widespread in England, it is more likely that the vineyards that disappeared in the Late Middle Ages succumbed rather to the prevailing socio-economic trends than to cooling summers.

A look across the Channel provides the context for the situation of viticulture in the late medieval England. In continental Europe low grain prices in the late fourteenth and fifteenth century encouraged a diversification of crops, and in Germany wine-growing expanded into new territory towards the north and east. Viticulture also increased in fourteenth-century central and western France and northern Italy, as well as in the fifteenth-century on the territory of modern day Belgium.[223] The trend towards cooler summers can not have been instrumental in the end of the English vineyards, while wine-growing expanded in northern and eastern Germany and Belgium.

[221] Stone, Medieval agriculture, 236–243.

[222] Campbell, Eastern Norfolk, 38–39 compares the eastern Norfolk demesne agriculture in the first quarter of the fourteenth century with the agricultural conditions of the first quarter of the fifteenth century: over that period the labour input of the permanent staff, the *famuli*, was cut by 26.8%; hired labour input was even cut further.

[223] Slicher van Bath, Agrarian history of western Europe, 144 and ibid. Slicher van Bath sees a clear connection between low grain prices and the increase of viticulture ibid., 144, 216–217. When grain prices increased vineyards would partly be converted into arable land.

English viticulture in the Middle Ages is not comparable to its counterpart of recent decades due to the wine-growers' different attitude towards the production of good-quality wine. Hence, the flourishing vineyards of the High Middle Ages are no proof in themselves for the summers of the Medieval Climate Anomaly being as warm or warmer than recent summers. When wine-growing decayed in England in the fourteenth and fifteenth centuries, the cooling temperature trend of the Little Ice Age played only a minor role, and some vineyards also operated in England during the Little Ice Age.[224] Consequently the existence of medieval vineyards provides no argument against the hypothesis of anthropogenic climate change. Summer temperature dropped between 1300 and 1400 in England and wine quality must have been reduced, but what changed in a much more fundamental way were population size and density as well as the availability and cost of labour. It was these changes and the import of wine from Gascony, Spain and elsewhere[225] that contributed largely to the end of most of the English vineyards during the Late Middle Ages.

[224] The vineyards are mapped in Jäschke, Englands Weinwirtschaft, 355–359 up to the sixteenth century.

[225] Unwin, Wine and the vine, 203.

Chapter 7
Harvest Length – An Indicator of Late Summer Precipitation

7.1 The Harvest Length and Its Socio-economic Context

The duration of the grain harvest varies from year to year. It is influenced by weather, the amount of labour per cultivated acre and the harvest size. Since the aim is to gather in the corn as quickly and as dry as possible, it can be assumed that longer harvests were hindered by increased precipitation totals and frequency, which necessitated longer drying periods and increased harvest works. Even towards the end of the pre-industrial period, in the eighteenth and nineteenth centuries, a heavy, badly laid crop required twice the labour input of a light standing crop.[1] This chapter will investigate the link between harvest-time precipitation and harvest length from the mid-thirteenth century to the mid-fifteenth century in East Anglia. [2]

The harvest length is given more often than the harvest date in the manorial accounts, and information on harvest length is also available for a longer time and reaches into the mid-fifteenth century. The data on the duration have been primarily collected from longer runs of account rolls, the manors are listed in Table 7.1. Between 1256 and 1448, 31 manors render 822 data covering 164 years (Fig. 7.1), the manors are largely identical with the places supplying information on the harvest date (Fig. 5.1), only Bawburgh (Norfolk) and Eccles (Norfolk) reference the duration but not the date of the harvest. Since the harvest duration is an expression of the relationship between precipitation levels, labour supply and harvest size, and these factors did vary across East Anglia, the data have been normalized to allow comparison. The number of references to harvest length available per manor varies between 75 for Sedgeford and three for Fincham, but only six manors have series of

[1] Collins, Harvest technology, 465.

[2] Hallam, The climate of eastern England 1250–1350, 125–132 examined the link between weather, harvest date and length and the size of the barley harvest on the mainly barley growing manors of Norwich Cathedral Priory.

K. Pribyl, *Farming, Famine and Plague*, DOI 10.1007/978-3-319-55953-7_7

Table 7.1 Harvest duration: Norfolk and Suffolk manors

Landowner	Manors giving harvest length information
Norwich Cathedral Priory	Bawburgh, **Catton**, **Denham** (Suffolk), **Eaton**, **Gateley**, **Gnatingdon**, Great Cressingham, **Hemsby**, **Hindolveston**, **Hindringham**, **Martham**, **Monks' Grange**, **Newton**, **North Elmham**, **Plumstead**, Scratby, **Sedgeford**, **Taverham**, **Thornham**, Worstead
St Giles's Hospital	Calthorpe, Costessy, Cringleford
St Benet's of Hulme	Flegg
Bury St Edmunds	Hinderclay (Suffolk), Redgrave (Suffolk)
Castle Acre Priory	Kempstone
Le Strange	Hunstanton
Other landowners	Eccles, Fincham, Akenham (Suffolk)

Prior's manors of Norwich Cathedral are *bold*

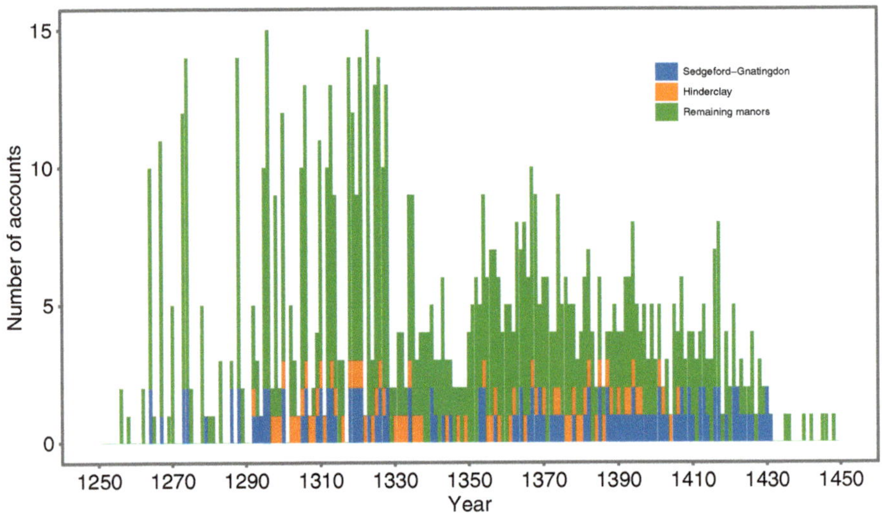

Fig. 7.1 East Anglian harvest length: number of accounts per year, 1256–1448

fewer than ten harvest lengths.[3] The most reliable and continuous information comes from some of the prior's manors of Norwich Cathedral, especially from Sedgeford and Gnatingdon in northwest Norfolk, and from the manor of Bury St Edmunds Abbey at Hinderclay (Suffolk).

[3] Bawburgh, Denham, Eccles, Fincham, Gateley, and Newton do not exceed ten data points. The mean value of the Denham series has been computed excluding the extreme years of 1258 and 1314.

Table 7.2 Harvest length and the Great Pestilence

	Mean value 1256–1431	N	Mean value 1256–1349	N	Mean value 1350–1431	N
Sedgeford	36.88	75	36.12	26	37.29	49
Gnatingdon	36.20	59	35.54	28	36.81	31
Eaton	27.53	38	26.73	30	30.50	8
Hinderclay	33.41	73	34.21	38	32.54	35
Hindolveston	32.62	50	31.15	26	34.17	23
Martham	31.11	54	31.79	28	30.38	26
North Elmham	34.28	32	34.14	22	34.60	10
Plumstead	31.80	49	30.88	24	32.46	24
Taverham	29.70	53	28.42	31	31.50	22

Mean values of the entire study period and sub-periods pre-1350 and post-1350 for eight Norwich Cathedral Priory manors and one manor of the Abbey of Bury St Edmunds (Hinderclay, Suffolk). Due to the impact of the plague on the labour force and thus harvest length the year 1350 has been excluded from the analysis except for Hindolveston and Plumstead, Gnatingdon in 1361 is also excluded

Average harvest length differed from manor to manor. Some places tended to have generally rather short harvests, whereas for other locations the harvest was more protracted. Geographically close manors do not necessarily show similar characteristics. Since demesne size and work management varied, it is not advisable to group manors. An exception is constituted by Sedgeford and Gnatingdon, which were run by one sergeant until c. the mid-1370s (Appendix 2). The data of these two neighbouring manors in Table 7.2 demonstrate that their statistical properties are very close, and they also display a strong relationship on the annual level with respect to harvest date and duration. Therefore these two series can be combined into one: the Sedgeford-Gnatingdon series supplies data for 93 years and hence constitutes the longest and most continuous series of harvest durations.

Information about the duration of the harvest is listed at several places in the manorial accounts. For the whole period it can be found in the *autumpnus* chapter, and also in the sections for *liberatio famulorum*, *vadium* and *corrodium*. Until 1327–1328 the accounts of Norwich Cathedral Priory give the start and end date of the harvest in the dairy section. After the accounting reform 1354–1355 the duration is also mentioned in the works account. At the beginning of the study period the dates for the beginning and the end of the harvest are supplied in the records. Later the duration itself is added; before 1350 it is usually detailed to the day, after 1354–1355 it is mostly rounded to the full week. From the mid-1350s onwards the end date of the harvest is omitted. The works account, which was newly added to the Norwich Cathedral Priory *compoti* at this time, also records the number of days actually spent harvesting during the harvest period (Sect. 3.2).

The duration of the medieval – or pre-industrial grain harvest was determined not only by the frequency of precipitation but also by socio-economic factors. The amount of labour per acre under crop, the harvesting method used, the management and the size of the harvest differed from manor to manor and also on one manor over time. In this respect the Black Death 1348–1349 functions as a watershed. Although the harvest date being dependent on the phenological phase of the grain was unaffected by demographic upheaval, the harvest duration was not. Before 1350 the high pre-plague population density and large workforce accessible to the landowners either as customary labour dues or as cheap hired labour, acted to some degree as a protection against wet weather impacts. The longer the ripe or harvested corn is exposed to the weather, the more likely the quality and probably also the quantity of the harvest suffer. Before the Black Death the East Anglian reeves and bailiffs could organize additional harvest works, when confronted with the need for speedy cutting and carrying, or the necessity of extra turning, unbinding, drying and re-binding of the sheaves in wet years. After the demographic collapse in the Great Pestilence such a strategy became very expensive.[4]

According to Henry Knighton not enough harvesters could be assembled in 1349 because of the high mortality in the Great Pestilence, and also because a sharp drop occurred in the demand for grain, so that on many grain fields the crops were left standing: '[...] multe segetes perierunt in campis, pre defectu colectoris. [...] tanta habundancia erat omnis generis bladorum, quod ullus de eis quasi curauit. [...].'[5] In 1350 the situation had probably somewhat improved in respect to the labour supply, but normality could not return quickly. In fact pre-plague normality was never to return. In 1350 mowers, who are cutting the grain quicker than reapers, are named explicitly for the first time in the accounts of Gnatingdon (Appendix 4). In that year a very low amount of labour per acre was available for the harvest on this manor. The relation got adjusted to pre-plague levels in the following years – by downsizing the sown acreage (Fig. 2.3). Therefore the non-climatic element in the information on harvest length is much stronger in 1349 and 1350 than in other years, and manors in which harvests were outright anomalous are excluded for those years.[6]

Over the following decades major adjustments were made to the size of the demesne lands, while labour became increasingly scarce and expensive. The array of crops grown also varied locally and over time. The situation differed across the region, hence there were diverging trends in harvest length after 1350. Most manors tend to have longer harvests in the post-1350 period; merely Hinderclay and

[4] On the harvest cost on Norwich Cathedral Priory manors, see Slavin, Bread and ale, 95.

[5] Knighton, Chronicon, vol. 2, 100–101. Knighton's description is echoed by information from Glastonbury Abbey. On a manor the part of the harvest that was difficult to gather in and economically not essential (legumes), was left out in the field due to a 'shortage of men and women', Titow, Evidence of weather, 403.

[6] For the harvest 1349 the *compotus* from Thornham, NRO, DCN 60/37/20, merely gives a standardized harvest length and can therefore not be used. This leaves only the account from Hinderclay in Suffolk for 1349, the harvest is very long, but not anomalous, Hinderclay, CUL, Bacon 468. Even though many manorial accounts survive for Norwich Cathedral Priory for the harvest 1350 only the harvest-length data from Hindolveston, NRO, DCN 60/18/31, and Plumstead, NRO, DCN 60/29/26, as well as from the non-cathedral manors in Suffolk, Redgrave, CUL, Bacon 335 and Akenham 1349–1350 (at Raynham Hall) are not anomalous. During the plague wave 1361 the harvest length in Gnatingdon, NRO, LEST/IC 12, was also extremely long and is excluded from the analysis.

Martham do not follow this trend (Table 7.2). The extent of the increase of the mean harvest length varies from place to place. At Sedgeford, Gnatingdon and Martham, but also at North Elmham and Plumstead, the average harvest lengths are very close in both sub-periods. At Taverham, Eaton and Hindolveston the difference is about 3 days.

From the 1380s onwards the data of bigger manors such as Sedgeford, Gnatingdon, Martham and Hinderclay display a lower variability than before. Often the recorded harvest lengths in those years were average: for Sedgeford and Gnatingdon this average was 35 days or 5 weeks, for Martham, Hinderclay and the much smaller manor of Taverham 28 days or 4 weeks. When the warm and dry summer of the 1410s arrived, the Sedgeford-Gnatingdon harvest length dropped to 28 days. Rarely was a harvest length of any manor during that decade lower; henceforth the warm and dry conditions of the 1410s are not visible in the Martham data.[7] In the Norwich Cathedral Priory accounts of this decade a kind of 'minimum harvest' is listed for many manors: it is the aforesaid 28 days which generally include 20 actual work days.

The 'minimum harvest' was probably connected to the increasing trend of mowing instead of reaping a considerable percentage of the grain crops. Mowing corn, primarily barley and oats, saved time and labour – Norwich Cathedral Priory expected one mowing work to replace two and half reaping works – but also caused more waste. It was still very limited in Sedgeford in 1357,[8] but became common in Sedgeford and Gnatingdon by the 1360s, when c.30% of the harvest fields were mown. Mowing helped to offset the labour shortage, and thereby reduce its influence on the harvest duration in the post-plague environment of the 1360s (Great Pestilence 1348–1349, Second Pestilence 1361, see Sects. 6.2, 6.5, and Chap. 10). It was even more widespread in the following decades, when grain prices were lower (Appendix 3). Indications are, that in the period of the 'minimum harvest' after c.1410 Sedgeford and Gnatingdon relied heavily on mowing.[9] During the 1410s the harvests were average to late, but also short. The decade also witnessed a run of dry summers (Sect. 8.3), so that harvests involving much mowing could be completed very speedily. They probably were indeed 'minimum harvests'. The high reliance on mowing also explains the raised variability in the harvest-length data in the 1420s. When wet or very wet summers occurred in 1421, 1423 and 1428 the

[7] The information from Martham after 1381 has been excluded from the analysis, since these data are too inflexible to reflect weather conditions. The same applies to Worstead 1346 and 1347, NRO, DCN 60/39/12-13.

[8] Merely 3.4% of the crops were mown; NRO, LEST/IB 24.

[9] When between the mid-1380s and c.1410 the number of acres mown was listed in the Sedgeford accounts, they amounted to 40–50% of the cultivated area; in Gnatingdon the average was c.55%. From 1408 onwards only the works done but not the acreage cut by the mowers are recorded. The number of mowing works indicates even higher percentages of acres being mown. This percentage steadily augmented until both manors were farmed out in the early 1430s. The fall in productivity resulting from the use of mowing was not untypical for Norwich Cathedral Priory manors in the first quarter of the fifteenth century. Due to a considerable reduction in the use of labour, this was also experienced in Martham, see Campbell, Eastern Norfolk, 38–39.

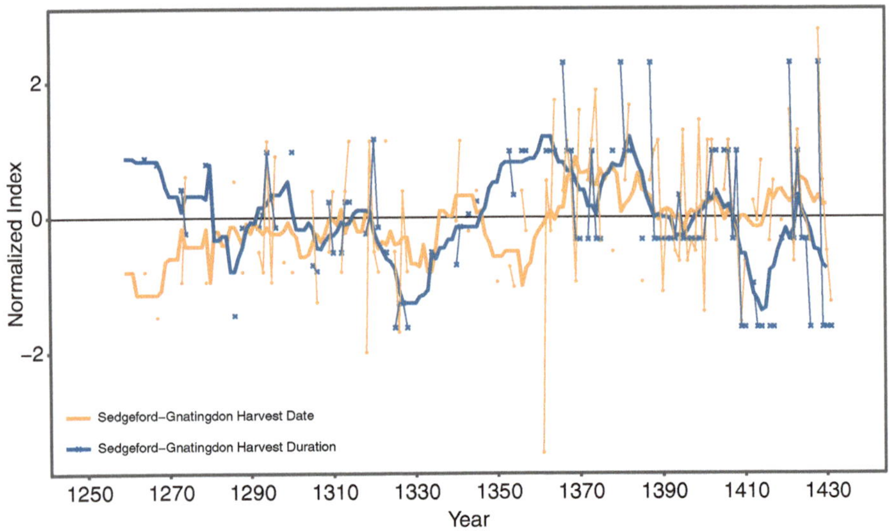

Fig. 7.2 Sedgeford-Gnatingdon harvest date and length: low frequency. Combined Sedgeford-Gnatingdon data on harvest date and length; break down of relationship between date and length before 1290, c.1350–1375 and c.1412–1425. Plotted is the 11 year running mean

mowing of corn was difficult or impossible and could not fulfil its potential in saving time and labour, so that the contrast between the wet summers with labour-intensive harvests and the normal or dry summers with short harvests is all the more striking (for Sedgeford and Gnatingdon see Fig. 7.2). For the development of the labour supply for harvesting on the manor of Gnatingdon, see Appendix 3.

Generally long and late harvests in East Anglia appear not to have ended as late as their counterparts in other English regions. The Winchester manors report in very bad years (1314, 1315, 1330) to have not yet finished the harvest by Michaelmas, 29 September, the official end of the harvest season. In East Anglia such cases were extremely rare, especially before 1350, when labour was abundant. In fact in Norfolk there was a noticeable reluctance to let a harvest extend beyond mid- or end of September of the Julian Calendar. This reluctance marked the harvests 1258, 1274, 1275, 1294, 1314, 1319, 1320, 1323, 1335, 1341, 1345, 1346 and for most manors even 1350. Before the arrival of the Black Death the only exception to this rule is Kempstone in 1330, a manor that was not part of the Norwich Cathedral Priory domain, but belonged to Castle Acre Priory. After 1350 there were many more cases when the harvest was finished shortly before or at Michaelmas, especially throughout the 1350s, 1360s, 1370s, parts of the 1380s and in the middle of the first decade of the fourteenth century. The cases when the harvest continued until October are 1330 (Kempstone), 1350 (Hindringham), 1382 (Gnatingdon), 1421 (Gnatingdon, Sedgeford), 1423 (Hindringham) and 1428 (Sedgeford). The tendency to finish harvesting before October reflects the climatic conditions of East

Anglia: it is the driest part of England. The chances to complete a harvest in late summer or early autumn are better in this region than elsewhere. However, the county also has a very high percentage of rain days: over 200 per year, shower activity is frequent,[10] consequently there is but little hope for a stable dry period in late September or October for bringing the harvest in. The plague wave afflicting Norfolk in 1420 probably played a crucial role in reducing the labour force and agricultural expertise – mowing is a more skilled harvesting method than reaping – as a result there were greater difficulties in keeping the harvest process compact and the loss of grain during bad harvests increased due to the extended exposure to the inclement weather.

7.2 The Relationship Between Harvest Date and Harvest Duration

On a decadal level the low frequencies of harvest date and duration are linked, that means early harvests were often short and late harvests tended to be long. Various factors play a role in this link. First, the weather conditions of the growing season, that lead to an early or late harvest, are unlikely to change suddenly with the start of the harvest. So if the harvest was late due to low temperatures often associated with raised precipitation levels, similar weather would often persist in harvest time. Second, in high summer the days are longer and allow for more work; grain cut in the morning also dries quicker during the higher summer temperatures. Third, in late summer the agriculturalists' wish to harvest quickly is countered by the cooler weather and the long dewy nights which hinder the corn from ripening properly, but consequently also prevent it from shedding.[11] Therefore in a late harvest haste is neither as essential nor as practical as in an early harvest.

In the Sedgeford-Gnatingdon series this relationship between harvest date and duration breaks down in three periods: before c.1290, 1350–1375 and c.1412–1425 (Fig. 7.2). The low density of data renders any analysis of the situation before c.1290 difficult. At the beginning of the period c.1350–1375 the harvest dates were rather early, but the harvests were long. In the 1360s then the trends were converging, harvests were average to late in timing, harvest duration was consistently long. Considering the timing and the degree of the disruption in the relationship between harvest date and length during the 1350s, it must be concluded that the socio-economic consequences of the demographic crisis – a decrease in the labour supply and the increase in wages – were important factors, but a high temperature variability and raised precipitation levels also played a role (Sect. 7.3 and Chap. 9). For the 1350s hardly any data on harvest works per acre are available, but the ratio of works

[10] Ogilvie, Farmer, Documenting the medieval climate, 128.

[11] Lisle, Observations in husbandry, 129.

per acre was comparatively high in the 1360s (Appendix 3).[12] This indicates on the one hand the wish to maximize the yields since grain prices were still elevated, and on the other hand higher precipitation levels, which led to an increase of work needed in the harvest. By the end of the 1360s the mowing of corn helped to strengthen the link between harvest date and length again.

As the percentage of corn mown instead of reaped increased in Sedgeford-Gnatingdon over the following decades, the relationship between harvest date and duration is restored. Only in the 1410s and early 1420s do the trends diverge considerably again. This was due on the one side to cold early springs which were delaying the start of the grain growth and so led to later harvests, and on the other side to the dry summers of the 1410s, when mowing could release its full potential as a means to save labour and time. These conditions ended with the wet summers in 1421 and 1423, and thus the relationship between date and duration was re-established.

It can be concluded that the Sedgeford-Gnatingdon data in the 1350s bear little relation to the actual rainfall levels around harvest time, in the 1360s they exaggerate rainfall to some degree, but understate it in the 1410s and the dry years of the 1420s. In post-1350 northwest Norfolk the way to stabilize the harvest length almost on a pre-plague level, was the preference of mowing over reaping. This reduced labour costs and work time, and also reinforced the link between duration of the harvest and the weather. Other manors might have pursued other adaptation strategies, and this explains the diverging trends of the harvest lengths on the manors listed in Table 7.2.

7.3 The Norfolk Precipitation Series

The usual calibration/verification approach as employed in the temperature reconstruction (Chap. 5) can not be used for analysing the influence of precipitation on the harvest duration, because summer rainfall is a more regional phenomenon than temperature and no detailed early modern Norfolk precipitation record is available for comparison. Additionally the eighteenth- and nineteenth-century farms were subjected to management practices which differed substantially from that of a medieval manor.[13]

Hence for verifying the value of the harvest-length data as an indicator of precipitation, it is useful to study the harvest-duration data from individual manors, and their link to the precipitation-sensitive tree-ring data, the precipitation index compiled by Ogilvie and Farmer for medieval England[14] and the yet unpublished summer precipitation index for medieval central Europe created by Pfister from information held in the Euro-Climhist database at the Oeschger Centre for Climate

[12] No data on harvest works is available for Sedgeford or Gnatingdon before the accounting reform 1354–1355. Therefore this comparison is limited to the period 1355–1431.

[13] Concerning harvest length and precipitation at Langham farm 1768–1867, see Sect. 4.1.

[14] Ogilvie, Farmer, Documenting the medieval climate, 124–128.

Research in Bern. The manors with sufficiently dense and continuous information allowing such a comparison are Sedgeford-Gnatingdon, Eaton, Hinderclay, Hindringham, Martham, Monk's Grange, Plumstead and Taverham (Fig. 5.1).

Two tree-ring width series, coming from southern England and from East Anglia, are a proxy for drought as defined by the Palmer Drought Severity Index (PDSI).[15] Both series are based on oak trees and capture the conditions during the months March-July, in particular the nature of precipitation. The oak data from southern England is highly correlated with northern France and also East Anglia, and both data series also correlate highly with each other. The overlap in the time-of-year for which the tree-ring and harvest-length data serve as a drought or precipitation proxy is very short, but the tree-ring data are the only available natural proxies for the English summer half year precipitation in the Middle Ages, therefore a comparison is informative. For some summers, which the chronicles, manorial accounts and in more recent times even instrumental observations describe as very wet and sometimes also as cold, as 1258, 1330, 1359 and 1816 (Sects. 6.3, 8.2 and 8.4), the precipitation levels are underestimated in the tree-ring data.

Generally the data on harvest length from the individual manors do not correlate well on the year to year level with the oak data and therefore with the implied precipitation levels; gap years strongly accentuate the problem. Between 1256 and 1349 correlations with the southern English data are highest, and amongst the manors with a sample size higher than 20 the correlation coefficient stands between 0.4 and 0.5 for Eaton, Hindringham, Monks' Grange and Hinderclay[16] and between 0.2 and 0.3 for Sedgeford-Gnatingdon, Hindolveston and Martham. After 1350 most of these manors have not sufficient harvest length information any more, but for those that do, correlations are low in the period 1350–1390. In the years 1391–1431 then the Sedgeford-Gnatingdon series and the oak data correlate again: $r = 0.36$ significant at $p<0.05$. For the period 1256–1349 the link between harvest length and precipitation is strongest on the manors close to Norwich and with easy access to the city's cheap labour, Eaton and Monks' Grange. Monks' Grange itself was run without customary tenants, so it can be assumed that the bulk of its harvest was done by hired labour. Eastern Norfolk, where Martham is situated, was also densely settled. A reasonable hypothesis might therefore be that on other manors the relationship between harvest length and precipitation on an annual level is weaker because of the management of labour and the constraints when relying on customary labour services.

In this light the poor correlations of harvest length and oak data 1350–1390 do not come as a surprise. First the Great Pestilence and then recurrent demographic crises disrupted the labour supply and the difficulty in finding or paying harvest workers disturbed the relationship of harvest length and precipitation. Between 1350 and 1370 the link in the low frequency of harvest date and length in Sedgeford-Gnatingdon breaks down completely, indicating that the harvest duration was to some degree driven by non-climatic factors. This is also the period when in

[15] Wilson et al., March–July precipitation reconstruction, 997–1017, Cooper et al., Hydroclimate variability, 1019–1039.

[16] The correlations are significant at $p<0.05$, the $r=0.49$ for Eaton even at $p<0.01$.

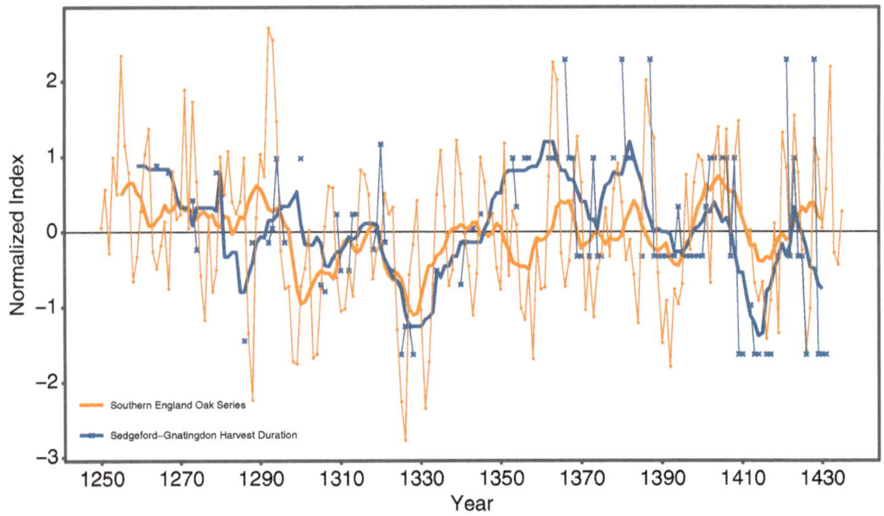

Fig. 7.3 Sedgeford-Gnatingdon harvest length and oak data: low frequency. Plotted is the 11 year running mean. Tree-ring data from southern England (Wilson et al., March–July precipitation reconstruction)

Gnatingdon the correlation between the number of harvest workers and sown acres is lowest, at 0.47 (Appendix 3). The improvement of the correlation after 1390 is probably due to the increased use of mowing in the harvesting process of Sedgeford and Gnatingdon, this overcame problems in the depressed labour market and strengthened the link between harvest length and weather.

Contrary to the annual correlations, the low frequency variability of the data on harvest length and the oak rings display similar trends and even levels for considerable periods of time on a decadal level. Both factors respond to precipitation variability. This is well visible in the Sedgeford-Gnatingdon series (Fig. 7.3). Between c.1310 and c.1340, the period when the data density is highest and the smoothed harvest data hence more reliable, the link in the low frequency between oak-ring and harvest-length data is also clear in the series from Eaton, Hinderclay, Hindringham, Martham, Monks' Grange and Taverham. Whereas in the decades after the Great Pestilence the link is disrupted, the smoothed Sedgeford-Gnatingdon curve shows similar trends as the oak data in the years 1370–1395, although the curves are offset. The relationship is strong 1395–1431, despite another offset between the data sets 1405–1420.

The strong relationship between harvest duration and tree-ring data breaks down in the following phases. For the years 1256 to c.1300 the lack of data on harvest duration renders an interpretation difficult, but the Sedgeford-Gnatingdon information is not contradictory to the southern English tree-ring data. Of much higher interest is the period 1350–1370, when the series of Eaton, Martham, Plumstead and Sedgeford-Gnatingdon and to a lesser degree also the other series show high precipitation levels during autumn, whereas the oak data indicate drier conditions around 1360 and wet conditions in the late 1360s. At the same time the low-frequency curves of harvest date and length in Sedgeford-Gnatingdon also diverge (Fig. 7.2). As mentioned above, the long-term impact of the first and second plague waves resulted in a labour shortage and a lengthened harvest process, but a high short-term climate variability contributed to the problem. As long as the harvests were dry and easy to gather in this appears not to have been a great problem. The harvests of the dry early 1350s were still only of average length and do not yet cause a substantial divergence between harvest-length and tree-ring data. After the second national plague epidemic in 1361 and in the wetter conditions of the 1360s the labour shortage was more evident. Also the harvest work force in Gnatingdon does not correlate well with the sown acreages between 1360–1361 and 1390–1391, though before the Black Death and after 1390–1391 there is a stronger link (Appendix 3). However, documentary sources independent from the East Anglian manorial accounts leave no doubt about the wet conditions during harvest time 1359, 1362 and 1365–1369.[17] The high rainfall levels at least from mid-summer onwards throughout harvest time 1365 and 1366 are not visible in the oak data, but combined with a lack of labour due to disease caused severe distress at harvest time (Sect. 8.4). The lack of labour lengthened the harvest period and hence enhances the rainfall signal, and since the rain started either too late in summer, or low temperatures and dull weather inhibited the tree-rings' sensitivity to rainfall levels, which was most likely the case in 1366, the precipitation is not reflected in the oak data. Also harvest-length information for the hot and dry summer 1361 is overshadowed by the Second Pestilence, thus it does not reflect the prevailing dry conditions. Hence 1350–1370 socio-economic factors prolonged the harvest duration, but the extent of the disruption of tree-ring and harvest-length data in the 1360s was increased because the tree-ring data do not show late summer or early autumn rainfall and are not always reflecting high rainfall levels in cold or dull years.

The 1410s are the last phase when harvest length and oak data deviate. It was undoubtedly a dry decade. Whereas the harvest lengths from Sedgeford-Gnatingdon imply extremely dry late summers and early autumns, the oak data show moderately dry conditions. Once more this offset coincides with another phase of disruption in the low frequency in harvest date and length (Fig. 7.2). This is connected to the now extensive use of mowing in the harvest process of Sedgeford-Gnatingdon, resulting in a very speedy harvest process, which overstates the prevailing dry conditions (Sect. 8.3).

A comparison of the harvest-length data from East Anglia and a tree-ring based summer drought reconstruction from central Germany over the last millennium by

[17] Ogilvie, Farmer, Documenting the medieval climate, 127 collected evidence for raised precipitation levels for August, September and October for all those listed years except 1369. For 1359 and 1368–1369 see also Sects. 6.3 and 8.4. Additionally the years 1363 and 1364 have high positive values in the oak data, indicating high levels of rainfall between March and July. However, no indications for increased rainfall at harvest 1363 or 1364 are given in the documentary data. There was flooding some time between spring sowing and harvest in the Winchester area in 1363, reflecting the high score of the oak data in this year. Titow, Le climat à travers les rôles de comptabilité, 320.

Büntgen et. al.[18] also reveals interesting overlaps in trends. The drought in the mid-1320s known from England and visible in the eastern England harvest data also affected Germany. As in the harvest-length data, high levels of summer precipitation indicated by increased growth occurred in the German trees between 1350 and 1370.[19]. The German drought index then gives low levels of precipitation in the early and mid-1370s,[20] but then relatively much summer rainfall must have occurred between c.1375 and the end of the 1380s. Again from the early 1370s to the early 1380s the harvest-length data from eastern England are remarkably close to these conditions. Both series show a drier phase in the early to mid-1390s. Whereas raised levels of precipitation marked some years in the first decade of the fifteenth century in East Anglia, the wetter years in central Germany are centred around 1400. Interesting are also the dry conditions in Germany in the second half of the 1410s and the spike of increased rainfall in the early 1420s. Both are reflected in the East Anglian data.

Apart from the tree-ring data from England and central Europe as a proxy for drought, the precipitation information from chronicles and other written records collated into indices – for medieval England by Ogilvie and Farmer, for medieval central Europe by Pfister – also offer the opportunity for a quantitative analysis of the harvest-length data.[21] Due to the non-continuous nature of the harvest-length series and the precipitation indices created from documentary data, some difficulties occur for statistical procedures. The use of the summer precipitation index for central Europe by Pfister is comparatively straightforward, because for about two thirds of the summers during the study period information is available. The problem faced is simply a reduction of years for which data are available from the Pfister indices and the harvest-length data for the purpose of correlation. The central European precipitation indices also contain a substantial amount of weather references from modern day Belgium and the Netherlands, and the Dutch coast is the same distance from eastern Norfolk as London, so some congruence between the Pfister indices and East Anglian harvest length can be expected. For the significant results of the Spearman's rank correlations see Table 7.3. These are moderate positive correlations, displaying a relationship between East Anglia and the central European data, indicating broad scale precipitation patterns during the European summer half year.

The situation for the Ogilvie and Farmer precipitation index for England is more complex, because it is not made on a seasonal, but on a monthly level. In this context gaps in the individual months, which are naturally quite frequent in the medieval period, result in problems when comparing the indices to a summer season or above monthly level precipitation series. Direct weather references in medieval narrative and administrative sources focus on extremes, so that it remains unclear if a missing value in the Ogilvie and Farmer series represents average rainfall levels or

[18] Büntgen et al., Tree-ring indicators of German summer drought over the last millennium.

[19] Ibid., 1007–1010.

[20] All the following references to the German drought index refer to, ibid., 1008.

[21] Ogilvie, Farmer, Documenting the medieval climate, 112–133. The unpublished summer precipitation index for central Europe is provided by Christian Pfister, University of Bern, OCCR, Switzerland.

Table 7.3 Spearman rank correlation coefficients between some Norfolk manors and the central European summer precipitation index

	Hinderclay	Martham 1256–1370	North Elmham	Monks' Grange	Sedgeford-Gnatingdon
rho	0.31	0.49	**0.62**	**0.50**	**0.58**

Bold typeface indicates correlations significant at p<0.01, otherwise p<0.05. The unpublished summer precipitation index for central Europe is provided by Pfister, University of Bern, Switzerland

if information on extraordinary rainfall has been lost over the centuries. Autumn weather, including September, was not of prime interest to the medieval chronicler,[22] hence there is a lower amount of information for August and September available than for June and July, which would generally weight a summer or harvest season index based on the monthly Ogilvie and Farmer indices towards early and high summer. However, due to the nature and preferences of the documentary sources the precipitation indices by Ogilvie and Farmer are much denser than the temperature indices. Documentary evidence for the fourteenth century is also sufficiently robust to assume that few major precipitation events escaped the historian's notice, so that Spearman's rank correlations between the harvest-length data and the Ogilvie precipitation index have been carried out. This has been done on a monthly scale for June, July, August and September and also for the mean index values for July–August, July–August–September and August–September.

The results show that it was not only the rainfall in August, the classical harvest month, that was essential for determining the harvest length, but also rainfall in July. The harvest lengths of Eaton and Plumstead even have a stronger correlation with the July precipitation, than with August rainfall. On most manors the link to the August precipitation exceeds or equals the one to the July precipitation, though there are often not enough data in the Ogilvie and Farmer index, to make the August correlation statistically significant. The combined Sedgeford-Gnatingdon series correlates at 0.49 with July rainfall and at 0.42 with August, these correlations are significant at p<0.01 and p<0.05 respectively. For these two manors on sandy soils the June precipitation was equally important as the July precipitation, although this is not a characteristic shared by many other manors. Rainfall in mid-summer precedes the onset of harvest, it can increase the bulk of the corn and heavy rain on the medieval crops, which were taller than modern varieties, can lay them flat. A bulky or badly laid harvest would take more time to complete. The bulking up of the grain due to rainfall in the months immediately preceding the harvest was probably of high importance at manors on soils which retained very little water, such as the sandy soils of Sedgeford, Gnatingdon,[23] Thornham and Great Cressingham. Mid-summer rainfall levels have lower correlations with the harvest length in the Norfolk

[22] Although autumn weather can have long term impacts, usually it has not such an immediate impact upon peoples' lives as severe weather during the growing season or cold weather during winter.

[23] See note 3 in Chap. 5.

Broads and on Cromer Ridge.[24] Hence correlations between harvest length and precipitation are for most manors equal for the periods July–August and July–August–September. For Sedgeford-Gnatingdon the correlation stands at 0.58 respectively 0.57, and both correlations are significant at $p<0.01$. The qualitative analysis of the harvest-length data shows extremely short or long harvests do fit independent English documentary evidence on precipitation during harvest time (Chap. 8), but the factor of the bulk of the harvest must have also played a role, and this role must have been more pronounced in middle-range harvest-length data, that means in years when the weather at harvest time was more or less normal.

The role played by the mid-summer precipitation, especially July, in the harvest duration explains the link in the low frequency variability between the harvest-length and the tree-ring data, which is after all proxy a for the March–July precipitation. The significance of the July rainfall also helps to illuminate the link between the individual series of East Anglian harvest lengths and the central European June–July–August precipitation indices by Pfister (Table 7.3). The comparison of the East Anglian harvest-duration series and the monthly rainfall indices for medieval England shows that the determinants for the harvest length were not just the precipitation during the harvest, but also the rainfall in the mid-summer period preceding the harvest, the East Anglian harvest length can be considered a proxy for July–August–September rainfall. Since the harvest process was slowed down by prolonged drying periods, it was more the frequency of the rainfall (i.e. rain days) than the absolute levels, that were decisive for the harvest length.

When combining the harvest-length data of individual manors into one East-Anglian harvest-length series, the Sedgeford-Gnatingdon data due to their high number and reliability especially after the Great Pestilence form the backbone of the regional series. They account for one third of the series before 1350, and for two thirds from 1350 onwards, except in years for which no parallel or no Sedgeford-Gnatingdon data are available. The resulting normalized East Anglian harvest length is shown in Fig. 7.4a together with the range of harvest lengths per year.

When comparing the low frequency in the East Anglian harvest-length series to the tree-ring data from southern England and East Anglia, a link in the low frequency with both tree-ring series is visible, but it is stronger with the East Anglian oak data (Fig. 7.4b). Indeed, on a decadal level harvest length and the East Anglian oak series move in parallel between c.1270 and the late 1290s and are in very close accordance between c.1290 and 1335. Then the curves diverge until c.1370. The offset between the two graphs narrows in the 1370s and 1380s and during those decades the trends in the East Anglian oak data are again reflected in the harvest-length data. In the late 1380s the curves converge closely until about 1430, when the bulk of the harvest-length data end. Only during the early 1390s and around 1420 are the values in the oak data somewhat higher than in the harvest data. After 1431 the data density in the harvest-length series is too low to allow a comparison. Also the correlation coefficients of the East Anglian harvest-length series with the precipitation indices by Pfister as well as Ogilvie and Farmer have a moderate link,

[24] Ibid.

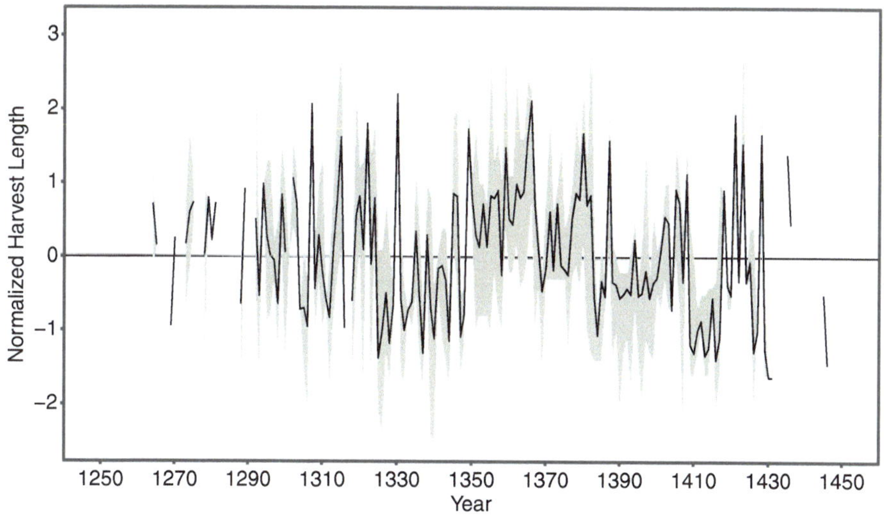

Fig. 7.4a East Anglian harvest-length series. Harvest-length series with shortest and longest harvest in any given year

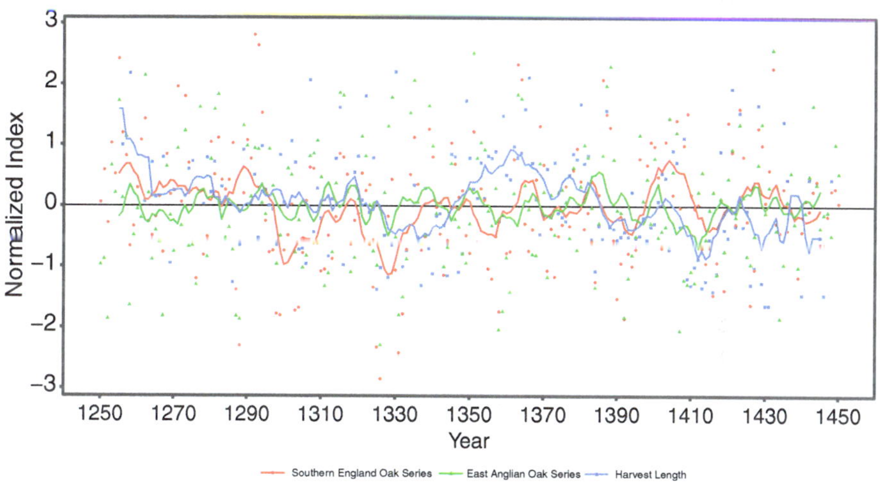

Fig. 7.4b East Anglian harvest-length series. Harvest-length series with tree-ring data: low frequency. Plotted are the 11-year running means of normalized data. Tree-ring data from southern England (Wilson et al., March–July precipitation reconstruction) and from East Anglia (Cooper et al., Hydroclimate variability)

they stand at 0.43 respectively 0.46 and are significant at $p<0.01$. Given these analyses the harvest-length series can serve as an East Anglia precipitation index for July–August–September. The normalized data were converted into indices on the −3 to +3 scale and are displayed in Fig. 7.5, indices of −2 or +2 and more extreme seasons are discussed in Chap. 8.

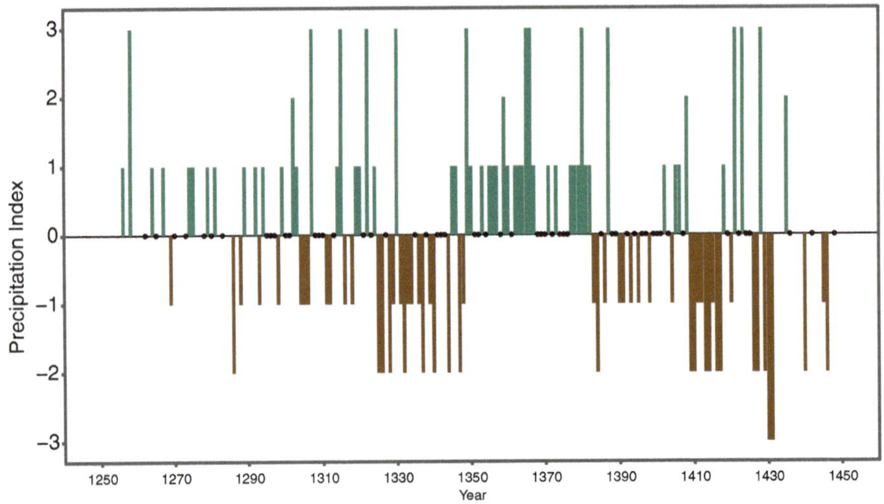

Fig. 7.5 East Anglian precipitation index: July–August–September. Plotted are indices −3 to +3. *Dots* are mark seasons with an index value of 0

The data for the 1260s, 1270s and 1280s are still comparably sparse, but indicate that those decades experienced average to wet conditions, apart from the second half of the 1280s, which appears as dry. The middle of the first decade of the fourteenth century was also dry. The rainfall levels were raised in some years of the agricultural crisis 1314–1323, but not continuously and apart from 1315 and 1322 not excessively. It followed a dry period between the mid-1320s and the mid-1340s. Apart from 1330 no very wet July-September period occurred in these decades. The increased rainfall levels at the beginning of the 1340s that culminate in flooding events in central Europe in 1342 and the Carpathian basin in 1343 appear merely as normal summers in the East Anglian harvest data. In the mid-1340s this calm picture of basically dry or average summers was disturbed by the wetter than average summers of 1345 and 1346. The situation had improved by 1347, but then the difficult years of the Great Pestilence followed. The Black Death overthrew the established balance between harvest workers and cultivated acreage. Also after 1350 the data were produced on a rougher scale and counted in weeks. Although parts of the early 1350s were very dry, these summers show up merely as average in the precipitation index. Summer precipitation after c.1355 was above average, parts of the late 1350s and the 1360s were very or even extremely wet. In the mid-1370s several subsequent summers saw average precipitation levels. However, the precipitation index switches back to wetter than average conditions in the late 1370s and only returns to indicate dry late summers by the mid-1380s. Another extremely wet July–September period occurred in 1387, but on the whole the late 1380s saw average rainfall. The 1390s again present the image of a calm decade with average to dry summers. In the first decade of the fifteenth century late summers were marred by raised precipitation levels in 1402, 1405 and 1406, and July–September 1408 appears as very wet. From 1409 onwards and throughout the 1410s harvest periods were almost consistently dry. Apart from the summers 1418 and 1419 no year saw even average rainfall conditions. At the beginning of the 1420s the previously rather stable conditions were upset and 1421 and 1423 had extremely high rainfall levels. In 1428 then a volcanic eruption can be connected to the very cold and wet summer. Apart from these years, however, the 1420s resemble the 1410s: the summers were dry. This also applies to 1430 and 1431. The few available data in the mid-1430s indicate wet or average and in the 1440s dry or average conditions.

Chapter 8
Harvest Length and Precipitation Extremes 1256–1448: Independent Evidence and Context

Independent documentary evidence for very high and low levels of rainfall in the East Anglian July–September precipitation index (Fig. 7.5) can clarify the meteorological conditions as well as contribute to an identification of risk factors for medieval agriculture in England. Due to the disruption caused by the Great Pestilence and the subsequent rise of mowing as a harvesting method, the analysis of the harvest length extremes is divided into the sub-periods 1256–1347 and 1348–1448. Many of the July–September periods with extreme precipitation levels have already been described in Sects. 6.2, 6.3, 6.4 and 6.5, since the temperature and rainfall extremes often coincided. This chapter uses the same sources as listed under Sect. 6.1. Together Chaps. 6 and 8 form a catalogue of climatological extreme spring and summer seasons.

8.1 Short Harvests 1256–1347

Short harvest occurred frequently during the first sub-period until 1347. The most striking examples are 1286, 1325–1326, 1328, 1332, 1337, 1340, 1344, and 1347.

8.1.1 Weather Conditions in the Second Half of the 1280s

The summers of the second half of the 1280s are known to have been dry.[1] In 1286 heavy thunderstorms occurred in the surroundings of Worcester at the beginning of May, indicating warmth.[2] Early summer was dry on the Winchester manors. The short harvest in East Anglia suggests a continuation of the dry weather into late

[1] Ogilvie, Farmer, Documenting the medieval climate, 125.

[2] Annales Prioratus de Wigornia, 492. The hail damaged houses, trees and crops.

K. Pribyl, *Farming, Famine and Plague*, DOI 10.1007/978-3-319-55953-7_8

summer and early autumn. On the continent the summer half year appears to have been unspectacular, the summer season in the Low Countries is categorized as II.

In 1288 follows another summer season during which dry conditions prevailed. The Winchester accounts report many drought impacts during the growing season and the harvest time was hot. The harvest was plentiful.[3] In East Anglia most places harvested quickly. The year was exceptional and several monastic annals refer to the very warm and dry conditions during summer time.[4] The Oseney Annals specify that no rain fell for five weeks around July:

> Anno eodem, principio Julii, scil. in festo Sanctorum Processi et Martiniani [2 July], incepit [calor] intollerabilis et siccitas magna invaluit, continue perseverans per v. septimanas, scil. usque ad festum Sancti Oswaldi [5 August]; ita quod interim nihil omnino pluebat.[5]

The heat and drought raised the risk for fires and the Worcester Annals record several fires (partly caused by lightning) which consumed churches, monasteries and even a fair.[6] The Dunstable Annals mention a raised level of mortality,[7] possibly due to gastrointestinal infections which occur more frequently when persistent dry weather endangers the supply of safe drinking water. The hot and dry summer was not limited to England, but stretched across the North Sea, where the Low Countries summer season has been classified as III.

8.1.2 Weather Conditions in the Mid-1320s

The warm and dry years 1325–1326 and 1328 have been discussed above, in Sect. 6.2.

8.1.3 Weather Conditions in 1332

The short harvest 1332 is part of the run of warm and dry years in the early 1330s and has therefore been considered in Sect. 6.2.

[3] Annales Prioratus de Dunstaplia, 342; Annales Prioratus de Wigornia, 495.

[4] Annales Prioratus de Dunstaplia, 341; Annales Prioratus de Wigornia, 495; Annales Monasterii de Oseneia, 315.

[5] Annales Monasterii de Oseneia, 315.

[6] Annales Prioratus de Wigornia, 495.

[7] Annales Prioratus de Dunstaplia, 341.

8.1.4 Weather Conditions in the Second Half of the 1330s

Dry summer weather favourable to grain growing did not only prevail in the first half of the 1330s, but it returned after the wet and cold year 1335. The springs and early summers 1336–1340 were not as hot as 1332 and 1333, they only saw average mean temperatures; they are described in Sect. 5.4. The growing seasons 1339 and 1340 followed on hard and long winters. However, these summers were mostly dry, for some years drought impacts are recorded in the East Anglian manorial accounts (Appendix 1). Most of these spring-summer seasons offered good conditions for grain cultivation and harvests were in general good during this period.[8] The summer half year 1339 was difficult towards the west of the country, and at Croxden Abbey the season was perceived as rather cold and wet,[9] the harvest of that summer was poor. Except in 1338 all harvests were at least shorter than average. Most likely the harvest of 1338, was of average length, because it was one of the best harvests between 1270 and 1429, the bulk of corn needed more work and time to be reaped and stacked.[10]

The short harvest 1337 in East Anglia comes after an average to warm growing season. In the Winchester accounts drought impacts for spring and early summer are recorded. By July the weather turned on the Winchester manors, rain came on and due to the rainfall the following harvest was long. This is not in accordance with the short East Anglian harvest, but for this year the weather diary of William Merle supplies detailed weather information that can shed light on this dichotomy.[11] May, June and the beginning of July were mostly dry and moderately warm, this was the season during which the Winchester manors had to cope with drought impacts. The second and third week of July, however, were wet and windy, as was also remarked by a Winchester manor, where rain and wind flattened the growing wheat. For August then, Merle noted very windy weather with occasional showers, the wind was very drying. This was the month of the East Anglian harvest and the weather would have allowed for a quick harvest. It seems that either further west and south the weather was wetter, or that the Winchester harvest lasted well into September. According to Merle this month was less windy, but showers were more frequent and 10–11 September were days of heavy rainfall. From then on until almost the end of the month there were moderate showers, so conditions for harvesting were much worse in September than in August. The Low Countries index was average and reflects the warm and dry part early in the summer as well as the wet second part.

Another short harvest follows in 1340. This year is extremely well documented. The manorial accounts from Sussex, Kinsbourne in Hertfordshire and of the estates

[8] Campbell, Nature as historical protagonist, 299.

[9] Lynam (ed.), Croxden Chronicle, x. For the sequence of wet and cool and dry and warm phases during the summer season in Merle's weather diary, see Sect. 5.4.

[10] Campbell, Nature as historical protagonist, 299. The harvest was so good that people remembered it fondly, see Sect. 5.4.

[11] Merle, Consideraciones temperiei pro 7 annis.

of the Bishopric of Winchester and Christ Church Canterbury agree on a hard and long winter, which was also dry at the end. The following delayed spring and the summer were dry and probably also hot. On the manors of the Bishopric of Winchester the dairy production was reduced, pastures and meadows suffered and for some mills the water levels dropped too low to continue working. The details on the meteorological conditions can be filled in with the help of Merle. He noted frost in the first week of April, afterwards warm and dry weather followed until mid-May, when rainy weather lasted for about a week. Then mostly hot and dry weather dominated until the beginning of July. After another wetter episode of about two wet weeks the heat and drought continued for about another month. Already at the end of August the first hoar frosts appeared. In East Anglia the April–July mean temperature was average, the speed of crop growth could never make up for the delayed growing season. The grain crops did not suffer from this weather, the grain price was very low in the agricultural year 1340–1341. The serene and hot conditions did not attract the attention of chroniclers. Reflecting this mixture of a hot late spring and summer and the long lasting cold conditions in spring as well as the early onset of hoarfrost in August, the index for the summer half year in the Low Countries is 6.

8.1.5 Weather Conditions in 1344

Even though the temperature of the summer season 1344 was only average in England and comparable (6) in the series from the Low Countries, it was a summer marked by dry weather. The manors of the Bishopric of Winchester submitted a good number of complaints concerning drought impacts and in Sussex a long dry spell defined the summer weather. Considering the short harvest in East Anglia, it is likely that these conditions lasted throughout August and into early September.

8.1.6 Weather Conditions in 1347

See Sect. 6.5 for the agricultural crisis in the late 1340s.

8.2 Long Harvests 1256–1347

Six long harvests fall in the period 1256–1347, they took place in 1258, 1302, 1307, 1315, 1322 and 1330.

8.2.1 Weather Conditions in 1258

The summer half year 1258 followed the massive eruption of the Samalas volcano in the Rinjani complex in Indonesia,[12] which caused severe rainfall anomalies and widespread famine in Europe and according to Matthew Paris and other chroniclers also in England.[13] Harvest time and some months before were rainy, as the long harvest in East Anglia and the complaints of some of the manors of the Bishopric of Winchester underline. The summer in the Low Countries stands at I.

8.2.2 Weather Conditions in 1302

For the late summer and autumn 1302, when a long harvest took place in East Anglia, hardly any documentary evidence from England is available. In late spring a dry spell affected some of the Winchester manors, and later some pastures along the Thames were flooded. The inundation and the long harvest in East Anglia indicate rain during harvest time, and in the Low Countries the summer was indeed very bad and is indexed as 2.

8.2.3 Weather Conditions in 1307

The year 1307 is part of the surprisingly poorly documented second half of the first decade of the fourteenth century. Not only in England, but also on the continent weather information is sparse. In the Low Countries the summer half year 1307 was average, but for the following two years neither a winter nor a summer index could be set. The manorial accounts of Norwich Cathedral Priory are missing for 1307 and mostly also for 1308. At a Somerset manor of the Bishopric of Winchester the weather was dry at some stage during late spring and summer. Also for Kinsbourne in Hertfordshire indications are that late winter and early spring were dry, but then this region also saw some severe weather. Fitting to these dry conditions, April–July mean temperature was high. However, no context can be provided for the very long harvest in East Anglia. In the light of the grain price that was rising in the year following, the harvest was not a good one, and that would be congruent with a long harvest troubled by rainfall.

[12] Lavigne et al., Source of the great A.D. 1257 mystery eruption, 16,742–16,747. Timmreck et al., Limited temperature response, L 21708, points out that, although the eruption was probably the largest in the last 7000 years, the drop in temperatures was limited. She concludes, that the temperature response to volcanic eruptions also depends on the size of the aerosol particles. In England, precipitation levels were already high in 1257, Britton, Meteorological chronology, 107.

[13] For a list of references, see Britton, Meteorological chronology, 107–109.

8.2.4 Weather Conditions in 1314 and 1315

The years 1314 and 1315 were studied in detail in the section concerning the agricultural crisis 1314–1323 (Sect. 6.4). It is enough here to state, that in 1314 the late and longer than average harvest in East Anglia was hindered by rain (Appendix 1), and that the harvest was probably not very long simply because it was already so very late – onset was between mid- to late August – and in East Anglia before 1350 there was the strong tendency to have the harvest completed before October. During the height of the Great Famine the harvest 1315 was then extremely long.

8.2.5 Weather Conditions in 1322

The long and wet harvest 1322 comes towards the end of the agricultural crisis 1314–1323 and has been described in Sect. 6.4.

8.2.6 Weather Conditions in 1330

For the cold spring and summer and the deplorably wet harvest (Appendix 1) 1330 see Sect. 6.3.

8.3 Short Harvests 1348–1448

After 1350, and especially after the accounting reform 1354–1355 at Norwich Cathedral Priory, the harvest length in the accounts is usually rounded to the full week. When looking out for very short harvests, it becomes evident that these are now masked by harvests lasting normally for a minimum of four weeks. This applies to all Norwich Cathedral Priory manors. Consequently, places such as Taverham, where harvests were relatively short before the Black Death, are less sensitive to the signal of drier seasons after 1350. The manor of Martham takes this development to the extreme and provides extended runs of 'minimum harvests' with very few deviations in the post-1380 period, so that its harvest length clearly no longer contains a climatological signal, therefore these data were not used in this study. The information from Sedgeford-Gnatingdon in northwestern Norfolk shows average harvest length at five weeks, short harvests at four weeks, long ones at six weeks and extremely long harvests at seven weeks. The harvests on the manors Hinderclay and Redgrave of the Abbey of Bury St Edmunds functioned in a similar step system. Data from other manors moves comparatively freely.

The 1350s and 1360s were rather turbulent with respect to harvest lengths. Labour shortage and raised rainfall levels resulted in elevated harvest lengths for some decades after 1350. Harvests that would count as short over the whole period 1256–1448 do not reappear before the 1380s. They are 1384, 1409, the 1410s, parts of the 1420s and the early 1430s, 1440 and 1446.

With the low grain prices from the mid-1370s onwards and the growing scarcity of labour, demesne farming became less profitable and shortly after 1380 accounting measures had to be tightened. The year 1382 marks the beginning of relatively frequent weather references or remarks on impacts of extreme weather in the Norwich Cathedral Priory accounts; this is probably linked to preferences of the new prior, Alexander de Totyngton (elected 14 April 1382), or a new administration established under him, for keeping tighter control on the expenses of bailiffs and reeves and the demesne's profitability (Appendix 1). The increased use of mowing instead of reaping the corn in Sedgeford-Gnatingdon after c.1390 and the reluctance to hire itinerant harvest workers (Appendix 3) was also part of the measures. Other landlords ended the direct management of demesne land or leased ever more land.

8.3.1 Dry Conditions in the Mid-1380s

A sequence of dry summers characterizes the mid-1380s, the dry conditions culminated in 1384. Already in the year 1383 the harvest in East Anglia was short and the summer half year is defined as 8 in respect to temperature in the Low Countries. The Winchester manorial accounts refer to a long drought between c. May and end of September 1383. It was also dry in Sussex. In Norfolk an epidemic took hold (Chap. 10).

The meteorological pattern of 1384 was very similar. According to the Westminster Chronicle a severe drought lasted until the first week of September, rivers and wells dried up and cattle died for want of water on the pastures.

> Preterea in ista estate tanta erat siccitas ita ut fluvii et fontes perhenni cursu de terra scaturientes, immo (quod magis mirabile videbantur) eciam putei altissimi siccarentur; et duravit ista siccitas usque festum Nativitatis beate Marie; [...]. Grosso vero animalia in ista estate quamplurima pro aque penuria perierunt.[14]

The dry conditions must have stretched at least from Norfolk, where the legumes were lost (Appendix 1), across Hertfordshire (Kinsbourne) and into Sussex. The Westminster Chronicle states that in September prolonged rain set in,[15] as is confirmed by the accounts from the Bishopric of Winchester. However, by that time the short harvest in East Anglia was probably already over. The summer half year in the Low Countries was warm (7).

---

[14] Chronicon Westmonasteriense 1381–1394, 88.

[15] Ibid.

The summer of 1385 followed the pattern laid out in the two previous summer seasons. Drought prevailed between May and September, even though the water levels dropped not as much as in the preceding year.[16] Around early summer it was dry in Sussex, Hertfordshire and Norfolk (Appendix 1). The summer in the Low Countries was very warm (8).

In 1386 the dry conditions were weakened in Norfolk. In this county (Appendix 1) and on a Buckinghamshire manor of Christ Church Canterbury there were drought impacts, but on the Winchester manors and in Kinsbourne harvest time was rainy, bringing probably also the index for the Low Countries' summer half year down to 6. The East Anglian harvest being short, it must have been finished before the rainfall set in.

8.3.2 Dry Conditions 1409–1420

The 1410s are well-known for their dry conditions during the summer season.[17] Except for 1418, the harvests between 1409 and 1420 were at least shorter than average or very short, and hence align themselves with other information on dry conditions for the late summer and early autumn period. The 1410s are indeed the decade of the 'minimum harvest' of four weeks. The extent to which a 'minimum harvest' was fixed is visible e.g. in the data of the works accounts of Gnatingdon. There, once people had gotten used to the dry weather, even the work units performed for harvesting show standardization. For Gnatingdon this begins in 1413, when the works for reaping were quantified at 240. This number, though usually highly flexible, is repeated in every single surviving account until 1419, namely in 1416, 1417 and 1419. For 1416 and 1417 even the works used for mowing the grain do not differ and stand in both years at 180.

The data on harvest lengths in East Anglia and weather information available in the Pipe Roll of the Bishopric of Winchester agree on very dry summers for 1409 (Sect. 6.2), 1410, 1413, 1414, 1415, 1416, 1419 and 1420. In all those years except 1414 this is supported by recorded direct references to drought or more commonly reports on drought impacts in the Norwich Cathedral Priory manorial accounts (Appendix 1). During this drought decade the springs and early summers of 1415 and 1419 were excessively dry; in 1415 drought prevailed from mid-March to the end of October on the Winchester manors. In Sussex a short period of wet weather was recorded for summer, potentially for harvest time. In 1419 the grain crops were partly of stunted growth and flattened by the drought in the Winchester area, the crops were in such a bad state that the harvest took longer than the weather alone would have indicated. In Sussex a spell of wet weather accompanied the harvest. In

[16] Ibid., 120.

[17] Ogilvie, Farmer, Documenting the medieval climate, 128.

1420 the heat and drought were excessive on the continent, causing famine, in England the weather was not as extreme, and even though the wheat harvest was almost a fifth below average on the Winchester estates, prices did not react strongly.[18] The same dry conditions that affected England during that decade and which are reflected in the East Anglian precipitation index and the information in the manorial accounts were also felt in the Low Countries, where the spring-summer 1409 and the summers 1412, 1413, 1414, 1417, 1419 and 1420 were dry.[19]

The references to drought impacts in the Norwich Cathedral Priory material (Appendix 1) focus on peas or vetches perishing in the fields, and legumes, oats, barley or rye being weak or having failed; a lack of hay or the replacement of hay as animal fodder are also mentioned. Often the number of fallow ploughings was low, although only in extreme cases can this be taken as an indicator of drought.[20] Not all of the references contain the addition of the (meteorological) cause, but these impacts are typical drought impacts and often relate to dry weather in late spring or early summer. The references to the lack of hay mostly come from northwest Norfolk and indicate drought, but they must be considered in the light of additional information, because still today 'the capacity for producing good growth of grass is not high in this area of low rainfall and often excessively drained soils'.[21]

The weather conditions in the summer half years 1411, 1412, 1415, 1416, 1417 and 1418 are more complex and are described below.

The year 1411 was not part of the drought cycle. Neither the chronicles, nor the accounts of the Bishopric of Winchester or those from Sussex refer to the weather of the growing season at all. The reconstructed April–July mean temperature was low, but the harvest was shorter than normally. In northwestern Norfolk not enough hay was produced (Appendix 1). The index of the summer half year in the Low Countries was set at 4 and thereby confirms the cool reconstructed growing season temperature for East Anglia.

The picture shifts in 1412. It was not yet as strikingly dry as later in the decade, but indications from Hampshire, Sussex and Norfolk point to a partly dry summer. A Winchester manor reports that its wheat was first affected by mildew and then withered in the dry weather. The Norfolk rolls list impacts that are typically drought impacts, such as a poor pea harvest (Appendix 1). The growing season temperature was average, the summer half year in the Low Countries has been classified as 6. The harvest length in East Anglia was shorter than average, but not as short as during the drought years, indicating either some precipitation during the harvest or a bulkier harvest.

[18] Le Roy Ladurie, Histoire humaine et comparée du climat, 115–119, Titow, Le climat à travers les rôles de comptabilité, 337.

[19] Camenisch, Endless cold, 1062.

[20] The variability of the number of fallow ploughings between c.1360 (references appear after the accounting reform 1354–1355 in Norwich Cathedral Priory) and 1431 differed from manor to manor. A comparison over longer periods of time is difficult, because work organisation also influenced the number of fallow ploughings.

[21] Agricultural land classification, sheet 124 Kings Lynn, 10.

The growing seasons 1415 and 1416 started out wetter. Storminess and abundant rainfall prevailed in northern Hampshire in 1415 until mid-March and in 1416 throughout March and April.[22] In 1415 this wet episode was followed by a massive drought. Dry conditions also followed in 1416. The grain price rose in the agricultural years 1415–1416 and 1416–1417, in the latter year it reached a very high level. By spring 1417 the grain supplies must have been dangerously diminished, since the royal administration began to organize grain imports from Prussia. The cause of the scarcity and hence of the poor harvest in 1416 was named as 'heavy rains'.[23] The manorial accounts only record rainfall in March and April 1416 and refer to drought afterwards, therefore this precipitation early in the growing season must have been the 'heavy rainfall' ruining crops. Its timing in March and April meant that the ploughing for and the sowing of the spring grains was hindered, and that the young winter crops could easily be damaged. If the rainfall extended indeed to eastern England is not clear because the East Anglian manors report impacts, such as legumes lost on the fields, that are normally attributable to drought (Appendix 1).

In 1417 the harvest was very short in northwest Norfolk, but of average length in the rest of the county. This summer season was not marked by conditions as dry as in the previous summers 1415 and 1416: the Winchester manors did not complain about drought in 1417. In Norfolk late spring or early summer witnessed a dry time, because the peas were perishing on the well-draining soil of Gnatingdon (Appendix 1) and in Great Cressingham no vetches were sown this year, as they had failed in the two previous years.[24] The reconstructed April–July mean temperature was below average – probably due to a delayed onset of the growing season since the preceding winter had been hard, it was characterized by the Winchester manors as long and rich in snow. The summer half year in the Low Countries falls into category 6. Late in the harvest in northwest Norfolk rain set in (Appendix 1). On the sandy soils of this region the effect of the dry phases of 1417 was enhanced, leading to a very short harvest, which dominates the signal, even though on the whole precipitation levels were not as extremely low as in the previous years.

In 1418 part of the summer was dry according to the manorial accounts of the Bishopric of Winchester. Many impacts usually associated with drought are given in the Norfolk *compoti*, so probably the east of England witnessed similar conditions (Appendix 1). However, on the Winchester manors the significance of the dry conditions paled in comparison to the damage mildew did to the crops. It had taken hold before the onset of the dry weather: first the kernels swelled up with the mildew and then they desiccated in the drought. The Winchester manors did not suffer alone, Norfolk was also affected. Mildew alone would make harvesting difficult and explain the longer than average harvest, but hail also damaged the standing crops on some of the estates of the Bishopric of Winchester. In Norfolk an 'encroachment of water' is reported for harvest time, sheaves needed additional turning as well as opening and rebinding (Appendix 1). While the harvest was thus longer than aver-

[22] Titow, Le climat à travers les rôles de comptabilité, 335.

[23] Sharp, Famine and scarcity, 135.

[24] Great Cressingham NRO, MC 212-13.

age, the harvest date was early, fitting into the picture created by the Winchester and Norfolk rolls of dry and thus rather warm conditions preceding the harvest. Summer in the Low Countries is classified as 6.

Medieval agriculturalists did not merely subject themselves fatalistically to these adverse weather conditions. Crop selection was used for minimising risks, as at Great Cressingham, where after two years of failed vetches these were not sown at all in 1417. Taverham, which like Great Cressingham is situated on sandy, freely draining soils prone to drought,[25] experienced young pea plants withering on the fields in all years between 1415 and 1419, except in 1417. Over this period of time the sown acreage of peas was reduced year by year and after the loss of all peas in 1419, no peas were planted in the following year. This policy was revised in 1421, when wetter conditions reigned.[26]

Summer half years in England with as little precipitation as in the 1410s are caused by the blocking action of high pressure systems over northwest Europe, which also lead to an increase in English summer temperatures. However, chronicles or administrative sources in England do not mention higher temperatures for the summer half years of the 1410s; hardly any weather information made its way into the dwindling number of chronicles and the manorial accounts concentrate on the drought conditions and the hard winters. The drought focus in the records of agricultural activities is not surprising, since it is the lack of rain, not the summer heat, that threatened the harvest success and the water supply. Also the dry conditions are most visible in the countryside, where they interfere with crop growing and hay production. Most chroniclers were town dwellers, and as such they were concerned about storms, heavy rainfall and flooding, which had an immediate impact on their lives, but as long as the water levels did not fall below a certain threshold, dry weather was unproblematic. Thus only one narrative source, the Annals of Ulster from Ireland, records summer heat for the British Isles in this decade.[27] Also in the Low Countries the indices for the summer half years 1409–1419 oscillate between 4 and 6, whereas 1420 was extremely hot.

The indices for the Low Countries are substantiated by the reconstructed East Anglian April–July mean temperatures being about average in this decade. As the indices from across the North Sea, the English temperature reconstruction is also marked by a low interannual variability (Fig. 5.4). The combination of very dry weather with only average summer temperatures is unusual, and it is likely that this is actually not only due to the conditions in summer, but also to colder than average springs which delayed the onset of the growing season. The 1410s were not just a decade of dry summers, but also of hard winters. Winters that were at least partially cold are mentioned by the Winchester manors for 1410–1411, 1413–1414, 1416–

[25] Agricultural land classification, sheet 126 Norwich, 14.

[26] Acres sown with peas at Taverham between 1414 and 1421 (NRO, DCN 60/35/43–50): 1414 15, 1415 10, 1416 6, 1417 8, 1418 8, 1419 8, 1420 0, 1421 4; in 1419 the whole 8 were lost and whereas in 1414 peas occupied 18% of the sown acreage, by 1421 they had dropped to 5%.

[27] Most likely for 1420, but the text arrangement could also put the reference into 1419, see Annals of Ulster, vol. 3, 82–83. The annals are not contemporary for this period.

1417 and 1418–1419; according to that evidence the winters 1413–1414, probably 1416–1417 and again 1418–1419 were also long. For springtime 1419 a Winchester account states explicitly that additional fodder had to be given to the animals 'et tantum per ordinationem ballivi propter longam continuationem venti borialis qui suo fluctu frigidissimo herbas primitus viventes subito mortificuit et devastavit.'[28] Therefore even in northern Hampshire the onset of the growing season was very late, and it was interrupted when a cold north wind cut down the grass and herbage that had already shot up. Due to its geographic position Norfolk is particularly vulnerable to such cold early springs and the subsequent delayed onsets of the growing season (Sect. 5.3). Milder conditions albeit with much bad weather – strong winds and prolonged rainfall – affected southern England in 1415 until mid-March and in 1416 throughout March and April.

In the 1410s East Anglian reconstructed growing season mean temperatures slightly below average occurred in 1411, 1414, 1415 and 1417. Most of these growing seasons followed on from long and cold winters, merely 1415 does not fit this pattern. April–July mean temperatures above average fall to 1409, 1410, 1416, 1418 and 1420, in the Winchester material none of those years is associated with a very long and cold winter, although the winter 1419–1420 was colder than average in the Low Countries, but summer 1420 then also very hot.

The 1410s were a decade dominated more by continental than maritime climatic conditions: hard winters were followed by cold springs and dry summers. The growing season was shortened. Potentially lower summer temperatures went hand in hand with these cold springs. The 1410s are indeed one of the periods, where the relation in the low frequency of harvest date to harvest length is weakened in the Sedgeford-Gnatingdon series (Fig. 7.2). This is partly related to an overstating of the dry conditions in the harvest lengths because of the mowing of the corn, but this can not explain the total extent of the divergence. Hence a change in seasonal weather conditions – a cooling of the summers – is suggested by the disruption of the link in the low frequency of harvest date and duration. The later onset of the growing season, the average summer temperatures in parallel to the increase in spring/summer droughts and the high frequency of severe winters indicate a deterioration in climate that marks the beginning of the Little Ice Age. Hard winters and cool/dry summers would recur and become more pronounced towards the middle of the fifteenth century, they were typical for the Spörer Minimum.[29]

[28] '[...] and it is so much because of the order of the bailiff which he made due to the long continuing north wind which suddenly killed and devastated with its very cold current of air the early herbs and grasses.'

[29] Camenisch et al., Early Spörer Minimum.

8.3.3 Dry Conditions in the Second Half of the 1420s and the Early 1430s

During the 1420s the downturn in the output of historiographic writings and the end of demesne farming lead to a scarcity of documentary sources that could supply information on weather, so that a contextualisation of the proxy data from East Anglia becomes increasingly difficult. For 1426 and 1427 and 1429 to 1431 the manors of Norwich Cathedral Priory recorded again 'minimum harvests' of four weeks, indicating dry weather during harvest time. The low precipitation levels in the summer half year of 1426 are confirmed by the St Alban's Chronicle which refers to 1426 as the year when wells and brooks dried up.[30] The summer half year in the Low Countries was warm (7), and London, Sussex and potentially Great Yarmouth (eastern Norfolk) were visited by the plague (Chap. 10).[31] In Norfolk an early start of the harvest was followed by a short harvest.

On the whole the year 1427 was much wetter than 1426, in Sussex it was mild and rainy, in the Low Countries conditions were average. In late May a severe thunderstorm brought heavy rainfall to the London area and in July hay and growing grain were damaged by an abundance of rain, perhaps also during a thunderstorm, and in early September another thunderstorm struck terror into the people around Cambridge.[32] However, for August no thunderstorms during an otherwise relatively warm summer typified by a synoptic situation favouring the development of convective storms are recorded and the harvest length in East Anglia indicates a dry harvest season. In early October a rheumatic fever, called the 'Mure', broke out and killed many.[33]

During the next sequence of short harvests, 1429, 1430 and 1431, even the accounts of the Bishopric of Winchester, which were still being produced, remain silent. Consequently it is unlikely that those years witnessed climatic extremes. The year 1429 is overshadowed by the effects of the very wet and cold year 1428 which had led to harvest failure (Sect. 6.3). By 1429 dearth reigned in England and drew the attention of the few narrative sources that were still being produced. The summer half year had wet phases, in Sussex flooding occurred and the St Alban's Chronicle mentions heavy rain in early May.[34] However, conditions in eastern England seem then to have been calmer and the harvest was short. A further indication that part of this summer half year was dry comes from the account of Gnatingdon in 1429–1430, where hay was partly replaced by other fodder (Appendix 1). A small hay crop (produced in 1429), was a frequent experience in northwestern Norfolk in the dry 1410s. The reconstructed East Anglian mean April–July temperature was

[30] Chronicon Rerum Gestarum in Monasterio Sancti Albani, 10.

[31] Bean, Population and economic decline in England, 428, Brandon, Late medieval weather in Sussex, 4.

[32] Chronicon Rerum Gestarum in Monasterio Sancti Albani, 15–17

[33] Ibid., 19.

[34] Ibid., 36.

slightly below average and the summer stands at 5 in the Low Countries, making a mixed summer half year of rain and dry weather likely.

In 1430 and 1431 early harvest dates reflect warm springs and early summers, in fact it was very warm in 1431, and harvest durations were short, indicating dry conditions at harvest time. In 1430 the peas perished in Gnatingdon (Appendix 1), a problem usually associated with a dry period in later spring. On the continent the summers of the first half of the 1430s tended to be warm, standing at either 6 or 8 in the Low Countries; for 1431 evidence from this region shows a dry summer.[35] The early and short harvests in northwest Norfolk in 1430 and 1431 fit into this context.

8.3.4 *Weather Conditions in 1440*

The short harvest 1440 in East Anglia comes after the famine of the late 1430s. In Sussex the year was classified as 'normal', which, after the wet and cold conditions ruining the harvests in the preceding years, must have been a great relief. The price for grain fell after the harvest, so it must have passed without interference from rainy weather. In the Low Countries the summer stands at 4.

8.3.5 *Weather Conditions in 1446*

For 1446 is neither in the Low Countries nor in England enough information available to categorize the summer, but spring 1446 was wet across the North Sea.[36]

8.4 Long Harvests 1348–1448

Long harvests after 1348 were frequent and often came in clusters. During the Black Death, the harvest of 1349 was very long, primarily due to the socio-economic upheaval and labour shortage caused by the epidemic (see Sect. 6.5). This upheaval persisted into the 1350s, when harvest lengths of the various manors appear less coherent in a given year than normally. In the decades after the Black Death long harvests occurred especially in the 1360s, around 1380, in the first decade of the fifteenth century and in the 1420s.

[35] Camenisch, Endless cold, 1062.

[36] Ibid.

8.4.1 Weather Conditions in 1359

In 1359 a long harvest is not only reported from East Anglia, but also from the Winchester manors, where it is ascribed to rainy weather. A Norfolk manor, Hindolveston, also reports rain and ended its long harvest very late, only a week before Michaelmas (Appendix 1). Hardly any English information is available for spring, but Titow suspects dry conditions in the Winchester area, because a higher than usual number of pastures was reserved for the lord's animals. Although this would fit to the reconstructed above average April–July temperature, it is not in accordance with the extremely inclement summer in the Low Countries. Brandon classifies the year as wet on the basis of the information in the Sussex accounts, and rain and humidity hindered the hay harvest in the region of the Winchester manors.

Therefore England witnessed during spring at least average, if not rather warm and somewhat dry conditions, this resulted in a comparative lack of pasture in the Winchester area, and a harvest earlier than average in East Anglia. Eastern England is less vulnerable to rainfall impacts than other parts of England, and it is not by accident that the longest harvest in 1359 was observed by Hindolveston, a manor on the Cromer Ridge, exposed to the winds and situated on soils that tend to waterlogging.[37] Latest in summer, however, the situation must have shifted totally, creating a very wet and potentially cold season. In northern France and western Germany the persistent rain occurred between June and early September, probably the situation was similar in England; on the continent inundations and bad vine and wheat harvests followed.[38]

8.4.2 Weather Conditions in 1361

A sudden shortfall in the labour supply caused by the national outbreak of plague 1361, resulted in a very long harvest period in Gnatingdon. Consequently the harvest duration of this year does not reflect the precipitation scheme of summer 1361, which was actually very dry and hot (see Sect. 6.2).

8.4.3 Weather Conditions in 1365 and 1366

The grain harvest was generally longer in the 1360s than in the 1350s. Within these longer harvests the years 1365 and 1366 stand out. In 1365 the reconstructed growing season temperature was high, the preceding winter was mild, it is classified in

[37] For the soils, see note 2 in Chap. 5.

[38] Alexandre, Le climat en Europe, 488.

the Low Countries as 3, and the onset of the growing season was consequently early. The summer half year in the Low Countries, however, was cool (3). Both years are marked in the Winchester accounts as having wet and long harvests, in 1366 the phenomenon was even more widespread than in 1365. In 1365 the accounts of Christ Church Canterbury also show a very long harvest,[39] so the rain in late summer stretched over most of England. According to John of Reading the hay and corn were destroyed by rain at hay making time.[40] At Kinsbourne in Hertfordshire the hay crop was indeed poor. Even though problems with the food supply marked the year following the harvest – the export of grain, malt and ale and other victuals was forbidden by royal decree in 1366[41] – the grain price was actually falling. The unresponsive grain price might be linked to the outbreak of an epidemic among men and beasts, the 'Pokkes', in 1365, which led to the sudden death of many.[42] The name of the disease indicates an affliction of the skin, but was probably not smallpox.[43] The 'Pokkes' left another scar on the English cultural memory. One of the most important works of English literature in the Late Middle Ages, Piers Plowman by William Langland, which draws much of its material from the decades after 1350, lists the 'Pokkes' alongside the plague and other diseases which were brought forth by Nature (Kind) from the planets:

> Kynde came after hym, with many kene soores,
> As pokkes and pestilences – and much peple shente;
> So Kynde thorugh corrupcions kilde ful manye.[44]

---

[39] CRU, Bickersteth, Minister's account rolls of Christ Church Canterbury 1305–1386.

[40] Reading, Chronica, 166.

[41] Cook (Bickerdyke), Ale and beer, 113.

[42] Reading, Chronica, 167. Creighton, Epidemics in Britain, 453–455 dates the 'Pokkes' to 1366. He was not familiar with John of Reading and used Walsingham, Historia Anglicana, vol. 1, and the English Brut Chronicle. However, after being very close to Reading's Chronica for almost two decades the Historia Anglicana suddenly diverges at the end of 1364 and gives very little information for the following two years, hence a reference to the 'Pokkes' is missing. Creighton, Epidemics in Britain, 454 also assumes a note on 'lethargy' and 'flux' probably for 1362 or 1363 in Walsingham, Historia Anglicana, vol. 1, 298 might actually refer to the 'Pokkes'. The Brut does indeed give the 'Pokkes' under 1366 (e.g. Brut, 316). However – as explained by the editor of John of Reading – the Brut is for the years 1345–1367 'little more than a roughly abbreviated translation of such portions of Reading's narrative as it was thought would suffice for English readers', Reading, Chronica, 8–9. While the English readers were supposed to have an interest in the history of epidemics, Reading's paragraph about 'Pokkes' in 1365 is unfortunately misdated in the Brut to 1366. This error had long-term consequences; it not only misled Creighton, Epidemics in Britain, 453–455, but the widely popular Brut Chronicle is also closely connected to The Chronicle of London from 1089 [sic] to 1483, 66–67, and William Gregory's Chronicle of London, 87 which were both written in the fifteenth century and also list the 'Pokkes' erroneously under 1366.

[43] Creighton, Epidemics in Britain, 451–455 rules out smallpox for fourteenth-century England. Reading, Chronica, 167 also refers to the disease as 'Morbillae' which might indicate 'measles' or may merely be another term for a disease blemishing the skin. On the confusing use and meaning of 'mesles' and 'morbilli', see Creighton, Epidemics in Britain, 448–451.

[44] Langland, Piers Plowman, 254.

The year was generally unhealthy, even the birds did not escape and many were found dead in the fields.[45] It is likely that the 'Pokkes' came on early enough to impact on the harvesting process; the Winchester manors noted the lack or death of tenants and the consequently raised harvesting costs.

The meteorological conditions did not improve in the following year. The spring-summer temperature was average, but the Sussex manorial accounts indicate excessive rain during the growing season and harvest. More Winchester manors than in the previous year experienced rainy and long harvests, and due to the late start of the harvest in Norfolk, the harvest dragged on to the end of September, a rare occasion on the Norwich Cathedral Priory manors. Due to the difficult circumstances, established work routines were disrupted (Appendix 1). In Sedgeford the harvest extended to Michaelmas, but was classified as unfinished.[46] At Great Cressingham the harvest started with the help of the *famuli* at 6 August, but at the 15 August another *famulus* arrived, sent by the 'lord of Norwich' (the prior or the *magister celarii*), and joined the local *famuli* for the remaining time of the harvest for five whole weeks.[47] In Taverham that, what might just be a recording error – in the harvest account the harvest is said to have started at St Bartholomew (24 August), an impossibly late date for Taverham, but in the works account St Laurence (10 August) is given as the harvest date[48] – takes on a bad taste in the context of the weather conditions in 1366. For the clerk writing the account, St Bartholomew was obviously not inconceivable as a harvest date in that year. Unsurprisingly in the Low Countries the summer half year falls into the category 2. Further problems became evident at harvest time (Appendix 1): the harvests of barley, mixed grain and peas were poor at places in Norfolk, the hay harvest was insufficient in northwestern Norfolk and also at Kinsbourne, Hertfordshire. Although the causes of neither the crop failures nor the poor hay harvest are explicitly stated, these are probably the consequences of the prolonged wet conditions.[49] Notwithstanding the at least regionally poor grain harvest, the corn prices did only rise marginally in the year following the harvest.

---

[45] Reading, Chronica, 166–167.

[46] Sedgeford NRO, LEST/IB 27.

[47] Great Cressingham NRO, MC 212/05.

[48] Taverham NRO, DCN 60/35/35, the duration of the harvest amounts in the various places of the account roll to six weeks. Since 24 August (year day 236) is so far beyond any harvest date that can be expected for Taverham (mean value 214), it appears that 10 August is the real harvest date. This also fits well in the context of the information from Sedgeford and Great Cressingham in this year.

[49] Difficulties in mowing the meadows were also encountered in 1366 on a Winchester manor. The growing conditions for barley were far from ideal in 1366, as quality and quantity of the barley harvest are strongly related to the weather of the sowing time of the previous crop, e.g. spring 1365. Barley should be sown in a fine tilth and frost would be helpful in producing such, but the winter 1364–1365 was mild in the Low Countries and is defined as 3. In the year of the harvest itself dry weather in August would be required for a good barley harvest, this, however, was surely not the case in 1366. Although the greatest negative influence on the bean and pea harvest is a very high temperature in May to July of the harvest year, the second important negative factor is high rainfall levels in the previous summer, obviously impeding the quality of the seed. Consequently the seed used in spring 1366 was of low quality. For the link between weather and productivity of barley, beans and peas, see Hooker, Weather and crops, 118–121.

It appears that in the gloomy, wet and disease-ridden years 1365 and 1366 the lack of labour caused first by the two major plague waves (1348–1349 and 1361) and then by the 'Pokkes' began to be felt keenly during the rainy harvests. There had only been one complaint about missing or dead tenants in the harvest paragraphs of the Pipe Roll in the Bishopric of Winchester between 1362 and 1364, but in 1365 and 1366 these voices multiplied and ensuing problems manifested themselves also in the unusual events on the Norwich Cathedral Priory manors. The lack of labour and raised costs contributed to the excessive length of the grain harvest, but the landowners could not offset these raised costs with the help of raised grain prices. The fall in demand for grain associated with the epidemic started to expose the problems that would finally lead to an end of demesne farming in wide areas of England in the following decades.

8.4.4 The Calm 1370s and the Long Harvests Around 1380

After the turbulent 1360s, most of the 1370s appear as a period of drier and more stable conditions and offer an interesting dichotomy in the behaviour of the harvest dates and length, since the dates continued in their highly volatile behaviour in the first half of the decade. Until 1377 the Sedgeford-Gnatingdon complex in northwest Norfolk harvested normally for a period of five weeks or 35 days, other manors also have harvests of mostly stable duration. In short, in those years the fluctuation in the harvest lengths is too reduced to reflect the full scale of precipitation variability during harvest time. The relative standardisation of the harvest length is most likely linked to the impact of the Third Pestilence, 1368–1369. The loss of life must have left deep scars in the villages, considering that already before, in the mid-1360s, *penuria servientium*, the lack of servile labour, was a common complaint in the manorial accounts, despite Norfolk being one of the most densely settled parts of England at this time.[50] To counteract this lack of labour it is likely that the percentage of crops mown instead of reaped increased (Appendix 3).

In 1371 and 1373 harvests were slightly longer than average. In 1371 spring and early summer had been warm and the harvest length seems to have been rather an

[50] Probably as a consequence of the lack of labour the sown acreage collapsed 1369 in Gnatingdon and Sedgeford. It is the spring corn whose acreages shrunk considerably on both manors. In Sedgeford an astonishingly high number of harvest works were performed by itinerant cockers, they probably were meant to make up the shortfall in the local labour supply. (In 1366 and 1367 309 acres respectively 318 acres were harvested with the assistance of six respectively 24 works by cockers. Then in 1369 60 cockers helped harvesting 280 acres.) In Sedgeford wheat and peas also failed. These facts might help to explain why the harvest duration in northwestern Norfolk was average in 1369, although weather references for southern England indicate very wet conditions. More on the weather conditions 1368–1370 in Sect. 6.3. The sown acreage on both manors recovered – Gnatingdon was close to the average of acreages sown of the mid-1360s by 1373, Sedgeford showed signs of recovery by 1372 and reached the level of the mid-1360s by the mid-1370s.

indication of harvest success than rain. Grain prices fell considerably in autumn 1371 (Sect. 6.2). The conditions in 1373 were the opposite. The growing season in East Anglia was cold, the summer half year of the Low Countries average, 5. Towards central Europe the summer was warm and dry.[51] Unfortunately the Winchester manors are silent for this year, but weather conditions in Suffolk were very variable. A very wet spring was followed by a dry later growing season and probably a wet harvest time. Hertfordshire was also subject to a very wet spring. It appears that the wet conditions at harvest time stretched from Sussex to Norfolk.

Between 1377 and 1382 lies a period of longer harvests, the harvest of 1380 being extremely long. All these summer seasons were average or close to average in the Low Countries. In England, however, the grain harvests 1377–1379 were good or even abundant,[52] and the bulk of grain was the main factor in the longer than average harvests in East Anglia. In Sussex the weather was very favourable for farming 1377 and 1378; in 1377 there was some rain in the area of the Winchester manors during harvest time, but the rainfall must have been very limited, since the grain price is extremely low 1377–1378. Meteorological conditions deteriorated in 1379, in Sussex it was wet and stormy and in the Winchester area rain created problems from June onwards and damaged the growing corn. Probably East Anglia was also affected, and the grain price rose in consequence.

In 1380 rain remained a prominent feature, as the extremely long harvest in East Anglia indicates. Sussex experienced a wet harvest, and in Hertfordshire spring and early summer were also wet.

The harvests in 1381 and 1382 were longer than average. In 1381 the weather was dry in Sussex and in Kinsbourne indications point towards a dry spring, too. In June the Peasants Revolt took hold in Norfolk and the consequent upheaval might have had an influence upon the availability and willingness of labour during harvest, even though the revolt itself was subdued in late June. Weather was not good throughout the harvest and could slow the work down, in northwestern Norfolk a field of peas was damaged in a storm (Appendix1).

In 1382 the weather was detrimental to agriculture, the growing season had been cold and the subsequent late harvest in northwestern Norfolk continued until Michaelmas and even a few days longer. The problems were due to the great rains at harvest time which made the opening and rebinding of the sheaves necessary (Appendix 1). It is possible that the harvest lasting beyond Michaelmas made manorial accounts for the harvest period itself necessary for the Norwich Cathedral Priory manors at Hindringham and Plumstead. Here the normal accounts stop before the onset of the harvest, which is unusual. In the case of Plumstead the closing of the account seems to coincide with the start of the harvest,[53] the account for Hindringham was stopped with the official onset of the harvest period, but not necessarily the real harvest, 1 August.[54] The severe rain during harvest time in Norfolk is in accordance with information from Sussex, Kinsbourne in Hertfordshire and the Winchester

[51] Alexandre, Le climat en Europe, 508.

[52] Walsingham, Historia Anglicana, vol.1, 343, 389, 427.

[53] Plumstead NRO, DCN 60/29/35.

[54] Hindringham NRO, DCN 60/20/32.

manors, where the harvest was wet. The latter specify an abundance of rain not merely during the harvest, but also in the preceding months or weeks. This generally wet and cold weather is responsible for the low April–July mean temperature. The summer half year in the Low Countries is classified as 6, on the continent spring was warm and summer cold.[55]

8.4.5 Weather Conditions in 1387

Amidst the generally warm and dry summers of the mid-1380s falls the very long harvest of 1387, a year when the growing season temperature was slightly below average. In Sussex the harvest time was very rainy. At the same time, an epidemic broke out in southern England. The Westminster Chronicle states 'unde in mense Septembris secuta est magna mortalitas hominum'.[56] The resulting long harvest, due to the high levels of precipitation and probably also to the impact of the epidemic, also caused problems in the Winchester area. One Winchester manor faced higher costs in harvesting because the harvest took long. However, spring and summer up to the onset of the rainfall were dry and also warm. In Sedgeford oats are said to have suffered from drought, in Gnatingdon the vetches grew badly (Appendix 1). In line with this information, spells of very warm weather defined the summer half year 1387 in England.[57] The summer season in the Low Countries is classified as 4, the rainy weather from August onwards affects the summer season index.

8.4.6 Wet Conditions in the First Decade of the Fifteenth Century

Around 1400 rainfall levels increased, and between 1402 and 1408 the harvest length settled generally on a higher plateau, indicating raised precipitation levels at harvest time. Growing season temperatures in these years were frequently low, as in 1399, 1401, 1402 and 1406; only in 1400 was it very warm. Due to the time-saving effect of mowing, the frequently reduced size of directly managed demesne land and the battle of bailiffs and reeves to keep labour costs low, the rain was sometimes not sufficiently severe to prolong the harvest time in East Anglia, as in 1399 and 1401. Harvests longer than average did occur in 1402, 1405, 1406 and 1408.

In 1399 and 1401 cool growing seasons were followed by wet harvests. In East Anglia some places harvested until mid- and also late September. Rain is reported

---

[55] Alexandre, Le climat en Europe, 521.

[56] Chronicon Westmonasteriense 1381–1394, 204.

[57] Chronicon Westmonasteriense 1381–1394, 204.

from Norfolk itself (Appendix 1), the manors of the Bishopric of Winchester and Sussex. On the Winchester manors the situation was bad in both years, in East Anglia harvest duration remained controlled. In 1401 some meadows were flooded in summer on the Winchester manors, and widespread flooding caused problems in autumn and winter in the Cambridgeshire fenland.[58]

Although the growing season temperature was again low in East Anglia in 1402 and the harvest long, the sky must have been somewhat brighter at some point in spring or early summer, since peas and vetches were partly lost because of dry weather. However, the harvest period was rainy, even the rye crop suffered from too much water at one manor (Appendix 1). In the Winchester accounts only rain at hay harvest is mentioned, but an impression of the weather in the west of the British Isles can be gained from Thomas Walsingham, who describes a royal military expedition to Wales which was caught in continuous rain, mixed with hail and snow, and cold around the second part of August and the beginning of September 1402. Indeed the weather was so bad, that the interference of magic was apprehended.[59]

Then in 1405 the growing season probably still passed without many problems, because the reconstructed East Anglian April–July temperature was only slightly below average, and the Winchester manors report no adverse weather for this year. However, the East Anglian harvest was longer than average. Most manors finished in the last third of September and rain at harvest time made extra works necessary. On Cromer Ridge the rainfall led to a reduction of profits from the meadows (Appendix 1). The low-lying fenland in Cambridgeshire was again affected by inundation.

The nadir was reached in 1406. Whereas the Low Countries summer half years 1401–1405 fluctuate unspectacularly between 4 and 6, summer 1406 suddenly drops to 2. This is mirrored in the English April–July mean temperature; the harvest was also longer than average and was finished in late September. Again extra works were needed for drying the grain and hay meadows were affected by the high water table on Cromer Ridge. Flooding is reported probably for autumn 1406 and the following winter from eastern Norfolk[60] (Appendix 1) and Cambridgeshire. Conditions on the Winchester manors were similar: the hay was weak because of the rain and there was a delay in the ripening of the grain crops and consequently a late harvest.

Relief regarding the weather conditions was offered by the year 1407. Growing season temperature was average, the East Anglian harvest of normal duration. Conditions during spring and summer were actually dry, as drought is mentioned in the Norfolk accounts (Appendix 1), and is confirmed by the little information that is

[58] The information about flooding in Cambridgeshire comes from Wisbech and was collected by Stone, Medieval agriculture, 123.

[59] Annales Ricardi Secundi et Henrici Quarti, 343 and a similar text in Walsingham, Historia Anglicana, vol. 2, 250–251.

[60] The accounts for Martham, NRO, NRS 5907, 20 D2, and Flegg, NRO, DN/EST 09/12, refer to the whole year1406–1407, but since other evidence points to a dry growing season in 1407, it can be concluded that the inundation occurred in autumn 1406 or the winter 1406–1407.

available from the Winchester manors.[61] A severe plague wave took hold particularly in the west of England (Chap. 10).

The dry summer 1407 and the extremely cold and long winter 1407–1408 – in England the cold and frost lasted well into March[62] – indicate a general shift in the large scale circulation pattern. 1407–1408 was the first severe winter since a long time (the last significantly colder than average winter (7) in the Low Countries was 1398–1399) and in England it was followed by an unremarkable and probably even serene spring and early summer, since April–July temperatures were average despite the delayed onset of the growing season. In summer, however, the weather must have changed, the East Anglian harvest was long and the chronicles show, that late summer and early autumn were dominated by high winds, extremely heavy rainfall and subsequent inundations. At the beginning of September river floods afflicted northern England, and for 7 September 1408 Walsingham describes an extreme rainfall event and the Continuatio Eulogii a flash flood in Ware, Hertfordshire.[63] For northwestern Norfolk accounts mention the loss of oats on the fields due to a tempest of wind at harvest time (Appendix 1). Oats were cut towards the end of the harvest and since cutting had started in northwestern Norfolk already in early August, it is likely that Walsingham, the Continuatio Eulogii and the accounts refer to the same storm. A storm also crushed the corn crops and legumes in two Hampshire manors of the Bishopric of Winchester. The Low Countries summer half year index reflects the wet and stormy harvest season and stands at 4. In England this summer marks the end of the wet cycle in the first decade of the fifteenth century.

8.4.7 *Weather Conditions in 1421*

In 1421 the people in Norfolk suffered from a late and extremely long harvest. Since 1421 was also an very cold growing season, the year has been considered in Sect. 6.3, but the impact of the rainfall on the harvest shall be given here.

In the Norwich region it was possible to finish harvesting in mid-September, but in northwest Norfolk the work was not completed before the first week of October. These extreme conditions are explained in the accounts, in Sedgeford there was an overflow of water, that affected the sheaves on the fields, and Plumstead reports 'watery weather' (Appendix 1). On all four manors of Norwich Cathedral Priory for

---

[61] The Winchester roll for 1407 is missing – maybe due to the plague in that region – hence not much information can be gained from that area, see Beveridge, The Winchester rolls and their dating, 100.

[62] The St. Albans Chronicle, 1406–1420, 27 (probably a work by Thomas Walsingham). An almost identical text is given in Walsingham, Historia Anglicana, vol. 2, 277.

[63] Walsingham, Historia Anglicana, vol. 2, 279, Continuatio Eulogii, vol. 3, 413. The flooded meadows in eastern Norfolk (Flegg NRO, DN EST 09/13 1408–1409) might also fall to autumn 1408 (Appendix 1), though this is not explicitly stated.

which the *compoti* survive – the two other ones being Gnatingdon and Taverham – a high number of additional works was needed for turning, opening, drying and rebinding the sheaves (Appendix 1). Gnatingdon pleaded 203 additional works, but in the audit merely 146 were approved. The average harvest works needed at Gnatingdon in the six harvests, that can be accounted for 1409–1420, was 487.75 works; in 1421 the number needed rose to 803 works.[64] Once more these conditions are mirrored in the Pipe Rolls of the Bishopric of Winchester, apart from long rainfall at harvest time, the destruction of the grain in the fields by the wet weather and even the sprouting of wheat in the field before it could be carried off are mentioned. The harvest in Sussex was also very wet. Flooding is reported for the manor of Flegg in Norfolk (Appendix 1). On 19 November 1421 the Dutch North Sea coast was submerged by an extreme storm surge[65]; it appears that the low pressure system responsible was not surprising in late 1421 given the wet conditions over the previous months.

8.4.8 Weather Conditions in 1423

After an interruption in 1422, the wet conditions returned already in 1423,[66] though they were not as extreme as 1421. The growing season temperature was very low and the East Anglian harvest very long. Most manors did not complete the harvest before the end of September, Hindringham was still at work around mid-October. There is ample evidence from the Winchester manors, Sussex and East Anglia for the rainfall at harvest time: for the harvest three out of four surviving Norwich Cathedral Priory rolls state additional works for turning and re-turning the sheaves, the rolls of Martham and Sedgeford identify 'rainy weather' and 'watery weather' as the cause of these extra-works (Appendix 1). A Winchester account echoes the Norfolk records and in Sussex the harvest was also wet. A version of the Chronicle of London describes the growing season and harvest time 1423 in the following words:

> Rex Henricus Sextus [1422–1423]: Anno p° Also the same yere in somer tide was great plente of al maner cornes and fruytes: but a litle before Midsomer there bigan to falle moch reyne, which contynued lasse or more every day as for the moost partie; howsoever the wynde stode unto viij daies bifore Christmas, so that men myght not gadre ynne there, and namely the codde corne, and yet was there plente of corne ynough.[67]

---

[64] Gnatingdon NRO, LEST/IC 40.

[65] Gottschalk, Stormvloeden, vol. 2, 51–100. Brooks, Glasspoole, British floods and droughts, 96 mention sea floods also in England on 19 November 1421, but cite no evidence. Flooding is mentioned for England in 1421 in Sussex, see Brandon, Late medieval weather in Sussex, 5 and in eastern Norfolk, manorial account for Flegg NRO, DN/EST 09–16, but the flooding of meadows in the Flegg district was a frequent occurrence and it is unclear if the referenced flooding was due to the storm surge in November 1421 pushing sea water up the river valleys or if it was due to rainfall.

[66] Brandon, Late medieval weather in Sussex, 4 also rates the harvest of 1422 as very wet.

[67] Chronicle of London from 1089 [sic] to 1483, 111. The weather reference comes from another manuscript published in the same edition, Cottonian MS. Julius B. I., and is given ibid., 165. It is not entirely clear, of this version of the chronicle is contemporary.

The text confirms the reports of the account rolls about a long and late harvest and specifies rainfall from shortly before Midsummer until almost Christmas. Before Midsummer the weather conditions must have been more favourable to agriculture, because there was a plenty of corn and fruits. The spring-summer months before the end of June can hence neither have been particularly cold nor could they have caused a harvest date as late as 1423. Indeed in Sussex part of the summer was dry. On some Winchester manors the dry conditions were persistent enough for some time between May and late June to lower the return rates for barley, pulses and even the more drought resistant wheat. In Hindringham itself mixed grain and peas are classified as weak in 1423, but the reason is unfortunately not given (Appendix 1).

Apart from the cold and wet post-June summer time an extremely cold and long preceding winter contributed to the late harvest in England: the winter half year 1422–1423 stands at 8 in the Low Countries and the season being very rich in snow, also left its traces in the English sources. One Winchester manor had to supply additional fodder to its livestock because of the long continuation of snow. Although this might refer to snowed up winter pastures, it seems more likely to be linked to a delay in grass growth in early spring. Late winter and early spring were the hungry gap for the livestock, the point when the fodder supplies ran out and the opening of pastures was potentially retarded. In Sussex, too, the winter was hard. Norfolk, being comparatively open to cold northerly winds, is even more likely to have suffered a delay in the onset of the growing season. It is remarkable, and probably indicative of the low population numbers of the time, that neither the problematic harvest of 1421 nor that of 1423 caused any rise in the grain price, and in 1423 there remained, according to the Chronicle of London, a good amount of marketable corn.

8.4.9 Weather Conditions in 1428

Although even the Norwich Cathedral Priory accounts thin out in the 1420s, in 1428 the final late and long harvest in the series up to 1431 can be distinguished. Without doubt it was a catastrophic wet and cold year, it has been described in Sect. 6.3.

8.4.10 Weather Conditions in 1435

By 1430 the stream of chronicles and administrative sources that illuminated the previous centuries has become but a trickle, so that no information is available concerning the meteorological conditions of the summer half year 1435 in England. Most winters of the first half of this decade were extremely hard and the summers generally dry and warm in Europe north of the Alps. In the Low Countries, 1435 was an average summer. In Canterbury an unspecified epidemic took hold in that year.[68]

8.4.11 Weather Conditions in the Second Half of the 1430s

The East Anglian manorial accounts which supply the proxy data for the spring-summer temperature reconstruction and the information on the harvest length for the precipitation index have mostly ceased to be produced by the 1430s. This end of the supply of account rolls and the low output of historiographic writings in England at this time lead to a lack of detailed knowledge about the weather in England during this difficult period. In the Low Countries cold, wet and dull conditions started in the middle of the 1430s. The summer half year 1436 was extremely cold (1) and the two subsequent summers remained on a temperature level slightly below average (4). The winters 1436–1437 and 1437–1438 were very cold (7–8). In England no weather information is available from the Pipe Roll of the Bishopric of Winchester for 1436 and 1437, in 1438 mildew destroyed the wheat crop. Rainfall levels were raised in the summers 1436 and 1438 in the Low Countries.[69] This is mirrored in England either in 1437 or 1438.[70] In England the grain price rose steeply in the years following the harvests 1437 and 1438, the most severe famine of the fifteenth century ensued , but conditions never approached the disaster of the Great Famine 1315–1317.[71] In 1438–1439 plague mixed with famine disease appeared in England and Scotland (see Chap. 10).

[68] Hatcher, Mortality, 29–30.

[69] Camenisch, Endless cold, 1062.

[70] Ingulph's chronicle of the Abbey of Croyland, 398. A similar text is in Latin in the Waltham Annals under 1437.

[71] Dyer, Standards of living, 268. For a description of the effect of the high grain prices, see McLaren (ed.), London Chronicle, 209–210 and Ingulph's chronicle of the Abbey of Croyland, 398.

Chapter 9
Climate and the Grain Price, 1264–1431

Bread and ale were staple foods in the Middle Ages. About 70–80% of the daily caloric requirements was derived from grain-based foods.[1] Whereas the demand for grain was comparatively stable from year to year, the supply and price of grain were variable.[2] Due to the dominance of bread, ale and pottage in the medieval diet, the grain price can be considered a 'socio-economic barometer'.[3] In this sense the long-term trend of the grain price is of less importance than its short-term fluctuation.[4] During the pre-industrial period meteorological conditions during the growing and harvest season, when the crops were especially vulnerable to the weather, were of critical importance for the harvest success and hence for the grain price. This was self-evident to all agriculturalists and the vulnerability and resilience of agriculture to weather specifically in England was described for the first time by William Merle around the year 1340 in his work 'De pronosticacione aeris'.[5] In the last chapter of this manuscript Merle made the explicit connection between weather, grain growing and grain price. He considered the impact of six mostly meteorological factors on grain growing: the lack of ploughing, abduction by floods, drought, cold, humidity and mildew.[6] When harvests failed, it would be difficult to supply land-locked regions in particular. Water transport of grain was comparatively cheap, but carts laden with grain could travel only about 16 km to the market to allow for a return journey to be made within the day. In times of high prices the distance could be doubled, but this was rarely exceeded.[7]

Merle's interest in weather and agricultural productivity marks only the starting point for the study of climate impacts in western and central Europe. In more recent

[1] Stone, Consumption of field crops, 11.

[2] Harvest quantity and grain price correlate highly in medieval England, Titow, Evidence on weather, 362.

[3] Bauernfeind, Woitek, Influence of climatic change on price fluctuations, 304.

[4] Hoskins, Harvest fluctuations, 1620–1759, 20.

[5] BLO, MS Digby 147, fols. 125r-138v.

[6] Merle, De pronosticacione aeris, BLO, MS Digby 147, fols. 136r-138v.

[7] Farmer, Two Wiltshire manors, 5–7.

K. Pribyl, *Farming, Famine and Plague*, DOI 10.1007/978-3-319-55953-7_9

times agrarian, economic and environmental historians have intensified the analysis of the influence of meteorological conditions on the grain price.[8] For England, studies have underlined the role of precipitation and temperature in harvest success.[9] A more refined model regarding the timing and impact of weather based on actual agricultural science has been developed for Switzerland by Pfister; however, as long as the influence of the high altitude of many Swiss regions is considered, the model would be applicable to western and central Europe, where climatic conditions are broadly comparable.[10] The factors impacting on grain growth cover the whole agricultural year; autumn rainfall and temperature, winter precipitation, spring as well as summer rains and temperature are decisive for the harvest success. In particular precipitation from winter to summer, and temperature in autumn and summer are important. In England's milder climate, the risk associated with winter duration is reduced while the risk stemming from precipitation is enhanced. Titow emphasized the influence of autumn and summer precipitation and winter severity on the harvest quantity in medieval English agriculture.[11]

The factors available for analysing the link between weather and grain price in this study, the April–July mean temperature and the July–September precipitation index, cover the whole growing season as well as the last stages of the grain ripening and the harvest; they encompass the summer temperature and precipitation which have been shown to be highly relevant for harvest quantity and quality in previous studies. The remaining weather factors influencing crop development, are impossible to assemble in a continuous series for late medieval England; conditions during autumn are seldom reported in medieval records.[12] Non-climatic factors also contributed to the harvest success and to the grain price formation, therefore a quantification of the impact of specific weather conditions on the grain price is rarely possible.

For the analysis, it is therefore necessary to concentrate on years that witnessed extreme weather conditions, i.e. extremely wet and cold or dry and warm growing seasons (Fig. 9.1), as well as on years that saw the highest grain prices or the sharpest rises in the first difference of the price (the movement of the grain price compared to the previous year) (Fig. 9.3). The English series of grain prices starts in 1264 and in this study the data provided by Phelps Brown and Hoskins and reworked by Munro are used.[13] The Munro grain price series refers to the agricultural or harvest year (from harvest to harvest) and is dated to the end of the agricultural

[8] Abel, Massenarmut, 35–37, Pfister, Fluctuations climatiques et prix céréaliers, Bauernfeind, Woitek, Influence of climatic change on price fluctuations, Scott et al., Grain prices in England, Brázdil, Durdáková, Effect of weather factors.

[9] On precipitation, see Hoskins, Harvest fluctuations, 1480–1619, 40–41, and on temperature Scott et al., Grain prices in England, 7, 11.

[10] Pfister, Fluctuations climatiques et prix céréaliers, 34–37.

[11] Titow, Evidence on weather, 362–364.

[12] Camenisch,, Endless cold, 1056.

[13] Munro, Revisions of the Phelps Brown and Hoskins 'Basket of Consumables' commodity price series.

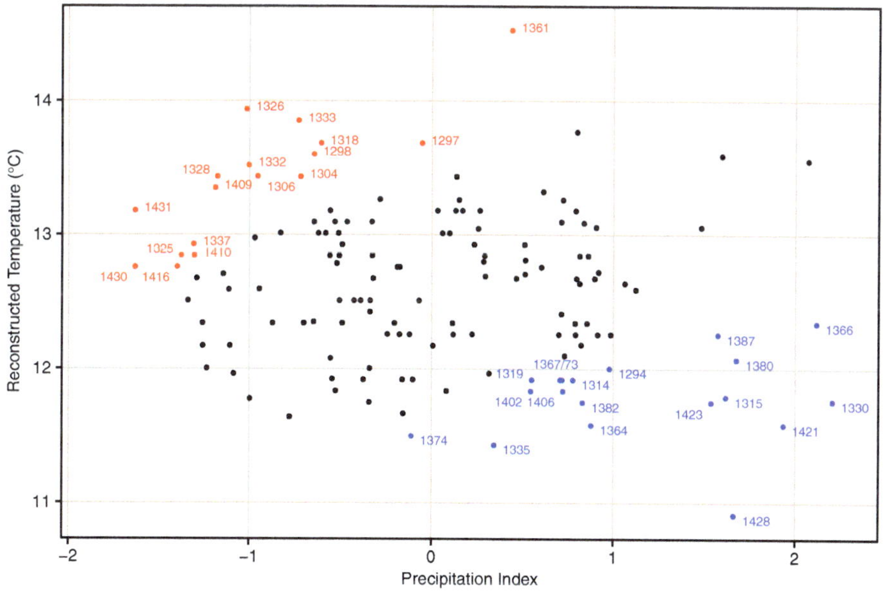

Fig. 9.1 Reconstructed April–July mean temperatures and July–September precipitation index, 1264–1431. Marked are extremely dry and warm years (*red*) and wet and cold years (*blue*). Precipitation data are normalized

year.[14] For ensuring that the prices coincide with the harvest and with the climate data the values have been advanced by one year. The prices for rye, barley and oats mirror the trends of the wheat price closely, albeit on a lower level, so that the wheat price is taken as representative of the variability of the overall grain price.[15]

With respect to long term trends, the climate data show that in late medieval England the climate was changing in such a manner as to interfere more frequently with the agricultural production. According to the reconstructed April–July mean temperature a cooling trend marked the time between the second half of the thirteenth century and the first half of the fifteenth century. In addition, the rainfall levels of July–September were particularly elevated in the 1360s, and variability in precipitation levels was extremely high in the 1420s. After c.1380 July–September rainfall decreased again, and drier conditions took hold in the fifteenth century. The cooling spring and summer temperatures resulted in shortened growing seasons. While this rarely posed a problem in East Anglia, an exception being 1428, it could be decisive in upland districts and in regions farther north. In the Lammermuir Hills in southern Scotland the altitude limit of settlement and oat cultivation dropped by c.190 m between 1250 and 1600; the falling summer season temperatures and

[14] On the dating of the various price and wage series for England, see Harrison, Grain price analysis, 136–137.

[15] This is also a phenomenon of early modern grain prices, see Hoskins, Harvest fluctuations,1480–1619, 40, Appleby, 871–880, Scott et al., Grain prices in England, 11.

Fig. 9.2 Stacking sheaves. Luttrell Psalter, Lincolnshire, circa 1325–1340 (British Library, Add. MS 42130, f. 173)

increasing wetness drove the cultivation limit to lower levels.[16] For the regions of intensive agriculture in southern, central and eastern England increased summer precipitation totals were more important than the long-term fall of spring-summer temperatures. The description of individual years of rainfall extremes (Chap. 8) has outlined the negative impact of very high rainfall during the growing season and harvest time. Frequent or abundant rain at harvest was not the only severe problem; over wide areas of England the soil tends towards water-logging, and so exacerbates the impact of high precipitation levels between sowing time and harvest. Wheat, the preferred and after the 1350s increasingly dominant bread grain, is particularly vulnerable to abundant rainfall. In an age before the widespread use of barns, raised precipitation levels in the post-harvest period were detrimental for the stacked sheaves (Fig. 9.2).[17]

The trend towards lower spring-summer temperatures and the raised precipitation levels in the decades after 1350 are borne out by the temporal distribution of years that lie in the extremely wet and cold spectrum (Fig. 9.1). Comparatively few of those years pre-date the second half of the fourteenth century: 1294, 1314, 1315, 1319, 1330 and 1335.[18] In the post-1350 period the cases of cold growing seasons and wet summers were much more frequent: 1364, 1366, 1367, 1373, 1374, 1380, 1382, 1387, 1402, 1406, 1421, 1423 and 1428. Before the middle of the fourteenth century such miserable springs and summers were always associated with either high prices or considerable rises in the grain price, and none of these summer half years were followed by low prices or prices lower than the previous year (Fig. 9.3).[19]

---

[16] Parry, Secular climate change, 9.

[17] Not all the harvest could be stored in barns. On the use of barns and granaries on the Norwich Cathedral Priory estates, where demesne barns were spacious enough for the demesne harvest, but also on barns in medieval England in general, see Slavin, Bread and ale, 119–139.

[18] The year 1322 was also very wet, but no data to reconstruct temperature are available; in the Low Countries the summer half year was very cold, van Engelen et al., A millennium of weather, winds and water in the Low Countries.

[19] An exception is the grain price after the harvest 1322, which dropped from the excessive level of the agricultural year 1321–1322, but remained very high – 1322-23 having the eighth highest price between 1264, the start of the price series, and 1431, the end of the April–July temperature recon-

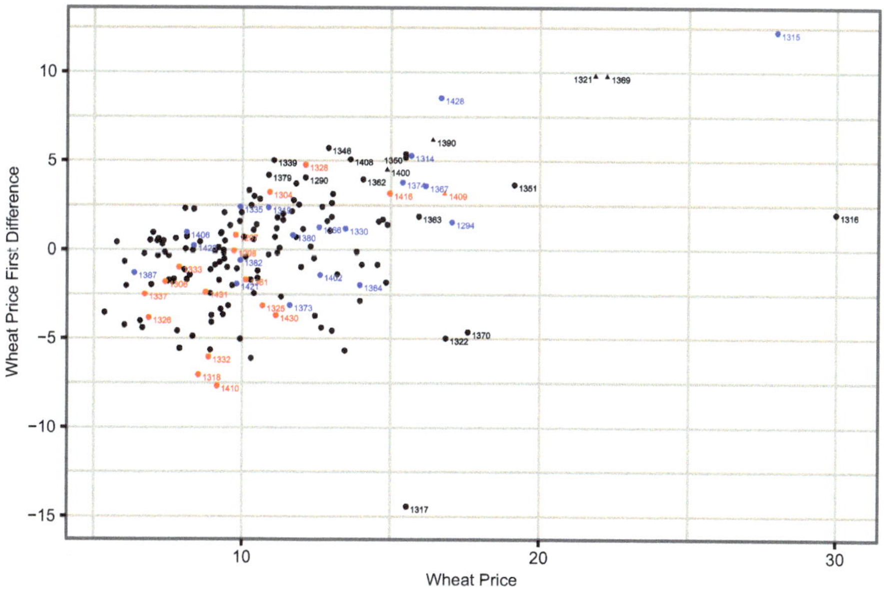

Fig. 9.3 Wheat price and first difference of the wheat price, 1264–1431. Marked are extremely dry and warm years (*red*), wet and cold years (*blue*) and years of epidemic disease in the high price segment (*triangles*). The epidemic disease was mostly plague, but in 1321 cattle numbers (including the draught oxen) all over England had just collapsed due to Cattle Plague. The precipitation and temperature extremes are taken from Fig. 9.1. The wheat price is from Munro, Revisions of the Phelps Brown and Hoskins 'Basket of consumables' commodity price series. Prices and first difference are advanced by one year and date to the harvest year. The year 1317 autocorrelated with the preceding year; the grain price in this years is high, because it follows famine conditions associated with even higher prices, but due a lack of seed-corn etc., the price had not yet normalized. To some extent this also applies to 1370, but this year was difficult for grain growing in the continent, where a dry spring was followed by summer rains (Le Roy Ladurie, Histoire humaine et comparée du climat, 76–77)

Since the grain crop was likely to suffer in years of rain and cold, such years left deep traces in chronicles and other written records. The risk associated with high amounts of rainfall in spring and summer was evident, and in his work 'De pronosticacione aeris' William Merle rated excessive rain as a frequent cause of dearth in England; it damaged the wheat in particular. He noted ten ways in which extreme 'humidity' or rain could ruin the wheat crop, which is more information than on all the other potential triggers of dearth combined.[20] Whereas Merle considered the meteorological factors involved in harvest failure and dearth from a rational angle,

struction. After the wet and cold summer half year 1335 grain prices rose, but did not exceed average. This is linked to the currency shortage and deflationary tendencies in the two decades before the Black Death, see note 63 in Chap. 2.

[20] Merle, De pronosticacione aeris, BLO, MS Digby 147, fols. 136v-137v. Excessive rainfall was also the commonest cause of harvest failure between 1480 and 1619, Hoskins, Harvest fluctuations, 1480–1619, 40–41.

rain and storms were also considered from a metaphysical point of view in the Middle Ages and Early Modern Period – as a malevolent force at work, disturbing the normal and regulated life. In the fifteenth century such a disturbance of the atmosphere was often perceived as the result of witchcraft.[21]

Despite the more frequent occurrence of wet and cold years in the second half of the study period the incidence of high prices or drastic price rises associated with them decreased and only occurred after the harvests in 1367, 1374 and 1428 (Fig. 9.3). The reconstructed low mean temperature in 1364 was caused by a long and very hard winter and a consequent late start of spring. It therefore had little influence on the harvest, although the winter sowing might well have suffered under the severe winter. In general the clustering of wet and cold summers without an impact on the grain price in the decades after the Great Pestilence is linked to the falling demand for corn due to the death rate in each of the passing epidemics. Many of those inclement years without a price response occurred shortly after or coincided with outbreaks of widespread disease (for a list of plagues outbreaks, see Chap. 10). Already during the Great Pestilence there had been no immediate impact of the bad weather on the grain price (Sect. 6.5). The wet and cold 1366 came one year after England had faced the 'Pokkes' (Sect. 8.4), the inclement weather 1380 and 1382 coincided with a prolonged sequence of regional plague outbreaks 1379–1383 (1379 plague in northern England, 1382 plague in London, 1383 plague in Norfolk). The bad weather in 1402 followed closely on the national outbreak of plague in 1400, and 1406 occurred during the low level plague wave 1405–1406 which came into full force in 1407. The soggy years 1421 and 1423 were in the direct aftermath of the pestilence of 1420, in the north of England plague was still present in 1421, and at many ecclesiastical institutions – and by implication most likely also in the wider population – the levels of mortality remained elevated throughout the early 1420s.[22] These epidemics reduced the need for grain and perhaps more importantly caused a disturbance of the market, so that prices could not react adequately to the supply situation.

Many of the wet and cold spring–summers after 1350 fell within the following 25 years, a period commonly known as the 'Indian summer' of demesne farming, when bad weather and subsequent poor harvests played a role in the sustained high grain prices after the Great Pestilence, and thereby helped to prolong the profitability of demesne farming.[23] The East Anglian precipitation index and the April–July mean temperatures show that there was indeed a sequence of wet and partly cool springs and summers, centred around the mid-1360s. Additionally strong fluctuations in temperature and precipitation were common in this period, as the dry and

[21] Cohn, Europe's inner demons, 212–213, McLaren, London chronicles of the fifteenth century, 71–72.

[22] For Christ Church Priory at Canterbury, see Hatcher, Mortality, 30, for Durham Cathedral Priory, see Hatcher et al., Monastic mortality, 677–678, and for Norwich Cathedral Priory, see Noble, Norwich Cathedral Priory, 58–61.

[23] For example Mate, Agrarian economy after the Black Death, 348.

hot summer half years interspersed in those years demonstrate. Spring and summer 1361 and parts of 1369 were dry and hot. In 1356 heavy rainfall at harvest time followed a prolonged drought in spring and early summer. Extremes also occurred in the winter half year: winter 1363–1364 was exceptionally cold and long, and throughout the whole period 1355–1375 winter conditions in central Europe were highly variable.[24] In the Low Countries summer and winter indices also fluctuate strongly from year to year.[25] Indeed the 1360s are the period when the correlation between harvest date and length in Sedgeford-Gnatingdon breaks down, implying socio-economic problems, but also reflecting this high climatic variability (Fig. 7.2). The unpredictability of the weather and also the unpredictability of the agricultural production were plain to see for everyone living through this period, and even found their way in one of the greatest literary works of the times: 'The vision of Piers Plowman' by William Langland.[26] It is impossible to quantify exactly the impact of the 'wederes unresonable'[27] or the diminished work force in the post-Black Death era upon harvest quantity and quality, but the climatic factor clearly figured prominently in the mind of contemporaries and must have played a major role in keeping grain prices elevated. The high level of interannual temperature variability and the high precipitation totals that gave demesne farming a renewed lease of life and landowners an extra lease of profits, were keenly felt by the lower social orders. In terms of weather conditions the 'Indian summer' of demesne farming was less of an Indian summer than of a dull autumn.

Whereas wet and cold summer seasons were feared in late medieval England, drought and warmth on the other hand were not perceived as risk factors for agriculture neither in the Middle Ages nor in more recent times. Already in the fourteenth century Merle did not consider drought to be a cause of famine in England since it generally did not damage the wheat, although it could prove problematic for barley and legumes.[28] The vulnerability of the pastoral sector to drought was higher as low water levels, poor pastures and a bad hay crop were detrimental to livestock. With respect to agriculture the relaxed attitude towards drought persisted throughout the Little Ice Age, and in the eighteenth and nineteenth centuries it crystallized in a number of sayings such as 'Drought never bred dearth in England'.[29] Indeed many years occurred during the Late Middle Ages when warm growing seasons with below average rainfall from July to September actually resulted in a good harvest. Such springs and summers can be located in 1297, 1298, 1306, 1318, 1325, 1326, 1332, 1333, 1337, 1361, 1410, 1430 and 1431 (Fig. 9.1). These harvests were associated with a drop in wheat prices or at least stable prices (Fig. 9.3), which is evident

[24] Pfister et al., Winter severity, 101–102.

[25] van Engelen et al., A millennium of weather, winds and water in the Low Countries.

[26] Frank, 'Hungry gap', 231–233, Langland, Piers Plowman, 188–189.

[27] Langland, Piers Plowman, 188

[28] Merle, De pronosticacione aeris, BLO, MS Digby 147, fol. 136v.

[29] Inwards, Weather lore, 7.

throughout the 1330s and early 1340s.[30] Dry and warm growing seasons with the described characteristics were rare in the decades following 1350, but then became frequent again from about 1410 to the start of the fourth decade of the fifteenth century. However, cases when dry and warm weather conditions became sufficiently severe so as to become detrimental for crop growth do exist. Prices rose notably after the extremely dry and warm years 1304, 1328, 1409 and 1416 (Figs. 9.1 and 9.3): after the harvests 1409 and 1416 they reached a very high level. Both years, 1409 and 1416, are however very specific cases. In 1409 other causes, such as epidemic disease, might also be involved in price formation. The price rise after the harvest 1416 was probably linked more to the prolonged heavy rainfall in March and April and less to the summer weather (Sect. 8.3). Years, such as 1293, 1331 and 1415, when drought and warmth were less pronounced or the lack of rain centred on a time before the period covered by the precipitation index, July to September could also see prices rise. The drought 1331 followed so shortly on the wet and cold conditions in 1330 that this sudden change from one meteorological extreme to another and the complete absence of rain between March and June in southern England were beyond the coping capacity of the medieval agriculturalist. Nevertheless while the prices following the growing seasons 1293, 1331, 1409 and 1416 indicate dearth, extremely dry and warm spring-summer periods were rarely connected to outright famine conditions. A variety of factors contributed to the beneficial effect of dry and warm weather on harvest success and grain price in medieval England. Of prime importance was the role of dry weather during harvest time, which allowed for a rapid and easy harvest. Due to its low moisture content the grain could then be stored with a low risk of rotting. Additionally, since many medieval English farming regions were situated on loamy or clayey soils that retain water, a shorter period of dry weather would have had no significant negative impact upon grain growth. Even a prolonged dry period in England is rarely of the same severity as in more continental climates. Furthermore wheat is a drought resistant crop and even though rye, barley, oats and legumes are sensitive to dry conditions and at least in Norfolk occasions are known when the barley harvest was disappointing, while the wheat harvest was good, the suffering of rye and the spring corn from drought was rarely reflected in the price. In fact for causing a detrimental effect on the wheat crop, the dry and warm conditions had to coincide with a crucial phenological stage or had to be very persistent. Unsurprisingly, above-average temperatures and drought conditions only attracted the attention of medieval chroniclers in extreme cases, although manorial accounts, coming directly from the agricultural world, supply a good deal of information on prolonged periods of dry weather in spring and summer.

A number of grain price spikes remain in this analysis that can not be connected directly to precipitation and temperature extremes before the relevant harvest (Figs. 9.1 and 9.3). Excluding the high grain prices following the harvests in 1321, 1350 and 1351 – which were due to a variety of meteorological and other factors and occurred in a post-Cattle Plague or post-Great Pestilence environment (Sects. 6.4 and 6.5) – there remain the harvests 1363, 1369, 1390 and 1400 which

[30] Deflation and currency shortage also contributed to low prices, see note 63 in Chap. 2.

were followed by major grain price spikes. It is difficult to relate the spikes after the harvests 1369, 1390 and 1400 directly to the prevailing weather conditions during the grain growing season. The springs and summers of these years saw relatively high temperatures and the months July–September were marked by average to low rainfall levels. Such conditions are normally beneficial for the wheat crop, therefore it seems that the high prices following these harvests indicate problems not only attributable to weather but to other factors as well. In fact all three years were plague years, in northern Europe the disease was particularly virulent in late summer and autumn. It can be assumed that the epidemic either impacted directly on the grain price by disrupting the grain market[31] or indirectly by causing a rise in prices through a perturbation of the labour market, agricultural practices, transport management or business proceedings. Severe disruptions in the agricultural sector are indicated by the survival pattern of manorial accounts from Hinderclay in northern Suffolk. In general many *compoti* of this important manor of Bury St Edmunds Abbey date to the second half of the fourteenth century, but the accounts ending with the harvests in the years of regional or national epidemics in 1369, 1375, 1383, 1389 and 1391 are missing, even though often complete sequences of account rolls exist for the preceding and succeeding years.[32] Such a pattern of survival is not arbitrary, but rather it is likely that during years of epidemic disease these *compoti* were either never composed, or the collection of information and the calculations were difficult, so that the accounts were retained in audit or in the office of the responsible obedientiary and hence were never transferred to the archive. Part of the disturbance of agriculture and administration could have resulted from a labour shortage during the harvest or from a fear of contagion. The accounts from East Anglia that do survive for the plague years 1369, 1390 and 1400 give harvests of generally normal length, indicating average or considering the potential labour shortage even below average rainfall in high and late summer. Farther south rain came on during the harvest in 1369 and 1400 making a higher input of labour in the harvest necessary exactly at a time when plague had broken out. In 1400, manors of the Bishopric of Winchester complained about the prolixity of the harvest, harvesting was very expensive because many of the customary tenants had to be replaced with costly hired labour.[33] The temporal coincidence of a rainy harvest – that led to an increased need for labour input – and a plague wave – that led to a shortage of labour – resulted in rising grain prices and most likely a reduced quality of the harvested grain. However, the combination of a plague wave and an early dry harvest, as in 1390, could also cause a problem of labour shortage. Usually early harvests after warm growing seasons were quick for avoiding the passing of the grain from reap-ripe state to dead-ripe state. This pheno-stage transition is rapid in warm high summers.

[31] Epidemic disease as the trigger for the grain price 1369 is also suspected by Campbell, Nature as historical protagonist, 304.

[32] Accounts for years between 1360–1361 and 1395–1396, Hinderclay, CUL, Bacon 479–506. On the epidemics see Chap. 10. The outbreak 1389 in Cambridge came in September and killed primarily the young, Chronicon W estmonasteriense 1381–1394, 402 and Walsingham, Chronica maiora, 1376–1394, 867. However, the Hinderclay account for the plague year 1390 does survive.

[33] Titow, Le climat à travers les rôles de comptabilité, 330.

Harvesting grain crops by hand in dead-ripe state results in raised losses of grain due to shedding. However, quick harvests were labour intensive, and at a time of plague this posed a choice for landowners between either high labour costs or the loss of part of the grain. In this context it is interesting that the national plague outbreaks 1361 and 1375 which either started unusually early in the year or occured when population levels had not yet reached the nadir,[34] were not followed by price spikes similar to those after the harvests 1369, 1390 and 1400, although after the harvest 1375 the price remained on the comparatively high level which had been reached after the cold spring and summer 1374. The high prices reached after the harvest 1409 might not only be connected to drought and warmth, but also to dysentery and plague, which was prevalent in the northeast, in Newcastle-upon-Tyne.

When the plague arrived in England in the mid-fourteenth century, the disease itself became a factor in grain price development. The social disruption during the plague waves in 1369, 1390, 1400 and potentially 1409 impacted on the grain and labour markets and contributed to a sudden steep rise in grain prices in the year following these summers. Superimposed on these short-term events was the negative demographic trend initiated by the Great Pestilence 1348–1350 and continued by recurring plague waves, which lowered the demand for grain and increased wages. At the eve of the arrival of the Black Death the English had numbered about 4.5 to 6 million, by 1377 the population stood at 2.5 to 3 million.[35] The demographic development became mortality-driven and population numbers continued to fall into the fifteenth century.[36] The resulting lower demand for grain reduced the vulnerability of the English people to weather-induced subsistence crises. Hence most of the cases of grain prices not responding to wet and cold summer seasons coincided with minor epidemics or came in the wake of larger outbreaks, mostly in the period after 1375, by which time the most rapid contraction of the English population had come to an end.

The rise of resilience against the ruinous effects of bad weather on the agricultural production, which came with an easing of the population pressure on the existing resources, did not only affect the extremes of weather and grain price but also took effect during average conditions. For analysing the relationship between weather – primarily precipitation – and the grain price in a quantitative approach, it is useful not to operate with the actual grain price, but with the first difference in the grain price (Fig. 9.4). In the period 1264–1350 the Spearman rank correlation between the East Anglian precipitation index and the advanced first difference in the national wheat price stands at rho = 0.38 ($p < 0.01$, two-tailed test). Since high amounts and frequency of precipitation played a crucial role in driving the wheat price up, but some drought years also resulted in raised prices, the correlation between the precipitation index and first difference of the price rises to a moderately

[34] Cohn, The Black Death transformed, 185. In 1361 plague affected provincial England also at the harvest time, see Sect. 6.2 and Chap. 10, but compared to later epidemics population density was still relatively high, so that people who were suffering from plague could easily be replaced by other workers.

[35] Hatcher, Plague, 68.

[36] Bailey, Demographic decline in late medieval England, 1, 15–17.

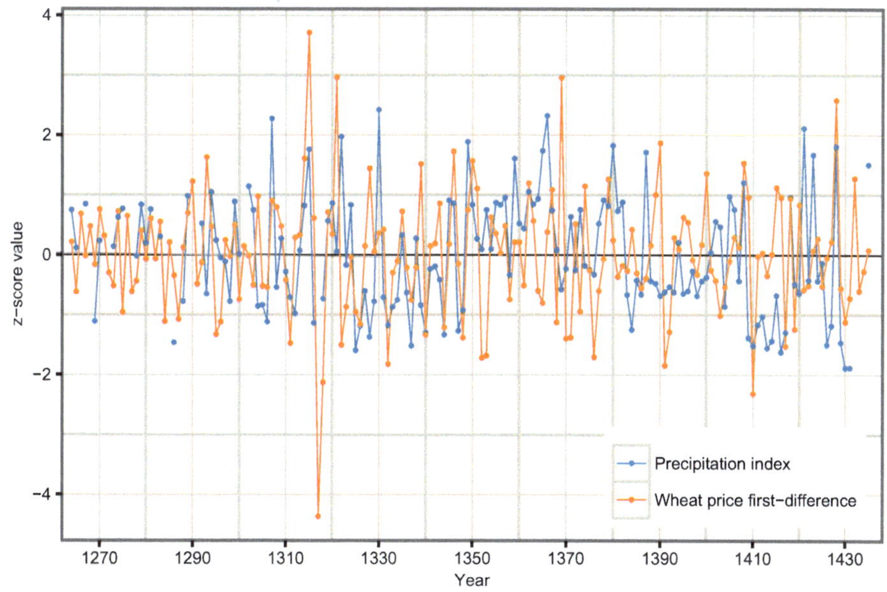

Fig. 9.4 July–September precipitation index and wheat price first difference, 1264–1431. The wheat price is provided by Munro, Revisions of the Phelps Brown and Hoskins 'Basket of consumables' commodity price series. It is advanced by one year and dates to the harvest year

high value (rho = 0.52, $p < 0.01$), when removing the known dry and warm years 1293, 1304, 1328, and 1331 and years when the price rise was largely due to (meteorological) events outside the growing season as 1321 and 1339 from the analysis. This suggests that in the pre-1350 period around a quarter of the variance in the first difference of the wheat price can be explained by the July to September precipitation alone. The link between precipitation index and first difference in the wheat price is close between c.1300 and 1350 and particularly so between 1330 and 1350 (Fig. 9.4). In the post-1350 period the correlation drops to rho = 0.11 (not significant at $p < 0.05$), and when excluding the price-driving drought and plague years 1369, 1390, 1400, 1409 and 1416 the correlation does not improve much and comes to rho = 0.22 (not significant at $p < 0.05$). The tree-ring data from southern and eastern England[37] correlate poorly with the advanced first difference in the wheat price before 1350, and the relationship is even considerably weaker after 1350. In the light of the high sensitivity of the grain price to meteorological conditions and particularly to extremes in the decades before the Great Pestilence, it is no coincidence that William Merle was observing the weather closely during this time, hoping to improve forecasting and to determine the effects of specific weather types on grain production and the grain price.[38]

[37] Wilson et al., March–July precipitation reconstruction, 997–1017, Cooper et al., Hydroclimate variability, 1019–1039 and on tree-ring data and harvest length, see Sect. 7.3.

[38] Merle, Consideraciones temperiei pro 7 annis and Merle, De pronosticacione aeris, BLO, MS Digby 147, fols.125r-138r.

With the demographic decline due to the plague waves, the vulnerability of the English population to weather-induced dearth and high grain prices also declined. As long as the weather was not marked by extremely raised rainfall levels combined with low temperatures, as in the late 1420s, or the miserable weather did not last longer than one season, as in the second half of the 1430s, fifteenth-century England was safe from famine, even though the cooling trend led to a gradual deterioration of the climatic parameters important to agriculture. In other words, for those who survived the plague waves, and who were part of the lower strata of society, living conditions improved to a large extent after the mid-fourteenth century. During the severe famine of the late 1430s, after all the worst famine of the fifteenth century, it would not be rumours about cannibalism that found their way into chronicles as in the time of the Great Famine roughly a century before, but the dry statement that in London people were reduced to consume bread not made of wheat, but of rye and barley, while in the country roots served as surrogate food.[39]

[39] Trokelowe, Annales, 95 for the Great Famine. On further contemporary and non contemporary references to cannibalism during the Great Famine and their credibility, see Marvin, Cannibalism as an aspect of famine, 73–84. For the second half of the 1430s, see McLaren (ed.), London Chronicle, 209–210 and Ingulph's chronicle of the Abbey of Croyland, 398. The comparatively low impact of this subsistence crisis is illustrated by Dyer, Standards of living, 268 who notes that there was no substantial increase of tenant mortality on midland manors 1437–1440, and many deaths in East Anglia were due to disease.

Chapter 10
Climate and the Plague, 1348–1500

The history of plague, *Yersinia pestis*, has captured the attention of generations of historians and medical scientists, and more recently of micro-biologists and climatologists. In Europe the Black Death 1347–1353 caused extremely high levels of mortality and recurrent plague waves would afflict the continent until the eighteenth century. Contemporaries of the plague as well as historians have searched for the causes and triggers of the numerous outbreaks. Famine and subsistence crises are supposed to have played an important role in the development of plague epidemics by malnutrition impacting the human immune system.[1] Scholars have also wondered about a potential climatic factor in the occurrence of pestilence. For the medieval mind plague was spread by miasma, corrupted or contaminated air, that entered the body through the pores. Contemporaries observed that certain weather conditions seemed to contribute to infection, and high temperatures were seen as favouring the disease. Warm and humid winds were considered risk factors and indeed plague outbreaks in continental Europe can partly be associated with short- or long-term rainfall or flooding.[2] In Italy fourteenth-century plague treatises already observed a seasonal pattern of outbreaks, these would occur in summer. In 1382 the Florentine doctor Francischino de Collignano linked plague waves to the heat of summer and declared autumn as too frigid for an outbreak.[3] In Italy the perception that plague tended to occur in summer persisted into the fifteenth century.[4] Another treatise, probably from Germany, observes that at least part of the local plague outbreaks in the First Pestilence began during the dog days,[5] that means in late summer between late July and late August which is taken to be the hottest time of the year. The idea of a causal connection between hot weather, miasma and the outbreaks of plague did reach the British Isles and was considered valid for the regional maritime

[1] Biraben, Les hommes et la peste, vol. 1, 147–154.

[2] Ibid., 134–139.

[3] Cohn, The Black Death transformed, 146.

[4] Ibid., 175–177.

[5] Horrox, The Black Death, 180.

K. Pribyl, *Farming, Famine and Plague*, DOI 10.1007/978-3-319-55953-7_10

climate. The widespread and lethal plague wave of 1390 is attributed by the Westminster Chronicle to the very hot summer and the consequent corruption of the air:

> Pestilencia: Circa principium mensis Junii immensus calor succrevit et duravit fere usque mensem Septembris. Qua de causa propter aeris corrupcionem magnam mortalitas hominum causabatur. Continuata est ista epidemia in diversis partibus Anglie, quamvis universaliter non desevit, usque festum sancti Michaelis [...].[6]

Plague treatises from northern Italy and north of the Alps placed mortality peaks due to pestilence primarily into late summer and autumn. The findings of Biraben and Cohn on the seasonality of mortality peaks in plague waves confirm this pattern and consequently the link to temperature levels – southern European late medieval outbreaks were most virulent from early to about mid-summer, further north the climax shifted to late summer and autumn.[7] The time of peak mortality in Italy was therefore not only the warmest time of the year, but also the driest period, and for some plague years chronicles record expressively drought conditions.

The massive shock that the ‘Great Mortality’ posed to Europe around 1350 and the questions regarding its nature, rapid spread, high death rate and socio-economic consequences have been the focus of many historical studies.[8] However, after 1350 the plague became endemic in Europe and flared up regularly and frequently in recurring subsequent waves which slowed the recovery of population levels. In England these later plagues caused further stagnation or more likely even a depression in the demographic development 1370–1520 due to a high mortality level even though fertility levels were also relatively high.[9] In this mortality-driven demographic setting population numbers fell from about 2.5 to 3 million in 1377 to about 2 to 2.5 million in England by 1520.[10] On the British Isles in the Middle Ages measures and interventions to prevent plague and to contain outbreaks were only in their infancy, or not at all in place, so the climatic conditions before and during a major outbreak of the medieval period can contribute considerably to refine our knowledge on the epidemiology of the disease that contemporaries identified as *pestilentia* or ‘plague’ in north-western Europe.[11]

The dates of the individual plague waves in England 1350–1500 are provided in the works by Creighton, Bean, Shrewsbury, Hatcher and Rawcliffe.[12] English plague

---

[6] Chronicon Westmonasteriense 1381–1394, 438.

[7] Biraben, Les hommes et la peste, vol. 2, 40–41, Cohn, The Black Death transformed, 42.

[8] For example: Biraben, Les hommes et la peste, Benedictow, Black Death, Campbell, Great transition.

[9] Bailey, Demographic decline, 15–18.

[10] Hatcher, Plague, 68–69, Britnell, Commercialisation 1000–1500, 155–156, Bailey, Demographic decline, 1. On the regional level, estimates for Suffolk for example point to a population decrease of 20%, Bailey, Medieval Suffolk, 183–184.

[11] On the origin and use of the terms ‘pestilentia’ and ‘plague’ throughout Europe, see Benedictow, Black Death, 5.

[12] Creighton, Epidemics in Britain, Bean, Population and economic decline in England, Shrewsbury, History of bubonic plague, Hatcher, Plague and Rawcliffe, Urban bodies.

outbreaks and continental epidemics partly overlap, so infection across the Channel and the North Sea might play a role. During the fourteenth century plague waves in England were often of national or supra-regional scale and chronicles describe how from the Second Pestilence onwards often the young fell victim to the disease. The intervals between plague waves were about 5–10 or 12 years and Gottfried assumes that plague outbreaks depended not only on human immunity but also on specific climatic and ecological conditions and the state of the local rodent and insect populations.[13] In the fifteenth century epidemic outbreaks assumed an increasingly local character.[14] The majority of information in the available sources concern the outbreaks in southern and eastern England and the midlands, but the references are not plentiful and the lack of historiographical writings for the mid-fifteenth century poses also a problem for the research of plague. There are difficulties in defining the start and end of medieval epidemics, so listed here are the years of the mortality peaks which are most widely recognized; the dates have been verified in the original sources. Some of these plagues might have started on a lower level in the previous year and might also have come to an end only in the following year. Generally the presence of plague on a low endemic level is likely throughout the whole period between 1350 and 1500.[15]

The Black Death arrived in England in 1348 and ran its course across the British Isles until 1350. In 1361 it was followed by the Second Pestilence. This outbreak was also very severe and most likely caused the second greatest mortality crisis in late medieval England and Europe. Naturally it ranks far behind the Great Pestilence, but took a higher death toll than the Great Famine.[16] The Third Pestilence came soon afterwards in 1368–1369, the latter year of this epidemic seems to have been worse. The Fourth Pestilence prevailed in 1375, plague remained present on a lower level in the following years, especially in northern England. In 1383 a regional plague outbreak focused on Norfolk. The Fifth Pestilence caused a high death toll in 1390; 1391 was a year of famine disease mixed with plague in Norfolk, northern and western England, and 1393 saw a renewal of pestilence in autumn in Essex. The next wave on a supra-regional scale came in 1400. For the years 1405–1406 low level epidemic disease, potentially plague, is recorded.[17] In 1407 a severe plague outbreak followed, it is described as one of the worst since a considerable time. Town and country were hit, London suffered badly and the west of England was most affected.[18] The year 1413 is also recorded as a plague year, probably the disease was confined to south-east England.[19] In 1420 another plague raged through Norfolk, at

[13] Gottfried, Black Death, 131.

[14] Creighton, Epidemics in Britain, 226, Hatcher, Plague, 57.

[15] Creighton, Epidemics in Britain, 202–211, Gottfried, Epidemic disease, 126–154, Bolton, Looking for *Yersinia pestis,* 33–34.

[16] Campbell, Great transition, 315.

[17] Creighton, Epidemics in Britain, 220–221, Shrewsbury, History of bubonic plague, 143.

[18] Walsingham, Historia Anglicana, vol. 2, 276 and Continuatio Eulogii, vol. 3, 410.

[19] The short description of the plague wave in 1413 by Walsingham, Historia Anglicana, vol. 2, 297 has caused some confusion in the past. Shrewsbury, History of bubonic plague, 143 assumed that

least parts of Essex, the London area and Kent,[20] and in 1426 it flared up in London, east Sussex and potentially also in Great Yarmouth in eastern Norfolk.[21] The outbreak of 1434 was widespread, and plague returned in the wake of the failed harvests in the second half of the 1430s and was probably mixed with famine disease in 1438–1439, again the latter year seemingly had a higher mortality. Urban plagues particularly in London occurred also in the 1440s and the 1450s. The written records for this period are very sparse, but Gottfried identifies a 'great mortality' affecting London, Reading and Suffolk in 1452 as a plague, although the documentary sources remain silent on nature and cause of the mortality.[22] Another regional pestilence was detected by Hatcher in Kent and other counties in 1457–1458. National outbreaks of plague returned in 1464, which was followed by a flare up in 1466 in London, southern Lincolnshire, Cambridgeshire and probably Norfolk.[23] The severe

the text was referring to a plague affecting many Englishmen outside of England, and Rawcliffe, Urban bodies, 365 implies the description refers to 'many parts of England' which 'plures Angli' does not. Plague was present in England, but it seems to have been regionally confined to the southeast – it was the cause of many deaths at Christ Church Priory, Canterbury, Hatcher, Mortality, 30 – and seems to have been also in the east, in Colchester, Rawcliffe, Urban bodies, 365. However, the area around London appears to have escaped unscathed, as there is no significantly raised mortality in Westminster Abbey, Harvey, Living and dying, 122–123. Another – most likely regional or even local – plague outbreak took place in High Clere in northern Hampshire 1415, Titow, Le climat à travers les rôles de comptabilité, 334. The 1415 small-scale outbreak illustrates the endemic nature of the disease well.

[20] A plague outbreak affected parts of Norfolk, Annales Monasterii de Bermundeseia, 485. Colchester was suffering equally from the plague, Rawcliffe, Urban bodies, 305. The epidemic must have extended into the vicinity of London, because the worst mortality amongst the monks of Westminster Abbey 1390–1515 falls to 1420, although this extreme mortality is not explicitly attributed to plague in the abbey records, Harvey, Living and dying, 125–126. However, plague caused a number of deaths amongst the monks of Christ Church Priory, Canterbury, too, Hatcher, Mortality, 30. About 1419–1421 a severe plague epidemic swept through northern England, though start and end date of this outbreak are unclear, Creighton, Epidemics in Britain, 221–222 and Rawcliffe, Urban bodies, 305.

[21] On the plague in London, see Bean, Population and economic decline in England, 428. According to Brandon, Late medieval weather in Sussex, 4, the plague of 1426 also came to east Sussex. Rawcliffe, Urban bodies, 365 cites evidence which indicates that the outbreak 1426 reached Great Yarmouth.

[22] Gottfried, Epidemic disease, 40–41, 97–98. He describes an epidemic which was most prevalent in autumn 1452, but the mortality remained high into winter 1452–1453. He found no evidence for the epidemic affecting Kent, Hertfordshire, Norfolk and Yorkshire, so it was limited in its geographical scope. Since the mortality peak occurred in autumn, similar as in a plague outbreak, the epidemic is included in the analysis of plague waves, even though its nature is not clear.

[23] Ingulph's chronicle of the Abbey of Croyland, 443 is mentioning a plague for either 1465 or 1466. However, the paragraph is describing a summer and it is the last entry before January 1467, so the text implies the plague to prevail in a year before 1467, most likely in1466, in the country around Crowland Abbey, that means in southern Lincolnshire, Cambridgeshire and Norfolk. In any case this plague is not dated to 1467 in the chronicle as is assumed by Rawcliffe, Urban bodies, 369. This dating of the plague wave is confirmed by the evidence on mortality collected in Gottfried, Epidemic disease, 88, 90, 105. He identifies excessively high mortality levels in Norfolk in 1466, which he attributes to plague. For 1467 his data does not show a plague style impact, ibid., 88, 90, 100.

plague in 1464 ushered in another period of more frequent large-scale waves. They date to 1471 and 1478–1479, whereby 1478 was still an outbreak on regional level around London and Oxford, in 1479 the plague moved into East Anglia and probably further regions and even northwards to Hull[24]; these waves in the early and late 1470s were causing high mortality rates.[25] In the northeast of England plague was also very active and cost many lives in the 1470s, but it displayed a different temporal distribution: Hull was infected in 1473, 1477 and 1479.[26] After a prolonged gap

[24]The dating of the plague of the late 1470s has varied amongst historians, and the years of the epidemic have been given as 1478–1479, 1479 or 1479–1480. In any case in 1479 the plague caused extremely high mortality; the epidemic came to an end in the winter 1479–1480. Evidence in narrative sources for the plague for 1479 comes from London, Norwich, Southwell and probably Hull, see Creighton, Epidemics in Britain, 231–233 and Bean, Plague, 429. Gottfried, Epidemic disease, 45, 95–96 also lists this information and adds independent data on the high mortality in 1479 in East Anglia, as does Bailey, Medieval Suffolk, 183 for Suffolk. However, the information presented by Creighton, Epidemics in Britain, 231–233 is a good foundation to date the start of the plague to 1478, at least in and around London. The evidence comes from 'London chronicles' which are normally organized according to the mayors' year, which changed with the mayoral elections at Michaelmas (29 September). The Chronicle of the Grey Friars, 22 and The Customs of London (Arnold's chronicle), 37 report the plague under the 17th year of Edward IV (in London mayors' years: Michaelmas 1477-Michaelmas 1478) and following the death of Duke Clarence (18 February 1478), so the epidemic is set to the summer half year 1478. Fabyan, New chronicles, 666 under the 18th year of Edward IV 1478–1479 writes of the plague 'whiche beganne in the latter ende of Senii' [September] in the presedynge yere, and contynued in this yere tylle the begynnynge of Nouembre [November]'; considering the mayor's years it is likely that Fabyan would refer to the time from September to November 1478. Finally Grafton's Chronicle, vol. 2, 68 also brings the start of the plague to the 17th year of Edward IV (at the very end of this regnal year, that means c. early March 1478), again after Duke Clarence' death; it lasted 4 months. Neither Fabyan nor Grafton are contemporary, but notwithstanding the potential imprecision of the 'London chronicles', the high mortality, though not explicitly the plague, in the area around London in 1478 is confirmed by the findings in Harvey, Living and dying, 104. Her work shows that many inpatients were recorded in the infirmary of Westminster Abbey in 1478 and 1479 and that mortality was high in both years, 1478 being actually classified as a crisis mortality. (It seems that Christ Church priory in Canterbury escaped the plague wave 1478–1479 largely unscathed as no raised mortality is found there, Hatcher, Mortality, 26.) A serious plague outbreak in 1478 is also recorded in Oxford, Chance et al., Medieval Oxford. In the light of this information, the plague in the late 1470s is dated to 1478–1479, 1478 being a regional plague.

[25]Gottfried, Black Death, 133 estimates that 10–15% of England's population lost their lives in 1471, and in the late 1470s a mortality of about 20% is likely.

[26]Normally the years of plague in Hull are given as 1472, 1476 and 1478, see for example Creighton, Epidemics in Britain, 231–232. The information comes originally from a local historian: Tickell, Kingston upon Hull, 132, he took his information from the town records. Frequently medieval town records are organized according to mayor's years. In Hull the mayor and other officers were elected the day after Michaelmas, 30 September, and sworn in on St Luke's day, 18 October, ibid., 663. Events in a mayoral year are usually given under the start date (in dominical years) of the term in office, hence the dates of plague in Hull actually refer to the years 1472–1473, 1476–1477, 1478–1479. Since plagues frequently occur in the summer half year it can therefore be concluded that in the mayor's year 1472–1473 the plague took place in summer 1473, in the year 1476–1477 the epidemic came in 1477 and in 1478–1479 Hull suffered simultaneously with most of England in summer 1479. That Tickell, Kingston upon Hull, 132 is indeed giving the start dates (in domincal years) of mayoral years for the plague waves finds confirmation in his statement, that

in plague activity the fifteenth century then closes with the severe plague 1498–1500, it afflicted Scotland in 1498, in 1499–1500 England was hit and the death rate was very high in the latter year,[27] for a list of these supra-regional and regional English plague waves see Appendix. 6. Plague epidemics limited largely to the unhealthy urban environment of London, which occurred in 1381, 1382, 1387, 1433,[28] 1437, 1442–1445, 1449, 1454, 1467[29] and 1487[30] are not included in the analysis.[31]

In England the study of the influence of extreme climatic events on health has a long tradition and dates in its medieval form – with the typical emphasis on the astrometeorological approach – back to John Ashenden's 'Summa astrologiae

each plague wave also killed the mayor before the term was up in September, first John Whitfield (elected 1472, Edward IV 12), then John Richards (elected 1476, Edward IV 16) and last Thomas Alcock (elected 1478, Edward IV 18), the election years are given ibid., 673.

[27] In 1498 the plague also was in parts of northern England, Rawcliffe, Urban bodies, 371. In 1500 London and the south were badly afflicted, Harvey, Living and dying, 125 and a number of London chroniclers report the plague and the death of prominent victims at the end of the mayor's year 1499–1500, indicating a high mortality in 1500, Fabyan, New chronicles, 687, Vitellus A XVI, 232 and Wriothesley, Chronicle of England, vol. 1, 4.

[28] Plague appears to have been very active on the regional level in the first half of the 1430s. In 1431 the disease can be located on the estate of St Albans, Chronicon Rerum Gestarum in Monasterio Sancti Albani, vol. 1, 62; in Canterbury, Hatcher, Mortality, 30. Gottfried, Epidemic disease, 36–37 assembles further evidence hinting at the presence of plague in London in 1431. A major outbreak occurred in 1434. In 1435 an 'epidemia' raised the mortality at Christ Church Priory in Canterbury, Hatcher, Mortality, 29. Hatcher assumes that the disease might also have been plague, however, normally the monks name plague precisely and not do not use the general term 'epidemia' for it.

[29] The plague of the year 1467 and its geographical coverage is a controversial issue. However, the evidence for a plague outside London is very weak. The parliament adjourning from Westminster and deciding to meet at Reading, as mentioned by Creighton, Epidemics in Britain, 230 and Rawcliffe, Urban bodies, 369 indicates plague in London. It is worth taking a closer look at the two narrative sources that are used as evidence for a plague outside London in 1467. The first, Ingulph's chronicle of the Abbey of Croyland, 443, does not date the widespread plague in its vicinity to 1467, but most likely to 1466 (see note 23). The second source, Herryson, Abbreviata Cronica, 10, mentions a plague for 1467, but according to Gransden, Historical writing in England, vol. 2, 254, he bases much of his information for the years after 1460 on a London source, so this reference is probably also linked to an outbreak primarily in this city. The data on mortality assembled by Gottfried, Epidemic disease, 88, 90, 100 do not show a plague mortality pattern in respect to numbers and seasonality in East Anglia in 1467. Hatcher, Mortality, 30 identifies plague at Christ Church Priory in Canterbury in 1465 and 1467, but mortality was very low in both years. Therefore the outbreak 1467 is considered to be limited to London with a very limited mortality also in the southeast of England.

[30] For 1487 see Creighton, Epidemics in Britain, 285, Hatcher, Mortality, 28, 30 and Rawcliffe, Urban bodies, 370. Hatcher notes a plague outbreak at Christ Church Priory in Canterbury, Rawcliffe locates the disease in Oxford, where it appears to have been present also in 1486.

[31] List of plagues (excluding 1381 and 1387) in Bean, Population and economic decline in England, 428 and in Gottfried, Epidemic disease, 47–50. It is not clear if the diseases of 1381 and 1387 were plague. For the great mortality in Oxford and London in 1381, see Chronicon Westmonasteriense 1381–1394, 20 and the high mortality amongst the young in September 1387, ibid., 204. The Westminster Chronicle provides comprehensive references to epidemics and plagues in central, southern and eastern England.

judicialis de accidentibus mundi', and in its modern form back at least to Thomas Short's 1749 publication 'A general chronological history of the air, weather, seasons, meteors etc.'. Over recent decades research in past climate has resulted in high quality information on the climate in medieval Britain and continental Europe becoming available.[32] When studying the role of meteorological factors in English disease outbreaks, climate data from England or close by continental regions is needed. Unfortunately the April–July temperature reconstruction ends in 1431 and the precipitation index in 1448 and they only partly overlap with the period of the year which seems to have been decisive in the outbreak of plague in England, high and late summer. It is therefore useful to extend the temporal coverage by also using the temperature summer indices created by van Engelen et al. for the Low Countries.[33] These have recently been refined by Camenisch for the fifteenth century, who also added a precipitation index; unfortunately her indices are not continuous and therefore can only be used in a supplementary manner in this study.[34] They can contribute refined information, especially with regard to precipitation, as can also the Ogilvie and Farmer indices for England.[35] For continuous precipitation data, the tree-ring based reconstruction by Cooper et al. is used, the data come from East Anglia.[36] Finally the summer temperature reconstruction for Burgundy based on grape harvest dates by Chuine et al., will also be included, because it covers the period of the year that is decisive for plague outbreaks: summer. This series is complete from 1370 onwards.[37]

Combining the temperature and precipitation proxies it is possible to create a picture of the meteorological conditions of plague waves in England. Figure 10.1 shows temperature represented by the Low Countries index and the Burgundy reconstruction and the East Anglian tree ring based precipitation data. Figure 10.1 reveals the existence of two subsets of plagues. The first subset is in accordance with the theory that plague outbreaks are linked to subsistence crises, these waves occur during summer seasons of comparatively low or average temperatures, and partly also during wet conditions. To these cases belong the plague waves 1348–1350 and 1438–1439, which both coincided with or set in during subsistence crises.[38] The background of food shortages around the time of the Great Pestilence, which hit a virgin-soil population, has been outlined by Campbell[39] and in Sect. 6.5, a short summary of the conditions in the second half of the 1430s is given in Sect. 8.4. The outbreak 1405–1407 belongs also partly to this group; the weather was cool and wet 1405–1406. Even though the grain price was very low, grain quality in

[32] Brázdil et al., Historical climatology in Europe, 363–377, Jones et al., High-resolution palaeoclimatology of the last millennium, 4–21.

[33] van Engelen et al., A millennium of weather, winds and water in the Low Countries.

[34] Camenisch, Endless cold, 1062–1063.

[35] Ogilvie, Farmer, Documenting the medieval climate, 124–128.

[36] Cooper et al., Hydroclimate variability.

[37] Chuine et al., Grape ripening as a past climate indicator.

[38] Campbell, Ó Gráda, Harvest shortfalls, 869–872.

[39] Campbell, Nature as a historical protagonist, 300–302, idem, Physical shocks, 20–29.

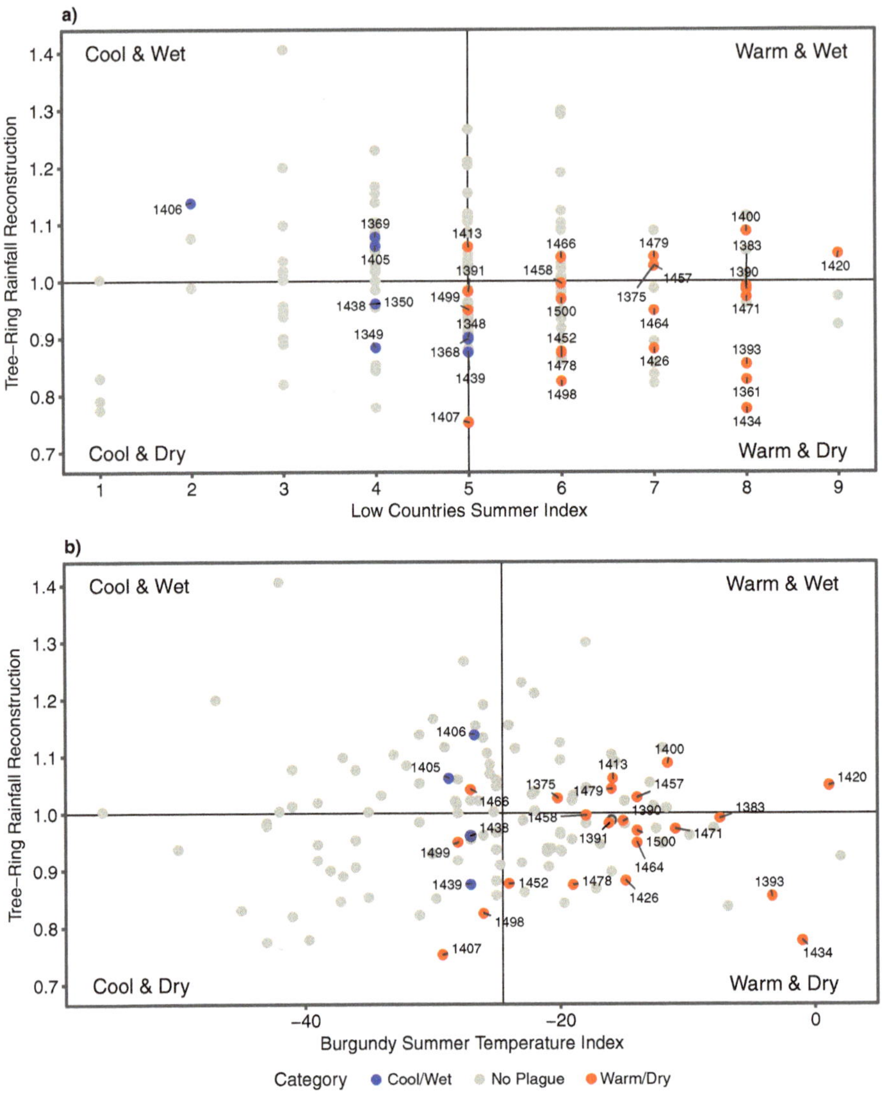

Fig. 10.1 Summer temperature and precipitation in plague years, 1348–1500. (**a**) Low Countries summer temperature index and East Anglian tree ring based precipitation reconstruction. (**b**) Burgundy vine harvest date summer temperature (April–August) reconstruction (from 1370 onwards) and East Anglian tree ring based precipitation reconstruction. The summer temperature for the Low Countries index is provided by van Engelen et al. (2001), the summer temperature reconstruction for Burgundy by Chuine et al. (2004), and the tree ring based rainfall reconstruction for East Anglia by Cooper et al. (2013)

both years is likely to have been negatively impacted by wet weather during harvest time, flooding was also common in those years. In 1407 the plague continued, but now mortality increased greatly particularly in London and in the west, even though the summer temperature was now average and the season also extremely dry (Fig. 10.1 and Sect. 8.3).

With these meteorological characteristics the year 1407 is leading towards the second subset of plague waves. This much larger category of plagues shows no direct connection to high grain prices or a lack of food supply, but all these epidemics occurred in summers which were markedly warmer than normally and also saw either below average or average rainfall levels. This group includes 1361, 1375, 1383, 1390, 1393, 1400, 1420, 1426; 1434, 1452, 1457–1458, 1464, 1466; 1471, 1478–1479, 1498[40] and 1500. Out of these 1361, 1393 and 1434 were not just extremely warm, but also extremely dry (Fig. 10.1). Generally it is possible that the summer months were even drier than the East Anglian precipitation proxy data indicates, since the tree-ring data is spanning a period longer than the actual summer season. Based on documentary sources, Ogilvie and Farmer identify also in 1375, 1383 and 1420 extraordinarily dry early to high summers, and indications also hint at a dry early summer time for 1390.[41] Chronicles describe 1464[42] as a time of drought. Further detail on precipitation can be added with the help of the fifteenth-century summer precipitation indices developed by Camenisch for the Low Countries, where the summers 1420, 1458, 1464, 1471, 1479 and 1498 were dry or very dry.[43] Very high summer temperatures during plague years in England are confirmed by narrative sources from this country for 1361,[44] 1375,[45] 1390[46] and 1464.[47] In the April–July temperature reconstruction for England, 1361 is the hottest year in the whole series (Sect. 6.2), 1390 is a warm year. Considering this information it is not surprising that a chronicle describes the plague 1361 as already starting in late March in London,[48] hence this outbreak diverged from the normal mortality pattern of the area north of the Alps which typically saw a peak in mortality in the time of late summer/autumn. The pestilence then came to Kent in July and moved into southern England west of London as well as into Suffolk and probably Norfolk in

[40] In 1498 the plague was in Scotland, normally the Scottish weather conditions are not well covered by the climate data used in Fig. 10.1. However, in the area of the southern North Sea, in East Anglia and the Low Countries, this year was still slightly above average in terms of temperature, and very dry. Further south, in Burgundy, 1498 was comparatively seen cooler. Also the years 1407, 1466 and 1499 were cooler than average in the Burgundy region.

[41] Ogilvie, Farmer, Documenting the medieval climate, 127–128.

[42] Short English chronicle, 80, Brief Latin chronicle, 180 and Herryson, Abbreviata Cronica, 8.

[43] Camenisch, Endless cold, 1062–1063.

[44] Titow, Le climat à travers les rôles de comptabilité, 319. For more evidence on high temperature, see Sect. 6.2.

[45] Walsingham, Historia Anglicana I, 319.

[46] Chronicon Westmonasteriense 1381–1394, 438.

[47] Brief Latin chronicle, 180.

[48] Continuation of Higden, Polychronicon, vol. 8, 360.

late summer and autumn.[49] The early onset of the plague wave in London is confirmed by Cohn, whose evidence based on the number of registered testaments displays a pattern of mortality more typical for southern than northern Europe for England's most populous city not only in 1361 but also in the very hot year 1375.[50]

Cases of plague that seem not to fit either the outbreaks associated with subsistence crises nor those taking root in hot and dry summer seasons are 1368–1369, 1391, 1413 and 1499. The meteorological conditions during the plague 1368 and 1369, with a high mortality in the latter year, were highly changeable (Sect. 6.3). Both years saw, at least in eastern England, periods of dry weather during spring and early summer and 1369 is a warm reconstructed April–July mean temperature, but in both years harvest times were marked by high rainfall levels. Flooding also occurred in 1369. In general the 1360s were a difficult decade for agriculture[51] and the wet harvests 1368 and 1369 contributed to the continuation of high grain prices. Therefore the plague in 1368–1369 can be placed in the group of epidemics connected with subsistence crises, but the dry and also warm spells in 1369 were factors in the high mortality of this year. The plagues 1391 and 1499 are primarily continuations of the severe outbreaks in the previous years. In 1391 a scarcity of foodstuffs coincided with the epidemic; again there are indications that the early summer was dry.[52] For the summer of 1413 a regional plague is recorded. Northwest European meteorological conditions appear average on first glance, however, even though neither the East Anglian tree ring reconstruction nor the harvest length based East Anglian precipitation index (Fig. 7.5) show drought, Ogilvie and Farmer give drought conditions in early summer, the records come from southern England.[53] Many drought impacts are known for Norfolk in the 1410s, and Camenisch shows a sequence of dry summers between 1412 and 1414 in the Low Countries.[54] Considering the dry weather in late spring and early summer 1413, high temperatures for this part of the summer season are also possible. The plague in 1413 falls therefore into the group of epidemics that were linked to warm and dry conditions, but neither these weather conditions nor the plague in this year were great in scale.

The influence of meteorological factors in the form of heat and drought on the occurrence of plague is apparent in Fig. 10.1. For the period 1348–1500 van Engelen

---

[49] For Kent: Chronicon Anonymi Cantuariensis, 212. The plague affected the harvesting process – probably late July and August – on a manor of Christ Church Canterbury, CRU, Bickersteth, Minister's account rolls of Christ Church Canterbury 1305–1386. Harvesting was also hindered on the manors of the Bishopric of Winchester, Titow, Le climat à travers les rôles de comptabilité, 320. In Suffolk peak mortality amongst adults (there is a lack of records for the mortality of infants) fell to the more typical time of late summer and autumn, Bailey, Medieval Suffolk, 183. In remote northwestern Norfolk the harvest was also unusually drawn out, which was likely a plague impact, Gnatingdon NRO, LEST/IC 12.

[50] Cohn, The Black Death transformed, 184–185.

[51] Mate, Agrarian economy after the Black Death, 344–349.

[52] Ogilvie, Farmer, Documenting the medieval climate, 128.

[53] Ogilvie, Farmer, Documenting the medieval climate, 128, data from Titow, Le climat à travers les rôles de comptabilité, 333.

[54] Camenisch, Endless cold, 1062.

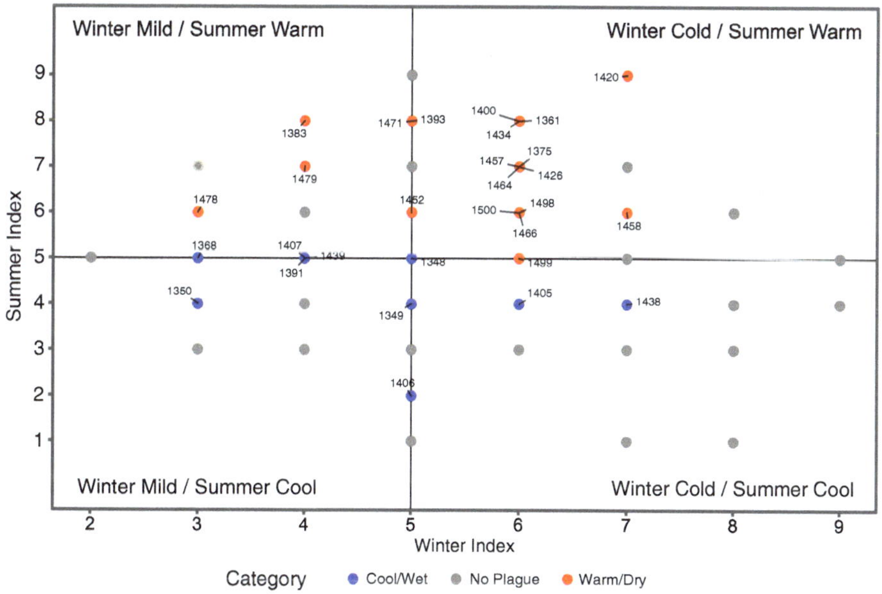

Fig. 10.2 Winter severity before and summer temperature in plague years, 1348–1500. The winter and summer temperature indices are provided by van Engelen et al. (2001)

et al. classify 26 summer seasons as being noticeable warmer than average (indices 7–9) in the Low Countries (Appendix 6).[55] Half of these warm seasons, 13 summers, saw supra-regional or regional plague waves in England. A further seven summer seasons were free of large-scale plague outbreaks, but such a supra-regional outbreak had occurred within 5 years previous to the hot summer, so the temporal interval was not sufficient for a renewal of the epidemic. Additionally some of those warm summers did witness local plague activity in England. It is likely that the Second Pestilence of 1361 had its roots in 1360.[56] In 1447 plague was present amongst the monks of Christ Church Priory in Canterbury[57] and for 1473 a plague has been recorded in the municipal documents of Hull.[58]

The importance of weather conditions in the occurrence of plague waves is not limited to the summer season. The temperature during the winter preceding a plague outbreak has also to be taken into account: not a single major plague outbreak in England between 1348 and 1500 is associated with a very cold winter on the van Engelen et al. index (8–9) (Fig. 10.2). Normally winters preceding a hot-summer plague witnessed slightly cool conditions (6), a few were average (5). Only the epidemic of the extremely hot summer 1420 came after a colder winter (7), and the

[55] van Engelen et al., A millennium of weather, winds and water in the Low Countries.

[56] Reading, Chronica, 147, Gransden, A fourteenth century chronicle, 275.

[57] Hatcher, Mortality, 30.

[58] See note 26.

1457–1458 plague spanned a comparable winter season. In addition, no hard winters preceded a pestilence during or shortly after subsistence crises, such winters were either average or even mild. Only the outbreak 1438 followed on a rather cold winter (7), but mortality seems to have been lower than in 1439. The end of the epidemic outbreaks in 1407 and 1434 coincided with an extremely cold winter. For a number of winters before either type of plague, 1369, 1383, 1390 and 1413, no index could be set for the Low Countries, this could be due to a simple lack of sources, but it could also indicate average or even mild winter weather which was so unspectacular that no writer felt the urge to comment; very cold conditions resulting in a freezing of the Dutch water channels could have not remained unrecorded in the Late Middle Ages. Even though data is sparse, it appears that mild winters, which were generally rare in the Late Middle Ages, were not favourable for major plague outbreaks either. Three plague years were associated with winters milder than average (3), but none with very mild conditions (see Fig. 10.2, no very mild winter (1) occured in the Low Countries 1350–1500). One of these plague years after a mild winter is 1350 which marks the end of the First Pestilence on the British Isles, and the two remaining years, 1368 and 1478, form the start of major biannual plagues, which had peak mortality in the subsequent summer.

Absolute temperatures of the winter before and the summer during the plague and temporal intervals between epidemic outbreaks were not the only determinants for plague. Years of high mortality due to pestilence were often summers when temperatures were markedly raised compared to the previous one to three summer seasons (Fig. 10.3 and Appendix 6). This temperature increase was often sudden

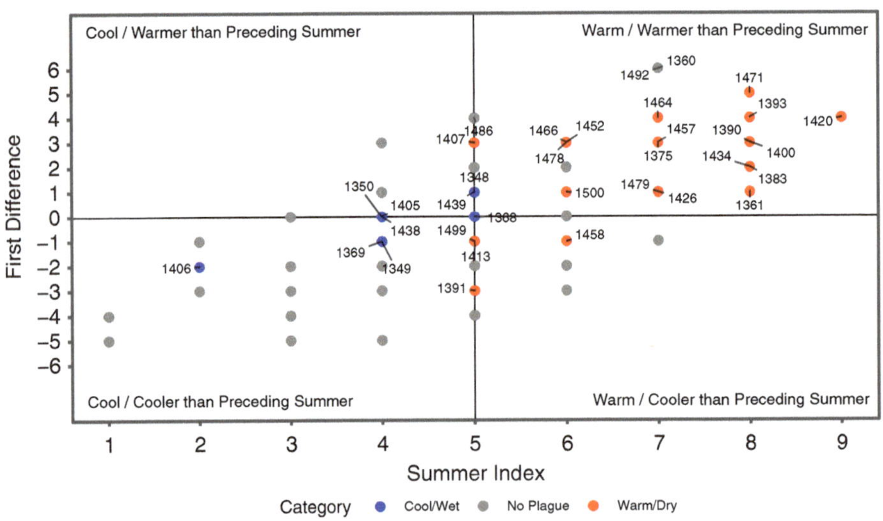

Fig. 10.3 Summer temperature and summer temperature first difference in plague years, 1348–1500. The summer temperature index is provided by van Engelen et al. (2001). First difference of summer conditions (compared to previous summer) is based on the summer temperature index from van Engelen et al.

and steep. Whereas for very warm summer seasons such a sudden increase of temperature can be expected as long as these hot years do not cluster, a sudden rise also marks plague years that saw only average temperatures such as 1407. When combining the two risk factors for plague, high summer temperatures (7–9) and a considerable rise in temperature compared to the preceding summer (first difference minimum 2), then 18 such years occurred in the Low Countries index between 1348 and 1500. The risk for plague in such summers was high: of these 18 summer seasons ten coincided with large-scale plague epidemics, most of these were supra-regional outbreaks (Fig. 10.3 and Appendix 6 category PP). Five further summers came too quickly after another large-scale outbreak, they fall within the 5 year limit. Two years, 1360 and 1492, were warm and had actually the extremely high first difference of 6, but saw only very limited (1360) or no plague activity (1492). Both years followed on summers that had been extremely cold. No other summers saw a rise in summer temperature as steep as from 1359 to 1360 and 1491 to 1492, and no major plague ever followed on or coincided with an extremely cold summer (1) (only the plague 1407 came after a very cold summer in 1406 (2) and the nature of the 1406 disease remains unclear). Hence it is likely that very cold summers, just as very cold winters, contributed to the absence of plague in the subsequent summer season, even if that happened to be warm (Figs. 10.2 and 10.3). In late medieval England only the summer 1495 fulfils the meteorological precursors for a widespread hot summer plague and is neither associated with a preceding very cold winter or summer, but was actually not a plague year. The risk for plague was also high in years when the meteorological precursors for plague were less pronounced, that means summers which were either warm (7–9), but not considerably warmer than the preceding years (first difference maximum 1), or years which were only warmer than average (6), but much warmer than previous summers (first difference minimum 3). There were twelve such summers in the study period, in six of these large-scale plague epidemics were recorded for England (Fig. 10.3, Appendix 6 category P). Half of the plagues in this category were of supra-regional and the other half of regional character, so on the whole they appear less devastating or at least less widespread than plagues coming in years which combined high summer temperatures with a high first difference. Supra-regional plagues in these summers also mostly had started in the preceding years.

The role played by warmer or rising temperatures in spreading the plague in England is also visible in the occurrence of plague waves in the neighbouring countries. In France, according to Le Roy Ladurie, even the spread of the first plague from the south towards the north after December 1347 and then during 1348 coincided with a prolonged run of winters that had been average or mild, and the summer half year 1348 was noticeably milder and drier than the run of wet and cold summers preceding it.[59] Plague also affected the more remote areas of northwestern Europe, as Scotland and northern England. Much less is known about the epidemics in these regions and some outbreaks were simply plagues that moved from southern

[59] Le Roy Ladurie, Histoire humaine et comparée du climat, 62–64. On the temperatures, see Pfister, Variations in the spring-summer climate, 71, Pfister et al., Winter severity, 101.

England northwards,[60] as most likely the Scottish plague of 1361–1362, which coincided with extremely high temperatures in England and the Low Countries. However, generally summer temperature in Scotland and northern England are not well correlated to those in the Low Countries let alone Burgundy,[61] but due to the North Atlantic Oscillation a certain congruence of meteorological conditions between Scotland and Scandinavia in July and August can be expected.[62] The temperature reconstruction for Europe by Luterbacher et al. from Roman times onwards also supplies data for individual geographical grid-squares,[63] and over northern Scandinavia the temperature reconstruction is mostly based on the local tree-ring information; therefore northern Scandinavian grid-squares are used as an indicator for the temperature in Scotland.[64]

Astonishingly often plague waves in the northern part of the British Isles do correspond to years of warm or at least average Scandinavian summer temperatures (of the period 1350–1500), most of these summer seasons were also warmer than the previous summers. This applies to the plagues 1374 and 1379 in northern England, the plague 1380 in Scotland, the epidemic of 1391 in York, the Scottish plague wave 1401-c.1403, the plague 1409 in Newcastle-upon-Tyne,[65] the northern English plagues c.1418–1421,[66] 1429, 1474 and 1477, the Scottish plague 1475, the general plague in all of England in 1479, the potential plague outbreak 1485 in York,[67] and the plague in Scotland and northern England 1498 and 1499. For most of these northern plague years the summer conditions in the Low Countries and southern England were neither warm nor marked by a rise in temperatures, so these warmer conditions did not extend to lower latitudes and so no plague outbreaks in southern

[60] The dates for outbreaks of pestilence in the northern parts of the British Isles come from Creighton, Epidemics in Britain, 233–236, 360–361 and Rawcliffe, Urban bodies, 362–371. In general fewer records are available than for England south of York.

[61] Jones, Hulme, The changing temperature of 'Central England', 178–183.

[62] Folland et al., The summer North Atlantic Oscillation, 1083–1085.

[63] Luterbacher et al., European summer temperatures since Roman times.

[64] Gridsquares used: 12.5E, 67.5 N and 17.5E, 67.5 N. The grid-square for Scotland is not employed, because no temperature proxy for the reconstruction is available for that region. The temperature reconstructed for Scotland (inferred from the temperature data and proxies from other European regions) shows, however, great similarity to the Scandinavian grid-reconstructions; the primary difference is a lower range of interannual variability in the Scottish reconstruction.

[65] This plague is sometimes dated to 1410, but the source from the spring 1410 indicates an epidemic in 1409 considering that most plagues have a mortality peak in late summer and early autumn. Bean, Plague, 430 also dates the epidemic to 1409. In Scandinavia the summer 1409 was warm, but only slightly warmer than 1408. However, in England 1409 is one of the warmest years in the April–July temperature reconstruction and also dry, see Sect. 6.2.

[66] Creighton, Epidemics in Britain, 221–222, Rawcliffe, Urban bodies, 365; the exact dates of this wave are however not clear, it may also have started in 1419. In the north 1418 was warm and also much warmer than 1417, 1419 was average and 1420 and 1421 temperatures were rising resulting in a warm 1421.

[67] The epidemic affecting York in June 1485 was classified as 'pestilence' by contemporaries, but might have been a early manifestation of the English Sweat that broke out in London in late September 1485, see Rawcliffe, Urban bodies, 370.

England were recorded. There are three exceptions: 1429 was in England and the Low Countries of average temperatures and therefore much warmer then the preceding extremely cold 1428. In the year 1479, when the plague was raging over most of England, it was also warm in the south. The years 1498–1499 were average after a colder period also in the Low Countries, and plague moved from Scotland southwards until it reached Oxford and London 1499–1500.

Plague waves accompanying subsistence crises also affected Scotland in the fifteenth century: the epidemic in 1439, if it was plague indeed, was connected with the severe famine of the late 1430s when English records also refer to plague mixed with famine disease, and in 1455 a plague epidemic again struck Scotland during a time of raised grain prices.[68]

Even though the data on the plagues of the northern part of the British Isles are less conclusive and the use of a remote temperature proxy cautions against too far reaching conclusions, it is evident, that the majority of northern plagues seems to be linked to summers that were considerably warmer than the preceding summer and at least on an average temperature level. The explanation for the synchrony of Scottish and Scandinavian summer trends would be a stable high pressure system between Scotland and Scandinavia. Such a high pressure system in summer would raise temperature and reduced rainfall over the northern North Atlantic, and shift storm systems and associated rainfall further south into continental Europe causing cooler and wetter summers there.

The use of climate data from areas as close as possible to the study area and precisely dated peak mortality in plague waves demonstrates, that the occurrence of severe plague waves in medieval England before human preventative measures were introduced was not random. On the contrary, plague, which was endemic in the country since 1350, was triggered into major outbreaks under specific meteorological conditions.

The climatic parameters of the first group of plague outbreaks, those during cool/wet conditions, do not come as a surprise: these epidemics coincided with or quickly followed subsistence crises and were often mixed with other diseases. Summer temperatures for those cases were either cool or average; precipitation levels often above average. In accordance with this impression of circulatory patterns dominated by westerlies, the winters preceding such an epidemic were mostly average or even slightly warmer than average. In all likelihood people became immunocompromized due to the malnutrition caused by food shortages. Partly the last year of such an epidemic which is also the year of the highest mortality in these outbreaks, is marked by average or even above average temperatures and dry spells as in 1369 and 1439, so the maritime conditions and westerly winds were less prevalent than in the preceding years. However, the connection between a cool/wet summer and the occurrence of plague waves is much weaker than that of warm/dry summer weather and plague. Between 1348 and 1500 plague outbreaks under cool/wet conditions are by far outnumbered by cool/wet summers that witnessed no plague epidemic (Fig. 10.1, in Appendix 6 only the actual plague years under cool/

[68] Grant, Independence and nationhood, 239.

wet conditions are given). For example in three summers of temperatures slightly below average and a first difference of 0 plague was recorded, but in total eight such summers occurred; two plague waves coincided with slightly cool summers (4) with a first difference of −1, but in total there were again eight summer seasons with these meteorological parameters. Since the cool/wet weather conditions were not themselves a major driver of plague outbreaks in late medieval England, other factors in these years must have been influential. All the plague outbreaks under cool/wet conditions were also associated with raised grain prices and/or harvest failure or low quality grain, which in turn would result in an increase in grain trade and population displacement in volume and distance as well as in an increase in poverty and malnutrition.

However, not only did the frequency of food shortages decrease with the falling English population density in the post-1350 period, but most plague waves did actually not accompany famines. In the years 1350–1500, when in the long-term summer season temperatures were falling at least until 1431 (Fig. 5.4) but probably well into the mid-fifteenth century, the typical summer for an English plague outbreak with high levels of mortality was warm, generally noticeably warmer than the average summer of that time, and it was dry. Often these epidemic seasons followed a period of one to three summers half years of cooler but not extremely cold conditions, but rarely would the grain price indicate a subsistence crisis in England. The winters preceding the plague wave were of slightly lower than average or of average temperature. When these meteorological patterns occurred and more than 5 years since the last major outbreak had passed, then the likelihood for pestilence to manifest itself on at least a regional if not on the national level was high in England.

Thus the classification of the plague as a disease of the heat by people who lived through it, is confirmed. However, the analysis of climate proxies of close proximity to the study region also reveals a more complex picture: plague was also embedded in a sequence of climatic conditions leading up to the hot and dry season outbreak, these conditions stretch back at least to the previous summer, but possibly even up to three summers preceding the plague. The most decisive meteorological precondition was the absence of very cold weather, especially the absence of very cold winters, after which no plague ever followed up to 1500 and which could stop a plague effectively. The identification of *Yersinia pestis* in European mass graves of supposed plague victims has proven that this bacterium played an important role in the high mortality events of the Late Middle Ages and beyond.[69] So how can the outlined medium-term meteorological pattern of 1–3 years before late medieval English hot weather plague waves be brought in accordance with the epidemiology of *Yersinia pestis*?

Recent research has broadened our view of plague. Plague is primarily an infectious disease of rodents transmitted to humans by rodent fleas, mostly by *Xenopsylla cheopis*, but in Europe also by *Nosopsyllus fasciatus*. In addition other ectoparasites can become vectors, amongst them the human flea, *Pulex irritans*, and lice, so that

[69] For a comprehensive review of this specialized field, see Bolton, Looking for *Yersinia pestis*, 18–25.

humans themselves can become hosts of the disease. Infection via insects vectors generally results in bubonic plague in humans. If pneumonic plague is present in cold weather, all vectors can be bypassed and be replaced by a direct human to human infection. It is possible that two plague outbreaks in fifteenth-century Iceland were primary pneumonic plague.[70] In fact all three components of the disease, bacteria, vectors and hosts, are subject to the influence of climate. Rodent flea abundance depends on temperature, rainfall and air humidity, fleas thrive under warm and moist conditions. With increasing temperature, flea development rates also rise until a critical threshold is reached at about 27 °C; immature stages are more readily affected by a shift to hotter and drier conditions than adult fleas. Rodent populations are subject to multiannual fluctuations which are linked to food availability, and which is in turn influenced by rainfall.[71]

In the region where the European plague wave 1347–1353 most likely originated, central Asia, plague is enzootic, and the Kyrgyz people considered the mass proliferation of rodents as a sign of the arrival of plague, sometimes a sudden collapse of the rodent population was also observed before the outbreak of the epidemic amongst humans. It has been observed that the mass proliferation of rodents is linked to the abundance of a wild plant (*Agriophyllum* genus) which serves as food and shelter.[72] In fact the main host species of plague in central Asia is the great gerbil (*Rhombomys opimus)* which is relatively asymptomatic to the disease. For this region of a dry continental climate with hot summers and cold winters, the epidemiology of plague and the rodent-flea-human interaction involved in plague outbreaks have been studied in detail and are very informative regarding timeframes and climate conditions involved in rodent und human plague outbreaks. In Kazakhstan the gerbil populations rise and fall in accordance over wide areas, climate fluctuations are behind this spatial synchrony influencing food availability, temperature stress, disease outbreak and parasites. Generally gerbils are more abundant in periods with increased vegetation, in the dry climate of central Asia this implies the shift to warmer and moister conditions.[73] The spread of plague requires a high abundance of rodent hosts and fleas, a high prevalence of plague amongst the gerbils therefore necessitates the number of gerbils as well as fleas to surpass a critical threshold. Flea populations in central Asia increase with warmer springs, which allows the reproduction of the fleas to start earlier, and higher rainfall particularly in summer; these conditions also further an increase of rodent numbers.[74] A high plague prevalence amongst the gerbils and a high flea burden are risk factors for

[70] Karlsson, Plague without rats, 276–284.

[71] Ben-Ari et al., Plague and climate.

[72] Nikanoroff, Rapport URSS, in: Jorge, R. (ed.), Les faunes régionales des rongeurs et des puces dans leurs rapports avec la peste, Paris 1928, 126–127, cited in: Audouin-Rouzeau, Les chemins de la peste, XIII 212.

[73] Kausrud et al., Climatically driven synchrony of gerbil populations, 1968.

[74] Stenseth et al., Plague dynamics are driven by climate variation, 13,110–13,113, Kausrud et al., Modeling the epidemiological history of plague in central Asia, 1–2.

human infection since the crowding of fleas forces them to seek out new hosts.[75] The relationship is so strong that human plague cases can be forecast with models using climate proxies for predicting sylvatic plague in central Asia.[76] The epidemiological pattern is complex and extends at least over several months: flea density, which is decisive for the timing and scale of human infections, starts to rise after summer rains resulting in a high flea burden in autumn. If the winter conditions are favourable to fleas, increased temperatures in the following spring cause an increased number of human infections. Hence relative to the increased rodent population and the raised prevalence of plague amongst the gerbils human infections occur with a lag of one year.[77] In fact a decline of the rodent population with a high flea burden contributes further to the spread of plague amongst the remaining rodents.[78]

In Europe rodent populations are also volatile and these fluctuations are equally influenced by climate variability. However, in Europe's much milder and wetter climate, and particularly in the maritime climate of the British Isles, the climatic triggers are not the same as for the arid regions of central Asia with their marked annual temperature range.[79] That plague, with its complex rat-flea-human relationship, adapts readily to regional climate is evident in its presence in climatologically and ecologically diverse areas around the world today. In China, for example, two plague foci exist: in the cold and arid environment in the north, plague outbreaks are favoured by warm and wet – but not too wet – conditions as in central Asia; in the tropical south-eastern plague region epidemics occur either in extremely wet years (due to dispersal of the rodents by floods) or in very dry years.[80] In Vietnam, plague was found to break out during the dry season.[81]

In Europe rodent population growth is affected by a variety of factors. The mice populations in particular – in England it is mostly the field vole (*Microtus agrestis*), and on the continent the common vole (*Microtus arvalis)* – experience growth cycles of 3–7 years.[82] Research on voles is extensive compared to other rodent spe-

[75] Samia et al., Dynamics of the plague-wildlife-human system in central Asia, 14,528.

[76] Kausrud et al., Modeling the epidemiological history of plague in central Asia, 5–10.

[77] Samia et al., Dynamics of the plague-wildlife-human system in central Asia, 14,531.

[78] Kausrud et al., Climatically driven synchrony of gerbil populations, 1968, Reijniers et al., Plague epizootic cycles in central Asia, 2.

[79] Thus the assertion by Semenza, J.C. and Menne. B., Climate change and infectious disease in Europe, 369 that 'warm, wet winters and springs increase rodent populations' is based on Kausrud, Climatically driven synchrony of gerbil populations, 1963–1968, and hence refers to central Asia, where the average winter temperature is −20 °C, but might not be valid in all regions of the geographically and climatologically diverse Europe. Therefore the use of this premise of warm and wet weather in winter and spring as factors contributing to rodent population maxima by Schmid et al., Climate-driven introduction of the Black Death, 1–5, is not helpful, even though the paper could not identify any climatological pattern that could be associated with plague or its potential wildlife reservoir in Europe.

[80] Xu et al., Nonlinear of climate on plague, 10,215–10,216.

[81] Cavanaugh, Marshall, The influence of climate on the seasonal prevalence of plague, 89–91.

[82] Elton, Periodic fluctuations in the number of animals, 140–142, Körner, Feld- Und Schermäuse in Solothurn 1538–1643, 442–443, Jacob, Tkadlec, Rodent outbreaks in Europe, 208–210.

cies of temperate Europe. The mechanism behind the vole population cycles are not clear, but they are linked to an interplay of climate, predation, starvation, parasites and disease. One of the factors affecting vole populations is the vegetation productivity especially in the form of hard mast, i.e. primarily beech mast and oak mast.[83] In England before the mid-twentieth century full beech mast years occurred on average every 5 years, only in full mast years is more mast produced by the trees than the rodents can actually consume. The mast provides food in autumn and winter and produces ground cover in spring. In northern Europe good beech mast years depend on sunshine and temperature during the summers 1–2 years previous to the mast year.[84]

Extreme climate conditions can cause a large scale synchronous effect on rodent populations in Europe.[85] This is illustrated by vole cycle maxima occurring in 1822 in France and Germany,[86] in 1891–1893 in Scotland, Norway, France and Thessaly[87] or the plagues in 2007 in Germany, western France and Spain.[88] The latter event was associated with a positive large-scale anomaly in plant productivity over wide areas of Europe in the preceding 9 months,[89] temperatures far above the average in central Europe and also higher than average throughout the continent including Spain and a severe spring drought in April in Germany.[90] Often such extensive and severe mouse plagues result in a sudden collapse of population numbers, also before modern control measures were practised.[91] It then takes years to rebuild numbers. Other rodent species also experience cycles. Particularly the black rat (*Rattus rattus*), an important host of *Yersinia pestis* since the start of the Third Pandemic in the late nineteenth century, will have reacted positively to hot and dry seasons, as long as it was present in Europe. The black rat is at home in warmer climates than Europe's temperate zones, where it could only survive in and around human habitation; its numbers in England were limited.

---

[83] Elton, Periodic fluctuations in the number of animals, 142–143, Heyman et al., Factors that drive hantavirus epidemics, 7–8, Imholt, et al., Klima, Nagetiere and Nagetier-assoziierte Krankheitserreger, 3, Zwolak et al., Advantages of masting in European beech, 754–756.

[84] Matthews, The influence of weather on the frequency of beech mast years, 111–114, Hacket-Pain et al., The influence of masting phenomenon on growth-climate relationships on trees, 324–327, Müller-Haubold et al., Climatic drivers of fruit masting, 93–94, Zwolak et al., Advantages of masting in European beech, 754–756.

[85] Imholt, et al., Klima, Nagetiere and Nagetier-assoziierte Krankheitserreger, 4.

[86] Anonymus, Landwirtschaftliche Berichte, 369–370, Gérard, Essai d'une faune historique, 229, Elton, Voles, mice and lemmings, 21, 62.

[87] Elton, Periodic fluctuations in the number of animals, 140–143, Elton, Voles, mice and lemmings, 34, 65, 145–150, 220.

[88] Jacob, Tkadlec, Rodent outbreaks in Europe, 214, Pinot, The interplay between seasonality and density, 10.

[89] Pinot, The interplay between seasonality and density, 10

[90] Löpmeier, Die agrarmeteorologische Situation im Jahr 2007, 50–55.

[91] Wolf, Uber die Feldmäuse, 62–67, Gatterer, Über die Verminderung der Feldmäuse, 67, Elton, Voles, mice and lemmings, 28, Körner, Feld- Und Schermäuse in Solothurn 1538–1643, 442–445, Farrell, Mouse plagues, 122.

Weather patterns that pose a risk factor for the development of vole population maxima in Europe north of the Alps are complex and the direct or indirect effects of specific meteorological parameters on rodent population dynamics are not yet always completely understood.[92] Of high importance are conditions during the winter preceding a massive increase of the vole population: mild but not too wet winters, to avoid flooding of burrows or high humidity in burrows which decrease burrow temperature, raise the outbreak risk.[93] Also winters with prolonged snow-cover increase the chance of rodent survival, because the snow provides cover as well as a temperature insulation for the burrows.[94] Winter air temperature must sink below −10 °C to seriously impact rodent survival.[95] If such mild or average but snowy winters are followed by dry and warm spring-summer seasons, and if these meteorological conditions coincide with the upward trend in the natural mouse cycle, the risk for a massive rise in rodent numbers is greatly increased. Dry and warm conditions have a positive influence on the survival rate of rodents and also tend to crack up the soil and therefore lead to more suitable burrowing conditions.[96] North of the Alps a spring or even summer drought after wetter years does not result in a rapid reduction of vegetation, so that during dry weather favourable conditions for a rodent plague persist. Considering the cyclical fluctuations of rodent population and the time necessary for reaching high population levels, it can be assumed that vole plague years were preceded by years which provided good conditions for vegetation growth, and therefore for high rodent fertility, and good rates of winter survival. This would imply the absence of further extremely cold winters and also of inclement summers and the presence of sufficient rainfall and warmth to further grass growth and vegetation cover. Summers of average temperature and of average or even wetter than average precipitation levels would fulfil these requirements. Recent research from Germany demonstrates the importance of the year preceding a common vole outbreak. Increased rainfall levels in spring raise the risk for a vole outbreak in the spring of the following year,[97] whereas increased precipitation in early autumn is correlated to autumn vole population peaks in the subsequent year.[98] Such autumn vole population peaks are also largely dependent on

---

[92] Imholt, et al., Klima, Nagetiere and Nagetier-assoziierte Krankheitserreger, 4.

[93] Imholt et al., Identification of weather parameters related to regional population outbreak risk of common voles, 557, Roth, Mäuse wieder auf dem Vormarsch,1.

[94] Wolf, Über die Feldmäuse, 28, Imholt et al., Identification of weather parameters related to regional population outbreak risk of common voles, 557, Esther et al., Correlations between weather conditions and common vole densities, 82, Imholt, et al., Klima, Nagetiere and Nagetier-assoziierte Krankheitserreger, 3.

[95] Roth, Mäuse wieder auf dem Vormarsch, 1.

[96] Clitomachus, Von Mäusen, 117, Gatterer, Über die Verminderung der Feldmäuse, 56, Saunders, Giles, Plagues of the house mouse, 241, Imholt et al., Identification of weather parameters related to regional population outbreak risk of common voles, 554, Esther et al., Correlations between weather conditions and common vole densities, 80–81, Roth, Mäuse wieder auf dem Vormarsch, 1.

[97] Esther et al., Correlations between weather conditions and common vole densities, 80.

[98] Imholt et al., Identification of weather parameters related to regional population outbreak risk of common voles, 555.

raised temperatures in the spring and lower rainfall in parts of the summer of the same year.[99] Hence, meteorological factors with a lag of one year influence central European common vole populations: raised rainfall levels in at least parts of the preceding summer half year increase the outbreak risk of voles in the present year. Such a sudden increase of rodent numbers in summer half years warmer, and therefore probably also drier, than the preceding seasons appears to be stable over time: it also occurs in a continuous record of numbers of voles and mice caught by the municipal mouse catcher in the Swiss town of Solothurn 1538–1643. The Solothurn series is the only long-term data series of historical rodent populations in Europe and also considerably longer than modern observations, it is also marked by mouse cycles of mostly 4–6 years.[100] The close relationship between meteorological conditions and common vole population maxima in central Europe can be employed to forecast vole outbreaks, and models for eastern Germany based on weather parameters are capable of predicting 60–85% of the outbreaks.[101]

The described pattern of European vole outbreaks in terms of frequency and meteorological conditions during and before the rodent population maxima is very similar to the weather preceding and accompanying medieval English hot-season bubonic plague outbreaks. The fact that English human pestilence outbreaks occurred in the Middle Ages frequently under meteorological conditions furthering also a mass increase in the main host of *Yersinia pestis*, the rodents, implies a causative link not merely correlation. The run up to a human plague year was marked by average or slightly cool temperatures, probably with somewhat raised levels of rainfall and would have provided good conditions for vegetation growth and contributed to an increase in rodent numbers. In this context the absence of extremely cold summers before hot-season pestilence is notable, these would have been summers with low vegetation growth or a shorter vegetation phase and therefore reduced food availability. Additionally no extremely cold winters precede any plague outbreaks in medieval England, most winters were average or slightly colder than average, a few were classified as cold; some were so non-descript that they were not graced with a reference by the chroniclers, which makes average to somewhat milder winter conditions likely. This sort of winter weather accords well with rodent survival rates in the cold season, and would ensure a strong basis of rodents of breeding age at the onset of the warm season. It is possible that the winters indicated by van Engelen et al. as slightly colder than average in the Low Countries saw prolonged snowcover which is favourable to rodent survival. With the onset of the warm season and dry weather rodents found optimal conditions and their numbers would have risen steeply, as indicated by the sixteenth to seventeenth-century mouse data from Switzerland; thus explaining the pattern of hot-weather human plague waves occurring in warm summers that were considerably warmer and drier than the previ-

---

[99] Esther et al., Correlations between weather conditions and common vole densities, 79–81.

[100] Körner, Feld- Und Schermäuse in Solothurn 1538–1643, 442–443.

[101] Imholt et al., Identification of weather parameters related to regional population outbreak risk of common voles, 554–558, Esther et al., Correlations between weather conditions and common vole densities, 78–82.

ous year(s). That rodent numbers had steadily risen towards the population peaks in warm and dry summers is potentially reflected in the presence of low level human plague in the year before, most likely this could be linked to already high but not yet exploding numbers of rodents after at least one good feeding season and with the benefit of beech and oak mast in autumn. The apparent minimum interval of about 5 years before the renewal of widespread human plague, even if hot and dry summer weather returned before that time, can not just be explained by lower human population density, but also has to take into account the problem of the rodent population to increase its numbers rapid enough and to reach a sufficiently high level of plague prevalence in its population after the last rodent population collapse which probably occurred around the time of the human plague wave. In light of the close connection between rodent population peaks and human plague, it is also not surprising that extremely cold winters could terminate a wave of bubonic plague, as long as pneumonic plague had not taken a hold; rodents are greatly decimated by extremely cold winters.[102] The plague outbreaks connected to subsistence crises were not coincidental with climatic factors furthering mass rodent proliferation. Hence they might find their explanation rather in lowered living standards and malnutrition, the increased trade volume of potentially rat-infested grain deliveries, and also in flooding which can dislocate rodents as well as humans.

The development of the necessary insect vector for transferring *Yersinia pestis* from the rodent host to humans is also fitting within the meteorological pattern accompanying late medieval English hot summer pestilence. Studies in central Asia have shown that rodent flea development lagged behind the rise of rodent populations, but that a warm spring, particularly after a wet summer would increase flea numbers. Similar mechanism might be at work in England, where even in dry periods air humidity rarely sinks much below 70%. Hence also during warm springs and summers with low precipitation levels air humidity remains close to optimum levels for flea development. Most plague waves north of the Alps are taking place in late summer and autumn, which is overlapping with the height of the flea season; in England August precipitation is also on average comparatively high. Therefore it is likely that with the weather pattern accompanying English human plague also the flea burden of the rodents increased. That rodent parasites became abundant towards the end of rodent population peaks has also been observed in continental Europe, when at the end of a vole plague dead mice were found covered in insects, particularly lice.[103]

The importance of temperature and precipitation levels are demonstrated by the plague-induced mortality peaks in 1361 and 1375. Both summer half years were

---

[102] The link between rodent population peaks and human epidemic disease do not necessarily imply the involvement of *Yersinia pestis*. Voles are also the host of other organisms harmful to man: the hantavirus and the *leptospira* bacteria endanger human health in years of mass mouse proliferation. However, these pathogens cause diseases that are not easily mistaken for bubonic plague as they are not as lethal as plague and do not lead to symptoms such as buboes or dark spots on the skin.

[103] Wolf, Über die Feldmäuse, 62–64.

very hot and dry, but 1361 extremely so. The mortality pattern in London shows the epidemic being most active in May to July in 1361 and in July and August in 1375, afterwards the number of deaths decreased considerably contrary to the normal mortality pattern north of the Alps.[104] It seems as if in these years the high spring temperatures triggered an earlier development of rodents and fleas, but then the very hot summer posed a problem for flea development and potentially also for rodent survival. The year 1361 was indeed one of the rare occasions in England, when drought conditions ruined the hay, fruit and grain crops. The drought began to impact on plant growth in May and with the subsequent increasing loss of vegetation the food supply of rodent populations must also have decreased, forcing fleas to look for new hosts in the following months.

At this point the question arises, why medieval chronicles in England did not report vole infestations and mouse plagues, when they were observing a human plague outbreak. In general the number of references to the mass proliferation of rodents in the medieval chronicles in Europe is very limited, only after c.1600 did such events receive more attention by the educated classes. Rodents were after all ubiquitous, in the town and in the country, and the vole cycles caused a constant fluctuation in the numbers of voles. However, except for the worst vole outbreaks, population peaks might have gone unnoticed, in the eighteenth century and even today tractates on agricultural and forest management lament the fact that even extremely bad vole plagues are 'invisible' until it is too late to prevent them and to mitigate the ensuing damage.[105] Normally rodent corpses were 'invisible', for example black rats often die out of human sight hidden in crevices or walls.[106] Also not every mouse cycle maxima made a veritable vole plague and not every vole plague was connected to *Yersinia pestis*. Voles can become a host of *Yersinia pestis*, even though they are largely asymptomatic to the disease just as the great gerbils of central Asia, but they were probably not the normal host of sylvatic plague in Europe; if this were the case plague would have demanded an even higher death toll, and the disease could only have been eradicated under even greater difficulty due to the high number and the extensive habitat of voles. But it is likely that there were non-urban plague reservoirs in medieval England, in East Anglia these reservoirs were mostly associated with marshland or heaths, the disease persisted as sylvatic plague[107] and the host species probably fluctuated in a way similar to the voles. In the fifteenth century these reservoirs included the marshlands along the North Sea coast between Great Yarmouth and Aldeburgh, the land on the river Waveney which marks the border between Norfolk and Suffolk, the higher ground between the wetlands of the rivers Yare and Bure to the east of Norwich and the land to the west of the city.[108] The East Anglian plague reservoir could be maintained without the black

[104] Cohn, The Black Death transformed, 184–185.

[105] Wolf, Über die Feldmäuse, 68–71, Gatterer, Über die Verminderung der Feldmäuse, 55, Saunders, Giles, Plagues of the house mouse, 241.

[106] Anonymus, Meußthurn, 57, Audouin-Rouzeau, Les chemins de la peste, XIII 177–180.

[107] Gottfried, Epidemic disease, 126–138.

[108] Ibid., 129–137.

rat and *Xenopsylla cheopis*, as in the sylvatic plague in the marshland of eastern Suffolk between 1906 and 1918, which involved primarily brown rats (*Rattus norvegicus*), but also rabbits, hares and ferrets. The disease affected humans in single or neighbouring households in 1906–1907, 1909–1910, 1911 and 1918; the presence of *Yersinia pestis* was confirmed by tests.[109] The reservoirs of sylvatic plague in East Anglia, the marshes and heathlands, were not remote areas devoid of human activity, on the contrary plenty of direct interaction between humans and rodents will have taken place in pastures, fields, woods and houses. Whereas the medieval chroniclers in England are largely silent on the subject of mass proliferation of rodents and human plague, European medieval and early modern plague tracts are more informative: they often do count the mass appearance of mice or rats and various insect species as a sign for plague, as they do also the abandonment of their habitation by rodents and other animals normally living underground.[110] This is not merely the description of putrefaction in the metaphysical sense, a symbolic upheaval of the established order of the world, but also a realistic observation of events indeed preceding the outbreak of pestilence amongst humans.[111]

The consideration of late medieval English mortality peaks in their climatological setting reveals the strong connection between meteorological factors and virulent human plague outbreaks. Scrupulous dating of the mortality peaks and the use of local and regional climate data are indispensable for identifying this pattern. Human plague-induced mortality peaks, most likely of bubonic plague, were not just defined by an affinity for hot and dry summer half years, but were also associated with a set of meteorological conditions preceding the epidemic by at least a year. Time lags of this scale are also observed in the much better researched central Asian plague region. The climatic conditions before and during English human plague are conditions also furthering population peaks in rodents, particularly mice. Since the connection between climate and human plague is strong, it can be concluded that *Yersinia pestis* had indeed found a host rodent species in late medieval England, the black rat in towns and other rodents in the marshes and heaths, and from this niche the disease broke forth into the wider environment when optimum conditions for a rodent population increase occurred. Therefore, at least for late medieval England, the recent theory that plague was frequently being re-imported from Asia in the centuries following the Black Death around 1350[112] is unlikely. This conclusion is supported by the recent finding of one identical genotype of *Yersinia pestis* in fourteenth-century skeletons from Manching-Pichl, southeastern Germany, and a seventeenth-century skeleton in Brandenburg, northern Germany, hence over this period of several hundred years one genotype of *Yersinia pestis*

---

[109] van Zwanenberg, Last epidemic of plague in England, 65–72, Black, Black, Plague in east Suffolk 1906–1918, 541–543.

[110] Sticker, Abhandlungen aus der Seuchengeschichte, vol.1/2, 126–132. Audouin-Rouzeau, Les chemins de la peste, XIII 34, 194–203, 215–234.

[111] Audouin-Rouzeau, Les chemins de la peste, XIII 235–239.

[112] Schmid et al., Climate-driven introduction of the Black Death, 2–5.

persisted in Germany.[113] Identical or highly similar genotypes of the bacterium have also been identified independently in Hereford in western England, in victims of the Great Pestilence in London and in fourteenth-century skeletons from Saint-Laurent-de-la-Cabrerisse in southwestern France. So far only one other *Yersinia pestis* genotype has been found in European plague victims, it comes from the Netherlands.[114] The discovery of the protracted continuation of one *Yersinia pestis* genotype in distant German regions between the fourteenth and seventeenth century suggests the existence of a wildlife reservoir of plague in the temperate climate zone of Europe for 300 years after 1350.[115]

[113] Seifert et al., Genotyping *Yersinia pestis* in historical plague, 2016, 3–7.

[114] Haensch et al., Distinct clones of *Yersinia pestis*, 4–6.

[115] Seifert et al., Genotyping *Yersinia pestis* in historical plague, 2016, 3–7.

Chapter 11
The Dance of Death – A Synthesis

The people of the Late Middle Ages experienced a time that was characterized as a Dance with Death. Mortality was high, and death could carry everyone unexpectedly to the grave, regardless of social position or age. In fact this Dance of Death was performed according to strict rules. The rules for the two major movements – famine and plague – have been carved out in this work. The common people could not escape either of these dangers, and whereas the well-off classes could buy their way out of hunger, they still had to face the plague. In a pre-industrial society climate is one of the major forces that creates the setting for famine and plague, hence it is one of the major forces forming the patterns of mortality in the Dance of Death.

The interest in the environmental conditions that would lead to a rise in mortality is older that the motif of the Dance of Death itself. For England the oldest systematic study of the link between weather and dearth survives from the first half of the fourteenth century. William Merle's 'De pronosticacione aeris' written c.1340 and his weather observations 'Consideraciones temperiei pro 7 annis' for the years 1337 to 1344 bear witness to his enquiring and practical mind. He worked in association with a group of Oxford scholars engaged in the *scientia astrorum*, which encompassed astronomy as well as astrology. Members of this groups were aiming at long-term forecasts of the weather with the help of the stars. Merle supplemented this astrometeorological work and the theoretical treatises of Ptolemy, Aristotle and Virgil with weather rules based on 'inferior signs' in the sky, 'farmers' rules' and the observation of nature, which could serve as the basis for short-term predictions.[1] Although the astrometeorological approach is most bewildering from the modern perspective, the scholars of the science of the stars and Merle had an empirical outlook and were part of the proto-scientific movement in Oxford.[2] Their aims included an identification of meteorological risk factors for the spectre of their times, famine, and a reduction in vulnerability by being able to predict the meteorological conditions that proved to be so detrimental to agriculture in England. Merle's high interest

[1] Thorndike, History of magic and experimental science, vol. 3, 143–145, Jenks, Astrometeorology, 194, Snedegar, Between scholasticism and folk wisdom, 31.

[2] Mortimer, William Merle's weather diary, 42–43, North, Cosmos, 290.

K. Pribyl, *Farming, Famine and Plague*, DOI 10.1007/978-3-319-55953-7_11

in famine and its mostly meteorological causes came at a time when the Great Famine of 1315–1317 was still fully within living memory. It had probably been the major event in Merle's childhood or youth, and that of his colleagues at Oxford. A phase of high interannual variability in spring and summer temperatures from the mid-1320s to the mid-1330s had just come to an end, and even though the 1330s were in general marked by dry summers, wet and cold summer seasons had repercussions for agriculture in 1330 and 1335. Then in 1338 incessant rains at winter sowing time and the subsequent cold and long winter ruined the winter corn, and dry weather in April 1339 damaged the spring sowing, so that the consequent dearth lasted from the harvest 1339 to the harvest 1340. Hence, just at the time when Merle is supposed to have composed his work on weather and subsistence crisis, the situation in the wider world demonstrates that Merle's and his colleagues' interest in the subject was far from purely academic or theoretical, but practical and immediate. The volatile and dangerous meteorological conditions from the 1320s to the 1340s affected a country that was still densely populated and consequently very vulnerable to any harvest shortfall, which also reminded people of the risk of a famine on the scale of the 1310s. Merle identified correctly excess humidity as the risk factor for agricultural production *per se* in England, and so described one of the major movements of the Dance of Death on the British Isles.

Merle's work resonated amongst the astronomers at Oxford. 'De pronosticacione aeris', including the chapter dedicated to weather and agriculture, was used by John Ashenden in the 'Summa astrologiae judicialis de accidentibus mundi' (completed in 1348), which in itself is devoted to the prognostication of general events such as storms, floods, droughts, earthquakes, war and famine.[3] The inclusion of his text in Ashenden's work guaranteed Merle a wide and long-lasting reception of his ideas, since the 'Summa anglicana' served as an astrological reference work in late medieval and early modern Europe. It was printed in Venice in 1489, and remained a textbook at the University of Vienna until the sixteenth century.[4]

The object of Merle's observation, recording and analysis – meteorological parameters such as the frequency and form of precipitation, temperature, wind and the occurrence of frosts – is of high interest for the climatologist and the environmental, agrarian or economic historian alike. For studying the impact of weather on the food supply and on human health regional climate data of at least annual but preferably seasonal resolution are essential, otherwise the detail in the meteorological parameters can be lost. Although the regional climate is inextricably linked to the global scale, it can diverge significantly from it, and it is the regional climate that shapes peoples' lives. Due the relative paucity of such data for England and the British Isles in the Middle Ages new data series had to be developed in this book.

The manorial accounts from East Anglia, mainly from Norwich Cathedral Priory, were so meticulously recorded and so well preserved, that those 600–700 year old parchments open an unprecedented window into the changing meteorological

[3] Thorndike, History of magic and experimental science, vol. 3, 329, Snedegar, Between scholasticism and folk wisdom, 30, Ashenden, Summa astrologiae judicialis de accidentibus mundi'.

[4] Meaden, Merle's weather diary and its motivation, 211.

conditions of late medieval England. The reconstruction of the mean temperature in the months April to July, which was achieved with the help of the grain harvest date, shows a cooling trend from the mid-thirteenth century until the third decade of the fifteenth century. Warmer and cooler periods alternated on a decadal level, many cooler times were often associated with problems in agricultural production as in the 1310s and early 1320s, or in the late 1340s, parts of the 1360s and 1370s, and the first decade of the fifteenth century. Interannual variability was equally difficult for crop growing and it was particularly pronounced from the later 1310s to the mid-1330s, and in the 1360s and the first half of the 1370s. Towards the end of the 1420s, in themselves a period of moderate variability, growing season temperatures again began to fluctuate strongly. In the July–September precipitation index, which is based on the harvest duration, periods of higher and lower rainfall levels also alternate. Wetter times were often times that witnessed difficulties in grain growing, they occurred from the mid-1310s to the mid-1320s, from the mid-1350s to the early 1370s, around 1380, in the first decade of the 1400s and in a number of years of the 1420s. Interannual variability was raised from about 1310 to 1330, in the late 1340s, in the first decade of the fifteenth century and in the 1420s. Superimposed on the interannual and decadal variability is a long-term trend of falling rainfall levels, however, due to the human factor in the harvest duration, the long-term trend of the precipitation index is not as meaningful as the short- or medium-term fluctuation. For evaluating the impact of extreme years in terms of temperature and precipitation, a catalogue of very hot or cold and dry or wet spring and summer seasons has been created (Chaps. 6 and 8), which also includes the available information on meteorological, socio-economic and agricultural conditions of these extremes from written records. That the most severe subsistence crises of the Late Middle Ages, those following the harvests in 1258, 1294, 1315–1317, and the dearth 1428 were linked to cold growing seasons, wet summer months and also to high interannual variability in temperature and rainfall is evident. Another phase of high precipitation and cool growing season temperatures with the potential to result in large-scale famine is masked by the Great Pestilence in 1348–1349. The two major famines of the fifteenth century, the famine of the second half of the 1430s and the famine of the early 1480s are equally the consequence of prolonged cool and wet weather conditions.

The data confirm the notion of the Late Middle Ages being not just a time of social, economic and cultural transition, but also of climate change. In England the shift in climatic conditions manifests itself in cooling summer season temperatures from the end of the Medieval Climate Anomaly towards the Little Ice Age; the intermediary period, the fourteenth century, was marked by a high variability of temperature and precipitation schemes. Climate and climate variability played an important role as a driver of late medieval change by influencing human mortality and hence demographic development via exercising control on agricultural success and failure, as well as providing the conditions needed for the outbreak of plague by driving the vegetation productivity and creating the average winters and the sudden shift to warm and dry springs and summers that were necessary for rodent population maxima and flea development.

The impact of climate, however, was not direct, but operated within a socio-economic, cultural and also environmental framework. The cool and wet summer half years ruining the grain harvest that caused such dearth, particularly in the mid-1310s, were to return even more often in the second half of the fourteenth century. With the diminished English population after 1350, however, harvests reduced by rain and cold translated less frequently into famine and high prices, although the suffering amongst the lower classes was not eradicated. In addition, coping strategies were improved over the Late Middle Ages, the international grain trade became more established and in times of crisis imports of grain were organised by the English crown or the mayor of London (1352, 1390, 1416, see Sects. 6.2, 6.5 and 8.3). With respect to epidemic disease before the introduction of *Yersinia pestis* into England, warm and dry summer half years or at least summers considerably warmer than previous years were not a major factor in mortality crises; the death-toll of gastrointestinal problems which frequently fall to summer was limited. Warm and dry summer months developed into a risk after 1350. Whereas the arrival of plague in England and its high mortality amongst a virgin population were not driven by regional climate factors in the form of warm and dry conditions, the spread of plague across Asia and Europe in the mid-fourteenth century was ultimately connected to climate change in the home region of *Yersinia pestis*, central Asia. In the sixteenth and seventeenth centuries, restrictions on the movement of people and quarantining in the case of an epidemic probably modified the spread of plague.

The significance of socio-economic factors for the shaping of climate impacts also becomes clear in the history of viticulture in England. At first glance a negative impact of cooling and increasingly wet conditions in the fourteenth century on wine-growing would appear to be the reason for the demise of English vineyards. However, taking the socio-economic setting into account, it becomes clear that English viticulture, which had never aimed at the regular production of good-quality wine, was much more efficiently ruined by the lack of labour and rising wages than by the deteriorating climate.

In England the secular trend of falling temperatures had less dramatic consequences for medieval society and agriculture than short-term and especially interannual variability of temperature and precipitation. Plague years are marked frequently by a sudden rise in summer temperatures and probably also a fall in precipitation levels compared to preceding summer seasons. With regard to agriculture, rapidly shifting meteorological patterns made reacting and adapting to new conditions impossible. During or shortly after the highly variable decades following 1350, the age of the 'Indian Summer' of demesne farming and maintained high grain prices while population levels had declined sharply, Piers Plowman laments not only about 'wederes unresonable', but continues:

> Wederwise shipmen and witty clerkes also
> Have no bileve to the lifte, ne to the loore of philosophres.
> Astronomiens alday in hir art faillen
> That whilom warned bifore what sholde falle after;
> Shipmen and shepherdes, that with ship and sheep wenten,
> Wisten by the walkne what sholde bitide

With unpredictable weather, agriculture too turned unpredictable and even experienced men could not assess the harvest in advance any longer, or plan a future course of action. Likewise, sailors and shephers now failed to read the signs of sky and nature.

> Tilieris that tiled the erthe tolden hir maistres
> By the seed that thei sewe whit thei selle myghte,
> And what to leve and to lyve by, the lond was so trewe;
> Now failleth the folk of the flood and of the lond bothe –
> Shepherdes and shipmen, and so do thise tilieris:
> Neither thei konneth ne knoweth oon cours bifore another.

The post-Black Death disillusionment with authorities included not only the religious and social elites, but was also directed at Merle's fellows, those scholars who were devoted to astrometeorology and who were unable to forecast the weather:

> Astronomyens also aren at hir wittes ende:
> Of that was calculed of the clem[a]t, the contrarie thei fynde.[5]

With the dramatic arrival of plague in Europe and England, the attention of those committed to the science of the stars, in reflecting the sensitivities and fears of the age, was turned towards epidemic disease and in particular plague. Plague was the movement *par excellence* in the Dance of Death. The regional climate data demonstrates that in England weather patterns had a decisive influence upon the occurrence of plague waves after the Great Pestilence. In the fourteenth and fifteenth centuries, at a time when public health measures were not yet developed, serene summer half years, especially those that were considerably warmer and drier than previous summers, potentially harboured death. At least some parts of the European society were aware – due to the close observation of their environment – of the role that heat played in the genesis of a plague wave, hence the frequent references to the subject. Plague waves formed not in every warm year, but at intervals. The occurrence of national and supra-regional plague waves instead of fragmented local disease outbreaks was set by feedback mechanisms and time delays, such as the replenishment of the vulnerable section of human society and the increase in rodent population numbers, which in turn was influenced by the climate-driven vegetation productivity and tree-mast cycles, the absence of hard winters and finally the weather conditions of the plague year.

The role of meteorological factors in combination with a vulnerable pool of people in a plague outbreak can be illustrated by the events of the 1480s. The early 1480s were a time of subsistence crisis and high grain prices,[6] climate indices from the Low Countries indicate low temperatures and wet summer half years. Nonetheless no major plague developed in England during or after the food shortages as it had done during the last famine in the outbreak 1438–1439. This is due to the severe plague wave shortly before the dearth in 1478–1479, the hard winter 1480–1481, and the fact that after the end of the food shortages, summer temperatures did not rise and remained on a low level throughout the 1480s.[7] With an

[5] Langland, Piers Plowman, 188–189.

[6] Hoskins, Harvest fluctuations, 1480–1619, 31.

[7] van Engelen et al., A millennium of weather, winds and water in the Low Countries.

absence of the meteorological factors involved in major plague waves, came an absence of large-scale plague waves themselves. The pool of vulnerable people remained though, and in the abysmally cold and rainy summer of 1485, it fell victim to a new vicissitude of nature: the English Sweat. When the weather improved shortly after, a plague outbreak of limited nature was registered in Canterbury, London and Oxford in 1487.[8]

The rules for plague in the Dance macabre remained valid throughout the fifteenth century, but towards the end of this period a strange lacuna of outbreaks can be noted. No widespread plague epidemic hit England again before the very last years of the century. The gap between the national outbreak 1478–1479 and renewed epidemic conditions in England from 1499 onwards is considerably longer than between most other major plagues. During the gap of the 1440s to the late 1450s the precursor of weather conditions is missing, it simply was rarely warm enough in those years and interannual temperature variability was comparatively low, but small-scale outbreaks in London or plagues confined to a single county were frequent.[9] However, in the late fifteenth century the meteorological patterns to trigger a plague outbreak were present in 1495 (Appendix 6).[10] This prolonged absence of plague, even though weather conditions were adequate for a disease outbreak, raises the possibility that either the aetiology of plague was changing at that time,[11] or that another pre-condition indispensable for a bigger outbreak, either the reservoir of vulnerable people or sufficient numbers of rodents, were not present. The findings on mortality and life expectancy in the monastic communities at Westminster Abbey, Christ Church Canterbury and Durham Cathedral Priory reveal a high mortality rate and low life expectancy in the late fifteenth century.[12] Further evidence such as probated wills indicate that the high death rate was not confined to the monasteries.[13] It is therefore conceivable that after the previous plague waves and other epidemics in the 1460s and 1470s, the famine of the early 1480s, the English Sweat 1485 and the

---

[8] See note 30 in Chap. 10.

[9] Low interannual variability in summer half year precipitation in England is clear in Cooper et al., Hydroclimate variability, 1026 and Wilson et al., March–July precipitation reconstruction, 1011. For the low summer half year temperatures and variability in the Low Countries, see van Engelen et al., A millennium of weather, winds and water in the Low Countries and Camenisch, Endless cold, 1062–1063. Also the data on vine harvests from Burgundy indicates cold summers, mostly temperature did not fluctuate much, Chuine et al., Grape ripening as a past climate indicator. The plagues in London are listed in Bean, Population and economic decline in England, 428 and Gottfried, Epidemic disease, 47–50 adds the provincial plagues.

[10] The weather during the summers 1490 and 1494 was approaching conditions that would have allowed a plague wave to develop (Appendix 6). The summer 1492 then bore many of the meteorological hallmarks of a plague year, it was warm and considerably warmer than the year before. However, 1491 had been a year of great trials, both the winter 1490–1491 and the summer 1491 had been extremely cold. A high mortality level prevailed at Westminster Abbey in 1491, see Harvey, Living and dying, 122–125. Medieval English major plagues did not follow shortly after extremely cold winters or just one year after an extremely cold summer, see Chap. 10.

[11] Gottfried, Black Death, 156.

[12] Hatcher et al., Monastic mortality, 674–685.

[13] Hatcher et al., Monastic mortality, 684–685.

high mortality in the extremely cold year 1491 the remaining pool of vulnerable people was simply too small to facilitate the development of a supra-regional English plague outbreak. In the mid-1490s a sequence of bountiful harvests may have helped to lower mortality in the southern monasteries and regions.[14]

Living through an age of climate change, the time of the onset of the Little Ice Age, took its toll on the people of the Late Middle Ages. In particular, extreme temperature and precipitation conditions, and periods of high interannual variability with the associated rapid cycling of weather patterns caused a considerable amount of environmental stress and so accelerated the rhythm of the Dance of Death. It is no coincidence that during those times of unstable climate the demographic shocks were pronounced. The period from the 1310s to the 1330s of frequently cool and wet conditions combined with a high variability in terms of temperature and precipitation in England, and the ensuing Great Famine and agrarian crisis have been described at length in this work, as has been the period of wet and cool summers in the late 1340s (Sects. 6.4 and 6.5). While reduced population pressure made the English more resilient to weather-induced harvest failures after 1350, the next period of mortality crisis came in the 1360s when the April to July temperature reconstruction for East Anglia, the summer and winter index for the Low Countries, as well as the central European winter conditions fluctuated most severely.[15] In this decade two supra-regional plague waves (1361, 1368–1369) and an outbreak of 'Pokkes' haunted England (1365). Early in 1362 a massive and memorable storm crossed the British Isles, and the winter 1363–1364 was one of the coldest of the last millennium in Europe.[16] The decade was punctuated by poor harvests in 1366 and 1367, the latter caused the price of grain to rise. Depending on the timing and severity of major plague waves, these could also contribute to a price spike for grain as after the harvest 1369. Another period of raised variability covers the years around 1400. Variability in the English growing season temperature was moderate, in the late summer precipitation it was high after 1400. The summer conditions in the Low Countries and in Burgundy fluctuated considerably. Two major plagues in southern England (1400, 1407) and a plague in the north (1409) occurred between 1400 and 1410, the summer half years in the middle of the decade were wet. After a quieter interval in the 1410s climatic variability increased considerably in the 1420s. The moderate interannual fluctuation of spring-summer temperatures in England throughout most of the 1420s surged at the end of the decade, and summer precipitation on the island was very variable. In the Low Countries the summers began warm in the 1420s and then deteriorated from year to year while fluctuating extremely, except in the early 1430s, to a nadir in the mid- and late 1430s. Burgundy

---

[14] Hoskins, Harvest fluctuations, 1480–1619. Harvests were good in 1492–1495 and in 1499.

[15] For England, see Sect. 5.2. For the Low Countries, see van Engelen et al., A millennium of weather, winds and water in the Low Countries, and for central European winter conditions, see Pfister et al., Winter severity, 101.

[16] This storm causes the storm surge of the St Marcellus Flood or Grote Mandrenke (Great Drowning of Men) on the Dutch and German North Sea coast, see Gottschalk, Stormvloeden, vol. 1, 368–378. For the winter 1363–1364, see Pfister et al., Winter severity, 101–102.

summers were also variable over this period. At the same time the winter variability was raised between the mid-1420s and the mid-1440s, while winters were on the whole also extremely cold, this was after all the time of the early Spörer Minimum.[17] Mortality crises in these two decades were frequent. During the 1420s a major plague (1420), a minor plague (1426)[18] and the 'Mure' (1427, a sort of rheumatic fever or influenza)[19] affected the English people. The summers 1421, 1423 and 1428 were wet and cold, and a dearth held England in its grip after the harvest 1428. The climate data for summer in the Low Countries and Burgundy show a calm start for the 1430s, a decade which then saw nonetheless quickly changing and extreme conditions in its middle years, and indeed the time was marked by two more major plague waves (1434, 1438–1439), a minor plague (1431), an unidentified epidemic disease in 1435[20] and the most severe famine of the fifteenth century which occurred in the second half of the 1430s. Compared to such a sequence of affliction, the 1440s and most of the 1450s appear almost benign. High climatic variability in the summer half year returned in the Low Countries and Burgundy around 1460, it came into full swing by the 1470s and stayed almost until the end of the century. From c.1470 onwards the demographic shocks were hard on each others heels. Whereas the 1460s still only witnessed one major plague (1464) and a regional plague (1466), the next decade was marked by two large-scale plagues (1471, 1478–1479), a severe epidemic of dysentery in the hot and dry summer of 1473,[21] and – unconnected to climate – the arrival of the 'French Pox' with the soldiers returning from the continent in 1475.[22] In the 1480s high variability in winter conditions was added to that of the summer seasons in the Low Countries. Renewed food shortages in the early 1480s were followed by the appearance of the English Sweat (1485) and a plague of limited scale in Canterbury, London and Oxford (1487). The freezing year 1491 was accompanied by a high mortality in the London region[23] and at the end of the 1490s plague returned to England (1499–1500). The consequent raised levels of mortality are reflected in the annual death rates at the monastic establishments at Westminster, Canterbury and Durham; in accordance with the evidence of epidemic disease from chronicles, monastic mortality crises were more frequent and severe after c.1460.[24] Life expectancy was falling in the southern mon-

---

[17] Camenisch et al., Early Spörer Minimum.

[18] See note 21 in Chap. 10.

[19] Chronicon Rerum Gestarum in Monasterio Sancti Albani, 19. According to Creighton, Epidemics in Britain, 225 and Gottfried, Black Death, 157 the disease might well have been influenza, which was prevalent around that time in France, the Netherlands and Spain.

[20] For 1431 and 1435 see note 28 in Chap. 10.

[21] Gottfried, Epidemic disease, 44, Rawcliffe, Urban bodies, 369.

[22] The 'French Pox' are the contemporary name for syphilis. Gottfried, Epidemic disease, 43–45. He finds no substantially raised death rate in the studied testamentary evidence in East Anglia and consequently doubts the epidemic nature of the 'French Pox'. On the spread of syphilis through Europe, see Gottfried, Black Death, 158–159.

[23] Harvey, Living and dying, 122–125.

[24] Hatcher, Mortality, 30, Harvey, Living and dying, 122–127, Hatcher et al., Monastic mortality, 676–678.

asteries particularly from the late 1420s onwards, when variability increased with regard to temperature and precipitation. Life expectancy stabilized on a lower level around 1450, and entered a veritable depression at Westminster Abbey between about 1460 and 1485[25], coinciding with the highly changeable weather conditions after c.1460.

Periods of increased short-term climate variability hence correlate with periods of frequent demographic shocks due to epidemic disease and subsistence crisis. The Danse macabre, the high death toll, went hand in hand with social change and this social change contributed largely to the end of the medieval world.

[25] Hatcher et al., Monastic mortality, 674–678.

Appendices

K. Pribyl, *Farming, Famine and Plague*, DOI 10.1007/978-3-319-55953-7

Appendix 1: Evidence on Weather in East Anglian Manorial Accounts 1256–1431

Appendix 1.1 Direct weather references and weather implied by the state of agriculture in Norfolk 1256–1431 (mainly Norwich Cathedral Priory accounts used)

Year	Period	Given period	Weather	Context	Place	Catalogue reference
1255–1256			Strong gale	(Re?-)roofing of the pea-barn and other houses after great wind	Sedgeford	NRO, DCN 60/33/01
1261–1262			Gale	Twigs brought down by wind	Hindolveston	NRO, DCN 60/18/02
1266–1267			Gale	Two ash-trees felled by wind	Hindolveston	NRO, DCN 60/18/04
1287–1288			Gale	Maple-trees and two ash-trees felled by wind	Hindolveston	NRO, DCN 60/18/09
1289–1290	Late spring, early summer		(Dry)	Peas were lacking	Framingham Parva	TNA, SC 9/925/33
1294–1295			Gale	An ash-tree and two maple-trees felled by wind	Hindolveston	NRO, DCN 60/18/11
1295–1296			Gale	Deadwood and twigs felled by wind	Eaton	NRO, DCN 60/08/09
			Gale	Deadwood, twigs, trees and an oak-tree felled by wind	Hindolveston	NRO, DCN 60/18/12
1299–1300	Late spring, early summer		(Dry)	Peas were lacking	Framingham Parva	TNA, SC 9/925/36
1302–1303	Spring and early summer		(Dry)	Peas and oats were lacking	Framingham Parva	TNA, SC 9/925/37
1304–1305	The long winter probably ended in March or April.		Long winter	Changes in the pea sowing rate due to long winter	Eaton	NRO, DCN 60/08/11A
			(Gale)	Twigs and a tree felled	Hindolveston	NRO, DCN 60/18/14A

1309–1310	Winter		Ice	A keeper had to watch the sluice because of ice. Sluice also repaired	Eaton	NRO, DCN 60/08/14
			(Gale)	A tree taken down that was felled	Eaton	NRO, DCN 60/08/14
1310–1311			?	Raised costs for threshing peas due to their severe weakness	Eaton	NRO, DCN 60/08/15 for pea harvest 1311
1311–1312	Late spring, early summer? (herbage growth)		Flooding	Herbage of Westfen meadow had to be saved from flooding.	Eaton	NRO, DCN 60/08/15
1312–1313	Tree could have desiccate in spring/ summer 1312 or 1313 or before. Probably 1312.		(Dry)	An ash-tree desiccated (sold at some time between autumn 1312 and autumn 1313.)	Hindolveston	NRO, DCN 60/18/18 probably for 1312
	April–June		Dry	Dry weather interfered with dairy production.	Sedgeford	NRO, DCN 60/33/18
1313–1314	April–late August		Flooding?	Oats flooded?	Hindolveston	NRO, DCN 60/18/19
		25 August–25 September	Rain and wind/ gale(s)	Problems encountered at harvest time due to rainy and stormy weather	Hindolveston	NRO, DCN 60/18/19
		11 August–8 September	Much rain	Problems encountered in harvest time due to much rainfall	Monks' Grange	NRO, DCN 60/26/15
			Gale	A tree felled by wind	Denham	NRO, DCN 60/07/06A
			Gale	Two trees felled by wind	Hindolveston	NRO, DCN 60/18/19

(continued)

Appendix 1.1 (continued)

Year	Period	Given period	Weather	Context	Place	Catalogue reference
1317–1318	Mid-spring – August		(Dry)	Lack of pasture during fallow time	Denham	NRO, DCN 60/07/07
	Time of spring or summer ploughing		(Dry)	Problems encountered by the ploughs due to the hard soil	Monks' Grange	NRO, DCN 60/26/16
1318–1319	Year when trees desiccated unclear, but probably spring/summer 1318.		(Dry)	Many maple trees and an ash-tree desiccated	Hindolveston	NRO, DCN 60/18/21 probably for 1818.
1320–1321			?	Larger scale sale of fallen wood or felled (by the forester) wood	Hindolveston	NRO, DCN 60/18/22
1322–1323	Year when trees desiccated unclear, probably 1322 or before.		(Dry)	Thirteen ash-trees desiccated	Hindolveston	NRO, DCN 60/18/23 probably for 1322 or before
			Gale	One tree felled by wind	Hindolveston	NRO, DCN 60/18/23
1324–1325	Spring and early summer		(Dry)	Peas and oats were lacking	Hindringham	NRO, DCN 60/20/22
	Late spring, early summer		(Dry)	Peas and beans were lacking	Framlingham, Suffolk	BL Add. Ch.16552 (Ridgard, Medieval Fram-lingham, 51–85.)
			?	Rye was lacking	Walsoken Popenhoe	TNA, SC 6/942/15
	April - August		Drought	Iron for plough due to great drought	Bircham	TNA, SC 6/930/5

	April-August	Summer	Drought	Digging for water for animals grazing on a pasture in summer	Framlingham, Suffolk	BL Add. Ch.16552 (Ridgard, Medieval Fram-lingham, 51–85.)
	April - August	Summer and harvest time	Drought	Iron for plough due to great drought in summer and harvest time	Framlingham, Suffolk	BL Add. Ch.16552 (Ridgard, Medieval Fram-lingham, 51–85.)
	August	Harvest time	Drought Low water levels	Drought and lack of water interfere with pasture leasing at harvest time	Framlingham, Suffolk	BL Add. Ch.16552 (Ridgard, Medieval Fram-lingham, 51–85.)
			Strong gale	Major repairs to the mill which was broken in a storm	Framlingham, Suffolk	BL Add. Ch.16552 (Ridgard, Medieval Fram-lingham, 51–85.)
1325–1326	Year when trees desiccated unclear, probably 1325 or 1326.		(Dry)	Massive sale of desiccated ash-trees – disallowed	Hindolveston	NRO, DCN 60/18/25 probably for 1325.
1326–1327			Gale	Three trees felled by wind	Taverham	NRO, DCN 62/01
1327–1328			Gale	A tree felled by wind	Hindolveston	NRO, DCN 60/20/28
1329–1330		15 August–c.10 October	Rain	Raised number of harvest works (including unbinding) due to very rainy weather	Kempstone	NRO, WIS 06
			Gale	An ash-tree felled by wind	Norwich Cathedral Priory gardener	NRO, DCN 1/11/1A
1333–1334			(Low water levels?)	Problems under 'malt' due to the sterility of the water	Gnatingdon	NRO, DCN 62/02
			Low water levels	Problems under 'malt' due to the lack of the water	Sedgeford	NRO, DCN 62/02

(continued)

Appendix 1.1 (continued)

Year	Period	Given period	Weather	Context	Place	Catalogue reference
1338–1339	Spring and early summer		(Dry)	Beans, peas and oats were lacking	Walsoken Popenhoe	TNA, SC 6/942/15
	April–August	Summer	Drought	Drought interferes with sale of hay and herbage in summer	Walsoken Popenhoe	TNA, SC 6/942/15
	April–August		?	Barley: weak seed corn	Sedgeford	NRO, LEST/IB 17 for state of the harvest 1339
1343–1344	April–August		?	Bread grain very weak	Sedgeford	NRO, LEST/IB 20 for harvest 1344.
1348–1349	Spring and early summer		(Lack of labour)	Oats perished in the field. (Plague year, crops are know to have remained unharvested due to lack of labour.)	Taverham	NRO, DCN 60/35/30 for harvest 1349
1352–1353	Spring and early summer		(Dry)	Oats were lacking	Shotesham	NRO, DN/EST 11/05
1354–1355		12 August–20 September	(Rain?)	Mixed grain (wheat and rye) perished at harvest time	Plumstead	NRO, DCN 60/29/29
	April–August		?	Mixed grain was weak	Plumstead	NRO, DCN 60/29/29A for mixed grain harvest 1355
1355–1356	Late spring, early summer		(Dry)	Peas were lacking	Sedgeford	NRO, LEST/IB 23
1356–1357	April–August		Dry	Only one fallow ploughing due to dry weather	Sedgeford	NRO, LEST/IB 24

1358–1359			?	Rye was lacking	Eaton	BLO, MS Rolls Norfolk 29
	April–August		(Dry?)	Very low number of fallow ploughing	Eaton	BLO, MS Rolls Norfolk 29
		3 August–21 September	Rain	Rainy weather during harvest time.	Hindolveston	NRO, DCN 60/18/36
1360–1361	August (plague year – disease also slowed down harvest process).	15 July–2 September	(Rain?)	Raised number of harvest works	Gnatingdon	NRO, LEST/IC 12
1361–1362	Probably the great storm c.15 January 1362.		Storm	Some timber trees blown down by great wind	Eaton	BLO, MS Rolls Norfolk 30
	April–August		(Dry?)	Very low number of fallow ploughing	Eaton	BLO, MS Rolls Norfolk 30
1362–1363	Late spring and early summer		(Dry?)	Substitute fodder given to animals due to lack of hay	Sedgeford	NRO, LEST/IB 26 for hay harvest 1363
	Spring and early summer		(Dry)	'Horsemeat', oats and barley were lacking	Gnatingdon	NRO, LEST/IC 13
	Spring and early summer		(Dry)	Barley was lacking.	Scratby	NRO, DCN 60/30/12
	Spring and early summer		(Dry)	Black peas, vetches and oats were lacking	Flegg	NRO, DN/EST 09/07
	Late spring, early summer		(Dry?)	Peas not bound in harvest	Gnatingdon	NRO, LEST/IC 13
1363–1364	Late spring, early summer		(Dry)	Peas were lacking	Great Cressingham	NRO, MC 212/04
	Late spring, early summer		(Dry)	Peas were lacking	Martham	NRO, NRS 5894, 20 D1
	Late spring to summer		?	Peas were weak	Hindringham	NRO, DCN 60/20/30

(continued)

Appendix 1.1 (continued)

Year	Period	Given period	Weather	Context	Place	Catalogue reference
1365–1366	Late spring, early summer		(Dry)	Substitute fodder given to animals instead of hay	Gnatingdon	NRO, LEST/IC 16 for hay harvest 1366
	Late spring, early summer		(Dry)	Substitute fodder given to animals instead of hay	Sedgeford	NRO, LEST/IB 28 for hay harvest 1366
	Spring and early summer		(Dry)	Mixed grain was lacking	Sedgeford	NRO, LEST/IB 27
	Spring and early summer		(Dry)	Peas and oats were lacking	Taverham	NRO, DCN 60/35/35
		10 August–post-September?	(Rain?)	Probably the harvest was unfinished by Michaelmas. Long harvest	Sedgeford	NRO, LEST/IB 27
		6 August–24 September	(Rain?)	An additional farm worker ordered to help in the harvest after it was already going on for two weeks. Long harvest	Great Cressingham	NRO, MC 212/05
1366–1367	Late spring, early summer		(Dry)	Black peas and vetches were lacking	Newton	NRO, DCN 60/28/06
	Spring and early summer		(Dry)	Oats were lacking	Taverham	NRO, DCN 60/35/36
	Spring		(Dry?)	Rye perished green	Gnatingdon	NRO, LEST/IC 17 for rye harvest 1367
	Spring		(Dry?)	Rye perished green	Taverham	NRO, DCN 60/35/37 for growing season 1367
			?	Rye was weak	Taverham	NRO, DCN 60/35/37 for rye harvest 1367

1367–1368	Late spring, early summer		(Dry?)	Peas perished green	Sedgeford	NRO, LEST/IC 30 for pea harvest 1368
	Late spring, early summer		(Dry)	Peas and beans were lacking.	Sedgeford	NRO, LEST/IB 29
	Late spring, early summer		(Dry?)	Substitute fodder given to animals instead of hay	Gnatingdon	NRO, LEST/IC 18 for hay harvest 1368
	April–August		(Dry?)	Low number of fallow ploughing	Taverham	NRO, DCN 60/35/37
1368–1369	Late spring, early summer		(Dry?)	Substitute fodder given to animals due to a lack of hay	Sedgeford	NRO, LEST IB 31 for hay harvest 1369
	Late spring, early summer		(Dry)	Peas were lacking	Sedgeford	NRO, LEST/IB 30
	April–August?		(Rain?)	Wheat was lacking	Sedgeford	NRO, LEST/IB 30
1369–1370	April–August		(Dry?)	Low number of fallow ploughing	Eaton	BLO, MS Rolls Norfolk 32
1373–1374	April–August?		(Rain?)	Wheat was lacking	Great Cressingham	NRO, MC 212/06
1375–1376	Spring and early summer		(Dry?)	Oats were lacking	Great Cressingham	NRO, MC 212/07
			?	Rye was lacking	Great Cressingham	NRO, MC 212/07
1376–1377	April–August		(Dry?)	Low number of fallow ploughing	Great Cressingham	NRO, MC 212/08
1377–1378	Spring and early summer		(Dry)	Peas and oats were lacking	Sedgeford	NRO, LEST/IB 35
	Spring and early summer		(Dry)	Oats were lacking	Ashby	NRO, DN/EST 09/01

(continued)

Appendix 1.1 (continued)

Year	Period	Given period	Weather	Context	Place	Catalogue reference
1380–1381		11 August–c.22 September	Strong wind	Ten acres of peas destroyed by strong wind at harvest time	Sedgeford	NRO, LEST/IB 38 for pea harvest 1381
1381–1382		20 August–c.1 October	Much rain	Raised number of harvest works (including binding and unbinding the sheaves) due to the very rainy weather	Gnatingdon	NRO, LEST/IC 20
		17 August–c.28 September	Rain	Raised number of harvest works (including binding and unbinding the sheaves) due to the rainy weather	Sedgeford	NRO, LEST/IB 38
1383–1384	Late spring, early summer		(Dry)	Most of the black peas perished green	Gnatingdon	NRO, LEST/IC 21 for pea harvest 1384
	Late spring, early summer		(Dry)	Most of the peas perished green	Sedgeford	NRO, LEST/IB 41
	Late spring, early summer		Dry	Peas perished green due to dry weather	Eaton	BLO, MS Rolls Norfolk 35 for pea harvest 1384
1384–1385	Late spring, early summer		(Dry)	Peas were lacking	Sedgeford	NRO, LEST/IB 41
	Late spring, early summer	Summer (growing season)	Dry	Peas were lacking due to dry weather in summer	Eaton	BLO, MS Rolls Norfolk 35
	Spring and early summer		(Dry)	'Horsemeat' was lacking	Gnatingdon	NRO, LEST/IC 21
	Spring and early summer		(Dry)	Oats and barley were lacking	Sedgeford	NRO, LEST/IB 41
	Spring and early summer		(Dry)	Barley was lacking	Eaton	BLO, MS Rolls Norfolk 35

			?	Rye was lacking	Sedgeford	NRO, LEST/IB 41
			?	Rye was lacking	Eaton	BLO, MS Rolls Norfolk 35
1385–1386	Late spring, early summer	Summer (growing season)	Dry	Peas were lacking due to the dry weather in summer	Sedgeford	NRO, LEST/IB 42 for pea harvest 1386.
1386–1387	Late spring, early summer		(Dry)	Vetches were lacking	Gnatingdon	NRO, LEST/IC 22
	Spring and early summer	Summer (growing season)	Dry	Problems with oats growing due to the dry weather in summer	Sedgeford	NRO, LEST/IB 42
1388–1389	Spring and early summer		Dry	Problems with oats growing due to the dry weather	Sedgeford	NRO, LEST/IB 43
	April-August?		(Rain?)	Wheat was lacking	Sedgeford	NRO, LEST/IB 43
1389–1390	Spring and early summer		(Dry)	Oats were lacking	Sedgeford	NRO, LEST/IB 44
	Spring to mid-summer		(Dry?)	Lack of straw	Gnatingdon	NRO, LEST/IC 24 for harvest 1390
1390–1391	Late spring, early summer	Summer (growing season)	(Dry)	Peas perished in summer	Martham	NRO, NRS 5900, 20 D1
1393–1394	Spring and early summer		(Dry)	Barley was lacking	Gnatingdon	NRO, LEST/IC 25
	April–August		?	Six acres of barley and six acres of oats can not be harvested, because grain too weak	Gnatingdon	NRO, LEST/IC 25
		3 August–c.8 September	Rain	Raised number of harvest works due to the rainy weather	Gnatingdon	NRO, LEST/IC 25

(continued)

Appendix 1.1 (continued)

Year	Period	Given period	Weather	Context	Place	Catalogue reference
1394–1395		16 August–c.13 September	Rain	Raised number of harvest works due to the rainy weather	Sedgeford	NRO, LEST/IB 48
			Pests	Oats damaged by rabbits and other animals	Plumstead	NRO, DCN 60/29/38 for oat harvest 1395
1395–1396	Late spring, early summer	Summer (growing season)	Dry	White peas perished in summer due to dry weather	Hindolveston	NRO, DCN 60/18/51 for pea harvest 1396
	Late spring, early summer		(Dry)	White peas were lacking	Plumstead	NRO, DCN 60/29/38
	Spring and early summer		(Dry)	Barley was lacking	Plumstead	NRO, DCN 60/29/38
			?	Rye was lacking	Plumstead	NRO, DCN 60/29/38
	April–August?		(Rain?)	Wheat was lacking	Plumstead	NRO, DCN 60/29/38
1396–1397	Late spring, early summer	Summer (growing season)	Drought	Vetches were lacking, five acres desiccated by drought in summer and could not be harvested	Gnatingdon	NRO, LEST/IC 27
	Late spring, early summer	Summer (growing season)	Dry	Peas were lacking due to dry weather in summer	Gnatingdon	NRO, LEST/IC 28 for pea harvest 1397
	Late spring, early summer	Summer (growing season)	Dry	Vetches perished due to dry weather in summer	Hindolveston	NRO, DCN 60/18/52 for vetches harvest 1397
	Spring and early summer		(Dry)	Black peas and oats were lacking	Hindolveston	NRO, DCN 60/18/51
	April-August?		(Rain?)	Wheat was lacking	Gnatingdon	NRO, LEST/IC 27

	end August–c.19 September	8 August–c.19 September	Rain	Raised number of harvest works (unbinding of sheaves, drying of peas) due to the rainy weather. Problems at the end of the harvest period	Hindolveston	NRO, DCN 60/18/51
	end of August–c.2 September	6 August–c.2 September	(Rain)	Raised number of harvest works. Problems at the end of the harvest period	Martham	NRO, NRS 5903, 20 D1
1397–1398	Late spring, early summer	Summer (growing season)	(Dry)	Vetches perished in summer	Hindolveston	NRO, DCN 60/18/52A for vetches harvest 1398
		18 August–c.15 September	Rain	Raised number of harvest works (unbinding of sheaves, drying of peas) due to the rainy weather	Hindolveston	NRO, DCN 60/18/52
		11 August–c.7 September	(Rain)	Raised number of harvest works	Martham	NRO, NRS 5904, 20 D2
1398–1399	April-August?		(Rain?)	Wheat was lacking	Plumstead	NRO, DCN 60/29/39
		24 August–c.28 September	Rain	Raised number of harvest works (unbinding of sheaves, drying of peas) due to the rainy weather	Hindolveston	NRO, DCN 60/18/52A
		10 August–c.14 September	Rain	Raised number of harvest works due to the rainy weather	Plumstead	NRO, DCN 60/29/39
		17 August–mid-September	Rain	Raised number of harvest works due to rainy weather	Sedgeford	NRO, LEST/IB 49
		10 August–7 September	Rain	Raised number of harvest works due to rainy weather	Costessey	NRO, NCR Case 24 Costessy 1377–1399

(continued)

Appendix 1.1 (continued)

Year	Period	Given period	Weather	Context	Place	Catalogue reference
1399–1400		28 July–c.1 September	Rain	Raised number of harvest works (including unbinding) due to rainy weather	Gnatingdon	NRO, LEST/IC 29
1400–1401		15 August–mid-September	Rain	Raised number of harvest works (including unbinding) due to rainy weather	Gnatingdon	NRO, LEST/IC 30
		15 August–mid-September	Rain	Raised number of harvest works (including unbinding) due to rainy weather	Sedgeford	NRO, LEST/IB 50
1401–1402	Late spring, early summer		Dry	Peas and vetches were lacking due to dry weather	Sedgeford	NRO, LEST-IB 52 for pea and vetches harvest 1402
		15 August–mid-September	Rain	Raised number of harvest works (including unbinding) due to rainy weather	Sedgeford	NRO, LEST/IB 51
			Heavy rain	Abundance of water overturns rye	Plumstead	NRO, DCN 60/29/40 for rye harvest 1402 or later
1402–1403	Late spring, early summer		(Dry)	Black peas were lacking	Plumstead	NRO, DCN 60/29/40
			?	Rye was lacking	Plumstead	NRO, DCN 60/29/40

1404–1405	Late spring, early summer		(Dry)	Peas were lacking	Gnatingdon	NRO, LEST/IC 31
	April–August?		Rain	Rainy weather interferes with pasture leasing	Hindolveston	NRO, DCN 60/18/55
		10 August–mid-September	Rain	Raised number of harvest works (including binding and unbinding the sheaves) due to the rainy weather	Gnatingdon	NRO, LEST/IC 31
		c.17 August–c.21 September	Rain	Raised number of harvest works (including binding and unbinding the sheaves) due to the rainy weather	Hindolveston	NRO, DCN 60/18/55
	Late August–early September		(Rain)	Raised number of harvest works (turning of peas)	Eaton	BLO, MS Rolls Norfolk 41
1405–1406	Winter half year?		Flooding	Pasture flooded	Hindolveston	NRO, DCN 60/18/56
		15 August–mid-September	Rain	Raised number of harvest works (including binding and unbinding the sheaves) due to the rainy weather	Gnatingdon	NRO, LEST/IC 32
		25 August–c.29 September	(Rain)	Raised number of harvest works (including binding and unbinding the sheaves)	Hindolveston	NRO, DCN 60/18/56

(continued)

Appendix 1.1 (continued)

Year	Period	Given period	Weather	Context	Place	Catalogue reference
1406–1407	Winter half year?		Flooding	Two pastures flooded	Flegg	NRO, DN EST 09/12
	Winter half year?		Flooding	Pastures flooded	Martham	NRO, NRS 5907, 20 D2
	Late spring, early summer		Dry	Peas were lacking	Flegg	NRO, DN EST 09/12
	Late spring, early summer	Summer (growing season)	Drought	Black peas and vetches perished in summer due to the great drought	Sedgeford	NRO, LEST/IB 54
	Late spring, early summer	Summer (growing season)	Drought	Black peas and vetches perished in summer due to the great drought	Sedgeford	NRO, LEST-IB 55 for pea and vetches harvest 1407
	Spring and early summer	Summer (growing season)	Dry	Sixteen acres of oats perished in summer due to the dry weather	Sedgeford	NRO, LEST-IB 55 for oats harvest 1407
	April-August		(Dry?)	Low number of fallow ploughing	Sedgeford	NRO, LEST/IB 54
	Probably drought 1407		Dry (1407)	Nineteen dry ash-trees sold. (Trees probably dried up in drought 1407 and sold as fuel in extremely cold and long winter 1407–1408.)	South Walsham	NRO, NCR Case 24 South Walsham 1399–1407, account for 1407–1408

1407–1408	Winter half year		(Hard winter)	Nineteen dry ash-trees sold. (Trees probably dried up in drought 1407 and sold as fuel in extremely cold and long winter 1407–1408.)	South Walsham	NRO, NCR Case 24 South Walsham 1399–1407, account for 1407–1408
		[5 August–mid-September]	Storm	Oats partly perished at harvest time in a storm	Gnatingdon	NRO, LEST/IC 34 for oat harvest 1408
		5 August–mid-September	Storm	Oats partly perished at harvest time in a storm	Sedgeford	NRO, LEST/IB 56 for oat harvest 1408
1408–1409	Winter half year?		Flooding	Three pastures flooded	Flegg	NRO, DN EST 09/13
	Late spring, early summer	Summer (growing season)	Lack of rain	Twenty acres of vetches perished in summer due to bad luck and lack of rain	Newton	NRO, DCN 60/28/08 for vetches harvest 1409
	Late spring, early summer	Summer (growing season)	Dry	Black peas and vetches perished in summer due to the dry weather	Sedgeford	NRO, LEST-IB 57 for pea and vetches harvest 1409
1409–1410	Late spring, early summer		(Dry)	White peas were lacking.	Newton	NRO, DCN 60/28/08
	Spring and early summer		(Dry)	Barley was lacking	Sedgeford	NRO, LEST/IB 57
			?	Rye was weak	Sedgeford	NRO, LEST/IB 57
1410–1411	Late spring and early summer		(Dry?)	Substitute fodder given to animals due to the lack of hay	Gnatingdon	NRO, LEST/IC 35 for hay harvest 1411
	July–August?		Wet	Problems with barley due to lack of dry weather the preceding summer?	Gnatingdon	NRO, LEST/IC 35 for barley harvest 1411

(continued)

Appendix 1.1 (continued)

Year	Period	Given period	Weather	Context	Place	Catalogue reference
1411–1412	Late spring, early summer		(Dry)	Peas were lacking	Gnatingdon	NRO, LEST/IC 35
	Late spring and early summer		(Dry?)	Substitute fodder given to animals due to the lack of hay	Gnatingdon	NRO, LEST/IC 36 for hay harvest 1412
	Spring and early summer		(Dry)	Oats were lacking	Hindolveston	NRO, DCN 60/18/59
			?	Rye was weak	Gnatingdon	NRO, LEST/IC 35
1412–1413	Winter half year?		Flooding	A pasture partly flooded	Martham	NRO, NRS 5909, 20 D2
	Winter half year?		Flooding	Fen flooded	Plumstead	NRO, DCN 60/29/42A
	Late spring, early summer		(Dry)	Peas were lacking	Martham	NRO, NRS 5909, 20 D2
1414–1415	Late spring, early summer		(Dry)	Vetches and peas perished in field	Great Cressingham	NRO, MC 212/12 for pea and vetches harvests 1415
	Late spring, early summer		(Dry?)	Substitute fodder given to animals instead of hay	Gnatingdon	NRO, LEST/IC 37 for hay harvest 1415
	Late spring, early summer		(Dry)	Peas perished in field	Martham	NRO, NRS 5911, 20 D2
	Late spring, early summer		(Dry)	Grey peas perished in field	Taverham	NRO, DCN 60/35/44
	Late spring, early summer		(Dry)	Peas perished in field	Taverham	NRO, DCN 60/35/45 for pea harvest 1415
	April–August		(Dry?)	Low number of fallow ploughing	Martham	NRO, NRS 5911, 20 D2
	April–August		Pests	Ten acres of wheat destroyed by rabbits	Gnatingdon	NRO, LEST/IC 37 for wheat harvest 1415

1415–1416	Winter half year?		Flooding	Fen flooded	Plumstead	NRO, DCN 60/29/43
	Late spring, early summer		(Dry)	Vetches perished in field	Great Cressingham	NRO, MC 212/12
	Late spring, early summer		(Dry?)	Substitute fodder given to animals instead of hay	Gnatingdon	NRO, LEST/IC 38 for hay harvest 1416
	Late spring, early summer		Dry	Peas perished in field due to dry weather	Plumstead	NRO, DCN 60/29/44 for pea harvest 1416
	Late spring, early summer		(Dry)	Peas perished in field	Taverham	NRO, DCN 60/35/46 for pea harvest 1416
1416–1417	Winter half year?		Flooding	Moor flooded	Plumstead	NRO, DCN 60/29/44
	Mid-spring		(Dry)	No vetches grown. (Given up after failure of vetches over the last years.)	Great Cressingham	NRO, MC 212/13
	Late spring, early summer		(Dry)	Peas perished in field	Gnatingdon	NRO, LEST/IC 38
		end of August–c.8 September	Rain	Raised number of harvest works due to rainy weather at the end of harvest	Gnatingdon	NRO, LEST/IC 38
1417–1418	Mid- to late spring		Mildew	Wheat destroyed in the field by mildew. (On the Winchester manors winter and spring corn were affected by mildew before drought struck.)	Martham	NRO, NRS 5913, 20 D3 for wheat harvested in 1418
	Late spring, early summer		(Dry)	Peas were weak and perished in field	Taverham	NRO, DCN 60/35/47
	Late spring, early summer	Summer (growing season)	(Dry)	All peas (eight acres) perished in field in summer time	Taverham	NRO, DCN 60/35/48 for pea harvest 1418
	Late spring, early summer		(Dry)	White peas perished in field	Plumstead	NRO, DCN 60/29/45 for pea harvest 1418

(continued)

Appendix 1.1 (continued)

Year	Period	Given period	Weather	Context	Place	Catalogue reference
	Late spring, early summer		(Dry?)	Substitute fodder given to animals instead of hay	Gnatingdon	NRO, LEST/IC 39 for hay harvest 1418
		27 July–c.31 August	Rain	Raised number of harvest works due to an abundance/a torrent of water at harvest time	Taverham	NRO, DCN 60/35/47
1418–1419	Mid-spring		(Dry)	No peas grown. (Given up after failure of peas over last years.)	Taverham	NRO, DCN 60/35/48
	Late spring, early summer	Summer (growing season)	(Dry)	Peas perished in the field in summer	Gnatingdon	NRO, LEST/IC 39
	Spring and early summer	Summer (growing season)	(Dry)	Peas and oats perished in the field in summer	Martham	NRO, NRS 5913, 20 D3
	Late spring, early summer		(Dry)	Black peas perished in field	Plumstead	NRO, DCN 60/29/45
	Late spring, early summer		(Dry)	Part of the peas perished in the fields	Plumstead	NRO, DCN 60/29/46 for pea harvest 1419
1419–1420	Mid-spring		(Dry)	No peas grown. (Given up after failure of peas over previous years.)	Taverham	NRO, DCN 60/35/49
	Late spring, early summer		(Dry)	Peas perished in the field	Martham	NRO, NRS 5914, 20 D2
	Late spring and early summer		(Dry?)	Substitute fodder given to animals instead of hay	Gnatingdon	NRO, LEST/IC 40 for hay harvest 1420

1420–1421		18 August–c.6 October	(Rain)	Greatly raised number of harvest works (turning, opening and drying sheaves)	Gnatingdon	NRO, LEST/IC 40
		11 August–c.15 September	Rain	Greatly raised number of harvest works (turning, opening and binding sheaves) due to watery weather	Plumstead	NRO, DCN 60/29/46A
		18 August–c.4 October	Flooding (due to rain and/or over-flowing of waterbodies)	Greatly raised number of harvest works (turning, opening and drying and rebinding the sheaves) due to overflowing water	Sedgeford	NRO, LEST/IB 63
		6 August–mid-September	(Rain)	Raised number of harvest works (turning, opening, drying and rebinding sheaves)	Taverham	NRO, DCN 60/35/50
	Winter or harvest time?		Flooding	Two pastures flooded	Flegg	NRO, DN EST 09/15
	Winter or harvest time?		Flooding?	Flooding?	Plumstead	NRO, DCN 60/29/46A
1421–1422	Winter half year?		Heavy? flooding	Two pastures flooded (*superinundatio*).	Flegg	NRO, DN EST 09/16
	Late spring, early summer		(Dry)	Peas perished in field	Hindringham	NRO, DCN 60/20/38 for pea harvest 1422
	Late spring, early summer	Summer (growing season)	(Dry)	Peas perished in field in summer.	Taverham	NRO, DCN 60/35/51 for pea harvest 1422
	Spring and early summer		(Dry)	Oats perished in the field	Martham	NRO, NRS 5915, 20 D2

(continued)

Appendix 1.1 (continued)

Year	Period	Given period	Weather	Context	Place	Catalogue reference
1422–1423	Winter half year?		Pests	Mixed grain (wheat and rye) destroyed in grange by mice and rats	Hindringham	NRO, DCN 60/20/38
	April–August		?	Peas were weak	Hindringham	NRO, DCN 60/20/38
			?	Rye was weak	Hindringham	NRO, DCN 60/20/38
		16 August–c.19 September	Rain	Raised number of harvest works (turning and turning again) due to rain	Martham	NRO, NRS 5916, 20 D3
		16 August–c.26 September	Rain	Raised number of harvest works (turning and turning again) due to watery weather	Sedgeford	NRO, LEST/IB 65
		12 August–end of September	(Rain)	Raised number of harvest works (turning and turning again)	Taverham	NRO, DCN 60/35/51
1423–1424	Late spring, early summer		(Dry)	Peas perished in field	Taverham	NRO, DCN 60/35/52
1427–1428		Mid-September–mid-October	Rain	Raised number of harvest works due to rainy weather at the end of harvest and the short days	Sedgeford	NRO, LEST/IB 68
1428–1429	Late spring, early summer		(Dry?)	Substitute fodder given to animals instead of hay	Gnatingdon	NRO, LEST/IC 42 for hay harvest 1429
1429–1430	Late spring, early summer		(Dry)	Peas and oats perished in field	Gnatingdon	NRO, LEST/IC 42
1430–1431	April-August		(Dry?)	Low number of fallow ploughing	Sedgeford	NRO, LEST/IB 70

Year: Given is the agricultural year 29 September to 29 September. Consequently 'winter' and 'winter half' year are identical with the years given, whereas 'spring', 'summer', 'harvest', 'growing season' etc. always refers to the year, when the agricultural year is closing. The documents are listed mostly under the agricultural year they are referring to, not under the year they were made. Exempt from this rule are the manorial accounts giving information on desiccated trees, because it is difficult to determine, when those trees succumbed to drought.
Period: Often the documents supply no date in connection with the weather event or the impact of adverse weather. The time period has therefore been set according to the period of the year during which the weather event had an effect on agriculture. These periods are mostly the time frame within which the event would have taken place, rarely would it have covered the whole period.
Given period: When the accounts refer to a season or date in connection with the weather event, this information is listed under 'Given period'. However, the medieval definition of the seasons is not identical with modern definition, so 'summer' referred to the (grain) growing season and included parts of the modern spring (usually May; if April was warm, also this month). Summer ended with the start of the grain harvest, therefore August was already classified as 'autumn' by the medieval people. Where necessary a specification of the medieval period is given under 'Period'. Dates given under 'Given period' are Old Style (Julian Calendar).
Weather: Direct weather references are supplied without brackets. Weather, which is strongly implied by the information in the accounts, but not directly mentioned, is set in brackets.
Places: Gnatingdon, Sedgeford and Taverham are situated on sandy soil; Taverham partly also on gravel. Great Cressingham is on the edge of the Breckland, an area of sandy soil and low rainfall. These villages are especially vulnerable to drought. Martham, the Flegg district and Plumstead are either close to the sea and/or had pastures in low-lying areas and/or river valleys. These were liable to flooding (due to high precipitation levels as well as to occasional sea floods).

Appendix 2: Administration and Managers of the Manors of Norwich Cathedral Priory with a Focus on Sedgeford and Gnatingdon

The Structure of Administration in Sedgeford and Gnatingdon

The manors of Sedgeford and Gnatingdon were not only geographically close, but were also managed by the same *serviens*/sergeant until at least 1368, but more likely until the mid-1370s or possibly even the mid-1380s.[1] After 1384 different sergeants or bailiffs were responsible for these manors. As long as the manor of Thornham was directly managed by Norwich Cathedral Priory, it was supervised by the same sergeant as Sedgeford and Gnatingdon.

The structure of administration varied between 1256 and 1431 (Append 2.1). Until c.1285 the sergeant was supported only by a reeve on each manor. Also a (rent) collector and a beadle (who often functioned also as a *messor*/hayward)[2] were occasionally mentioned in the accounts. In 1286 a keeper of the grange appeared for the first time in Gnatingdon. After this time the name of collectors and beadles were also more frequently noted. After 1295 no reeve was given any more in the Gnatingdon accounts. This system was maintained until the Black Death, but in the early 1350s a reeve appeared again in Gnatingdon. Parallel to the accounting reform of Norwich Cathedral Priory 1354–1355 the administrative system for Sedgeford, and Gnatingdon was modified. For Sedgeford, in addition to the sergeant and the reeve the *messor*/hayward is now listed instead of the collector. In Gnatingdon, too, the collector is replaced by the hayward. In the early 1380s the reeves disappear on both manors. Finally in 1391 the office of the sergeant was abolished in favour of that of the bailiff; this adjustment was universal on all Norwich Cathedral Priory manors.[3] In Sedgeford the bailiff was supported by a hayward until 1408 and after 1430, in Gnatingdon he rendered account alone.[4] Therefore the complexity of the administrative structure of these two northwestern manors was greatly reduced in the mid-1380s, when the position of reeve was dropped at Sedgeford, and again after 1392,

[1] In this chapter the end year of the accounts is used.

[2] Bennett, English manor, 178–80. The hayward was in charge of operations connected with sowing and harvesting the crops, as well as hay making. At harvest time he would assemble the reapers, supervise the gathering of the corn and check the size of the sheaves. A beadle acted as a village policeman.

[3] This was most likely only a change of name, not of function or duty, Bennett, English manor, 162.

[4] It is likely that under newly elected priors the management and organisation of the manors was reviewed. In any case it is striking that the last time a reeve is responsible for farming at Sedgeford was 1382, the year when prior Alexander de Totyngton was elected. The first available account of Sedgeford that fell fully under his control is for 1385 and no reeve is mentioned in it. It is also under Totyngton's priorate that the hayward at Gnatingdon drops out of the accounting procedure. When his successor Robert de Burnham was elected in 1407, the accounts 1407–1408 still show the management structure established under Totyngton, a bailiff and a hayward for Sedgeford. In the following accounting year, when Robert de Burnham could change the administrative structure, he did do so; the hayward disappears. Just after his resignation, at the end of the 1430s, a hayward contributed to rendering account again for Sedgeford.

when the hayward stopped being included in the accounting procedure in Gnatingdon (Appendix 2.1). This downscaling of the number of manorial officers coincided with the rise of explanations for failures in farming in the *compoti*, such as weather references. It appears as if Norwich Cathedral Priory reacted to the reduced profitability of demesne farming in the face of a falling population, low grain prices and rising wages with a reduction of personnel and associated costs. The lower level of mutual control by the various officers was compensated by stricter accounting procedures.

The Recruiting Pool and Career Paths for the Various Offices

Analysing the social position and the place of origin of the various officials helps to highlight the responsibilities as well as the system of control that was in place on the manors.

The sergeant of the thirteenth and first half of the fourteenth century was generally a non-local and free man,[5] his name was usually combined with his place of origin. Under him worked men of local origin, probably well-off and reliable customary tenants of Norwich Cathedral Priory.[6] This group of the rural society populated the offices of reeve, hayward, collector and beadle and must have known the land and the villagers well. To switch from the office of the collector to that of the beadle and vice versa was normal, even holding various offices successively in both manors was not uncommon, for example Walter Hert made his appearance in Sedgeford as well as Gnatingdon.

For establishing a further check on the probably closely cooperating sergeant and reeve, Norwich Cathedral Priory introduced the keeper of the grange in 1286 in Gnatingdon. This keeper usually came from another manor of the cathedral priory, listed in the *compoti* are men from Hemsby, Hindolveston, Catton, Trowse [Newton] and Eaton. Probably these keepers were trustworthy customary tenants of Norwich Cathedral Priory who were temporarily sent to other cathedral manors, where they had no connections of their own, to check the dealings of the local officers and especially to verify the quantity of the grain going in and out of the grange. It should be noted that the monks went to considerable lengths to ensure this additional level of control. Whereas Hindolveston is comparatively close to Sedgeford and Gnatingdon, Catton, Trowse [Newton] and Eaton are neighbouring Norwich and consequently a day's journey from northwestern Norfolk. Hemsby is even further away, it lies on the eastern Norfolk coast.

It is possible that the position of the keeper of the grange was a kind of starting position for a career in direct demesne farming at Norwich Cathedral Priory (although not for customary tenants). In 1321 Peter de Acle was keeper of the grange at Gnatingdon. If he can be identified with Peter de Ocle, which is quite possible, he had an interesting career before him. In 1327 Peter de Ocle was keeper of the grange at Martham and in 1334 sergeant of the manor of Taverham. Having proven his ability he returned to Sedgeford and Gnatingdon as sergeant and can be traced there between 1340 and 1345.

[5] Bennett, English manor, 163.

[6] Ibid., 169.

Generally the monks aimed at keeping the sergeant in office for a long time. Henry de Henham was sergeant for Sedgeford and Gnatingdon from at least 1295 to 1306, possibly even a few years longer.[7] Trusted men probably moved from one manor to another during their career at Norwich Cathedral Priory. Before he came to Sedgeford and Gnatingdon Henry de Henham is found as sergeant in North Elmham in 1283. Also the reeves were chosen for more than one year, many can be traced for two successive years or more, they could also be re-elected several times.[8]

In the years immediately following the Black Death this continuity of personnel was broken. On the local level the social chaos following the demographic collapse becomes apparent in demesne administration. The sergeants changed quickly: in 1350 to 1354 four sergeants can be traced for Sedgeford and Gnatingdon. In 1353 they were even administered by different sergeants.[9] It was during this time that the office of the reeve re-appeared at Gnatingdon, most likely in an attempt by the cathedral priory to support the new and possibly less qualified sergeants and regain higher profitability (Appendix 2.1).

From the 1360s onwards the cathedral priory managed to re-establish the policy of keeping talented officers, especially the sergeants, in place for a long time. It appears likely, though, that the social stratum, from which the sergeants and bailiffs were chosen, had changed. Whereas for the early sergeants their place of origin (usually a place at great distance from northwestern Norfolk) was employed in their name, this is rarely the case for the sergeants after 1370. With names merely giving 'atte Chyrche', 'Reed', 'atte Water' and 'Fornfeld' as specification, the sergeants and bailiffs running Sedgeford and Gnatingdon during their last decades as directly managed demesnes appear to have been locals.[10] Some definitely were.

John Howlyn started as hayward in Sedgeford in 1370 before becoming sergeant, first in Gnatingdon in 1385 and then in Sedgeford in 1390. He was then appointed Sedgeford's first bailiff and remained in office until at least 1396.[11] He is an example of the new opportunities opening up for the lower levels of the society in a time haunted by plague: before him haywards had never become sergeants. His story also exemplifies another new characteristic of the high offices, they were now allocated amongst 'dynasties'. John Howlyn Junior, most likely John Howlyn's son, was bai-

[7] Since the manorial accounts are not continuous, it is difficult to establish the exact periods in office of the various officials.

[8] See Walter de le Brok [Brook], respectively Walter atte Brok. Assuming that the name refers to the same man, he held the position of the reeve at Sedgeford several times between 1292 and 1318. He also was collector in 1313.

[9] Sedgeford was managed by William de Hindolveston. He had been sergeant in Plumstead the previous year and was to return there in 1354.

[10] During the thirteenth and fourteenth century short-lived place-name surnames were replaced by hereditary surnames in England. Therefore it is difficult to determine, if the later reeves and bailiffs on the Norwich Cathedral Priory manors were outsiders or came from the village. Around 1300 short-lived place names were still common as surnames.

[11] It is difficult to trace his ancestors, but they could be the 'Huelyn' family appearing in Sedgeford as collectors in 1312 (Galfred Huelyn) and 1319 (Elias, son of Richard Huelyn). Perhaps also Walter Hulm/Hulem, reeve in 1273 and collector in 1294, is/are connected to the Howlyn family, although 'Hulm' would more likely be a topographical reference, meaning 'islet'.

liff at Gnatingdon 1401–1406. Until c.1370 it was only for the minor positions of reeve, hayward and collector that the cathedral had drawn its personnel again and again from the same (local) families. Now this also occurred for the office of bailiff. The Howlyn family profited from the social upheaval and the new opportunities in the post Black Death society, however, possibly it was also finally brought down by plague. John Howlyn [Junior] was replaced as bailiff by John Reed in February of the accounting year 1405–1406; an unusual arrangement, because the monks avoided change of personnel during the accounting year. In 1405–1407 a plague wave swept over England (Chap. 10). Nonetheless, the family must have stayed around Sedgeford after 1406. In 1433 Richard Houlyn (Howlyn) was one of the men to whom the manor was farmed out.

The bailiff of Gnatingdon 1391 to c.1398, Richard Bryd, also was a local. His ancestor(s) John Bryd/Brid held offices in Gnatingdon in 1313 and 1328.

The local origin of sergeants and bailiffs would also explain the disappearance of the reeves after 1382, the sergeants and bailiffs now had local knowledge themselves. An exception in this respect is constituted by the bailiffs Thomas Fornfeld and Samson and Edmund Itringham (Itteringham). The latter probably came from the village of the same name about 50 km to the east of Sedgeford and Gnatingdon. However, they too appear to belong to the same family, at least it is unlikely to come across an identical place of origin amongst the important managers of the manorial economy of Norwich Cathedral Priory in only about 20 years by pure coincidence.

Again trusted men could move from manor to manor during their career. Samson [de] Itringham was bailiff in Gnatingdon in 1400 and bailiff in Sedgeford 1403–1417 (probably continuously). In 1401 he managed Martham at the east coast for a short while. In 1418 he became the *firmarius*/farmer of Hindolveston and in 1420 its bailiff. Then his trace is lost.

Thomas Fornfeld who was bailiff in Sedgeford from at least 1421 until 1429, held the same position at Hindolveston from 1406 to at least 1409. In 1437 he farmed Hindringham together with his son, who continued in this position without his father some years later.

'Dynasty building' can also be observed at the manor of Hindolveston, here Richard Fuller took over as bailiff in 1421 and was bailiff or farmer of Hindolveston throughout the 1420s. In 1433 he farmed the manor together with his son Galfred, who continued as farmer alone after 1436.

That experienced bailiffs of long standing were willing to become the first farmers of the manors they had served before or on other cathedral manors underlines two facts. First, these men, who knew the agrarian economy intimately, believed in the possibility of farming these manors profitably themselves. Second, the cathedral priory, who could not make direct demesne farming work any more, was eager to engage familiar and proven men as farmers, to ensure a regular flow of rents. During the 1420s and 1430s, when death once more – in the form of plague waves, other epidemics and famines – demanded a high toll, and expertise amongst men began to run thin, the cathedral priory must have been relieved to see its manors in the hands of its former employees. Therefore, where the cathedral priory failed in maintaining direct control of its manors, men whose fathers had been merely (well off) customary tenants or free men moved up the social leader and became yeomen.

Appendix 2.1 Administrative personnel at the manors of Sedgeford and Gnatingdon 1256–1431

	Sedgeford						Gnatingdon						
	Bailiff	**Sergeant**	**Reeve**	**Hayward**	**Collector**	**Beadle**	**Bailiff**	**Sergeant**	**Reeve**	**Hayward**	**Collector**	**Keeper of the grange**	**Beadle**
1256									Nicholas Newman		Serle Ruefd		
1264								Thomas de Dereham	Serle Ruefd				
1267									Nicholas le Newman				
1273		William Pulham	Walter Hulm		Walter Hert			William de Pulham	Robert Prophete				
1274		William Pulham	Walter Bond		Walter Hert			W[illiam] de Pulham	Will[iam] del Pic				
1279		Galfredus Hengham	Martin Suvel			[?]							
1286		Andrew Yhelverton	Galfredus			Walter Hert		Andrew de Yhelverton			Walter Hert	John de Hemsby	Gilbert le Newman
1287													
1288		Andrew Yhelverton	Galfredus [?]			Walter Hert		Andrew de Yhelverton	Gilbert le Newman			John de Hemsby	Walter Hert
1292		Radulf [?]	Walter de le Brok		Colyn								
1293									W[illiam] in le Pyt			John de Hindolveston	
1294			Colyn?		Walter Hulem								
1295		Henry de Henham	Elias Bustard		Roger Hert			Henry de Henham	Gilbert Newman			John de Hindolveston	

1296		Henry de Henham	Elias Bustard		Thomas Owypel			Henry de Henham			Robert Prophete	John de Hindolveston	
1297													
1298												John de Hindolveston (no function)	Roger Hert
1299													
1300		Henry de Henham	Walter atte Brok		Martin Suvel							John de Hindolveston	Roger Hert
1305								Henry de Henham				John de Hindolveston	Robert Newman
1306		Henry de Henham	Walter atte Brok		Galfredus Hirham			Henry de Henham				John de Hindolveston	Robert Newman
1309								[lost]				John de Catton	Robert Newman
1310		Hubert Lakenham	Robert Futur		Roger Godefrey			Hubert de Lakenham				John de Catton	Roger Hert
1311													
1312		Jacob de Dycleburg	Walter atte Brok		Galfredus Huelyn			Jacob de Dycleburg			Roger Hert	John le Karter	
1313		Jacob de Dycleburg	Walter atte Brok		Thomas Owypel							John de Catton	John Bryd
1314	Andrew de Yhelverton		Robert Tulte		Walter atte Brok								

(continued)

Appendix 2.1 (continued)

	Sedgeford					**Gnatingdon**					
	Bailiff	**Sergeant**	**Reeve**	**Hayward**	**Collector**	**Bailiff**	**Sergeant**	**Reeve**	**Hayward**	**Collector**	**Keeper of the grange**
1318	[?] de Elmham		Walter atte Brok		Radulf Palmer		John de Elmham			Roger Hert	W [lost]
1319		John de Elmham	Richard Tulch		Elyias son of Richard Huelyn					Roger Hert	William de Trowse
1320		John de Elmham	Thomas Tabbard		Reyner Hert		John de Elmham			John son of Radulf de Brunham	William de Trowse
1321		John de Elmham	Thomas Tabbard		Reyner Hert					Thomas Edolf	Peter de Acle
1322											
1323										John Newman	John Potter
1324											
1325							John de Elmham			John de Brunham	John Potter
1326		John de Elmham	Thomas Tabbard		Henry Perkyn		John de Elmham			Roger Hert	John Potter
1327										Thomas Tabbard	John Potter
1328		Roger Castr	Roger Hert		William Godfrey		Roger [no place of origin]			John Brid	William de Eaton

1334		Simon de Babrigge	Richard Brid		Richard Clonelek		Simon de Babrigge			Thomas Edolf	Peter Cotwot
1340		Peter de Ocle	Henry Caly		John Brunham		Peter de Ocle			Roger son of Roger Hert	Peter Cotwot
1341		Peter de Ocle	Henry Caly		Stephen Hert						
1342											
1343							Peter de Ocle			Roger Hert Junior	Peter Cotwot
1344											
1345		Peter de Ocle	John Berry		Roger son of Roger Hert						
1350			Alain Futuor		Richard Profete		John de Langley			William in le Pit	
1353		William de Hindolveston	Roger Reyner		Alain Futuor		Richard de Wimdham	William Prophete		William Prophete	
1354		Radulf Wirmegeye	Roger Reyner		Robert Edolf		Radulf Wymegeyer	William Dany		John Newman	
1356		Radulf Wirmegeyer	Roger Reyner	Thomas de Swaffham							
1357		Radulf Wymegeye	Gilbert Dagheman	Thomas de Swaffham							
1361								William Martyn		Galfredus Cloneleek	

(continued)

Appendix 2.1 (continued)

	Sedgeford				Gnatingdon			
	Bailiff	**Sergeant**	**Reeve**	**Hayward**	**Bailiff**	**Sergeant**	**Reeve**	**Hayward**
1362		John Foulton	Richard Profete	Gilbert Choselе				
1363						John de Foulton	Walter Etolf	John Heyward
1364		John de Foulton	Gilbert Dagheman	Roger Reyner		John de Foulton	Walter Etolf	John Heyward
1365								
1366		John de Foulton	Gilbert Dagheman	William Martyn				
1367		John de Foulton	Richard Profete	Gilbert Dagheman		John de Foulton	John Heyward	Gilbert de Shosele
1368		John de Foulton	Gilbert Dagheman	John Hayward		John de Foulton	John Newman	William Martyn
1369		John de Foulton	Walter atte Brook	Gilbert Dagheman			John Heyward	John Profete
1370		John de Foulton	Gilbert Dagheman	John Howlyn				
1371								
1372		John de Foulton	John Saad	Robert Elyot				
1373							Godfrey Harwyne	John Heyward
1374		William atte Chyrche	Robert Elyot	John Howlyn				
1375		William atte Chyrche	Richard Profete	Walter atte Brook				

1378		William atte Chyrche	Richard [lost]	Gilbert Dagheman				
1380		William atte Chyrche	John Saad	Gilbert Dagheman				
1381		William atte Chyrche	Roger Crisp	Gilbert Dagheman				
1382		William atte Chyrche	Robert Elyot	Gilbert Dagheman	?			
1385		Richard Diggart		Robert Elyot		John Howlyn		[lost]
1386								
1387		Richard Diggart		Stephen de Acre		John Howlyn		John Howlyn
1388						John Hyanm?		Ehoe? Wynde
1389		Richard Diggart		John Howlyn				
1390		John Howlyn		Robert Elyot				
1391					Richard Bryd			John Akke
1392	John Howlyn			Stephen de Acre				
1393	John Howlyn							
1394	John [no family name]			Martin Cornel	Richard Bryd			
1395	John Howlyn			John atte Water				

(continued)

Appendix 2.1 (continued)

	Sedgeford				**Gnatingdon**			
	Bailiff	**Sergeant**	**Reeve**	**Hayward**	**Bailiff**	**Sergeant**	**Reeve**	**Hayward**
1396					Richard Bryd			
1397					[lost]			
1398					Richard Bryd			
1399	William Goldyng			John Howlyn				
1400					Samson de Iteringham			
1401	William Goldyng			John Sarle	John Howlyn Junior			
1402	William Goldyng			Thomas Hytche				
1403	Samson Itringham			Thomas Hytche				
1404								
1405					John Howlyn Junior			
1406					John Howlyn Junior, February: John Reed			
1407	Samson Itringham			Roger Martyn	Jacob Reed			
1408	Samson Itringham			Roger Martyn				
1409	Samson Itringham				Jacob Reed			
1410	Samson Itringham							

1411								
1412	Samson Itringham				Jacob Reed			
1413	Samson Itringham				Jacob Reed			
1414	Samson Itringham							
1415								
1416	Samson Itringham				Jacob Reed			
1417	Samson Itringham				Edmund Iteringham			
1418								
1419					Thomas atte Water			
1420								
1421	Thomas Fornfeld				Thomas atte Water			
1422	Thomas Fornfeld				Thomas atte Water			
1423	Thomas Fornfeld							

(continued)

Appendix 2.1 (continued)

	Sedgeford				Gnatingdon			
	Bailiff	**Sergeant**	**Reeve**	**Hayward**	**Bailiff**	**Sergeant**	**Reeve**	**Hayward**
1424	Thomas Fornfeld							
1425	Thomas Fornfeld							
1426	Thomas Fornfeld							
1429	Thomas Fornfeld							
1430	Thomas atte Martyr			John Acre?	Robert Crisp			
1431	Robert Crilp [sic]			John Acre				

The year given refers to the closing of the account

Prominent Local Families

The most successful prominent local families, the Howlyns and the Bryds have already been discussed above. About a dozen other families formed the pool from which Norwich Cathedral Priory drew its reeves, haywards, collectors and beadles for Sedgeford and Gnatingdon. Amongst them are the names Hert, del Pit/ in le Pit, Newman, atte Brook, Profete, Futur, atte Water and Crisp (Appendix 2.1). Some served on both manors, some merely on one, a few were also connected to Thornham.

Two of the mentioned families were considerably less lucky than the Howlyns and Bryds. The Hert family, for which Walter Hert, Roger Hert, Roger Hert Junior (Roger's son) and Galfred Hert (in Thornham responsible for the summer account in 1349 and the keeper in 1349–1350) can be traced in administrative positions in Sedgeford, Gnatingdon and Thornham between 1273 and 1350, never reappeared after 1350. The same applies to the del/in le Pits, who were active in Gnatingdon and Thornham between 1274 and 1350. Other names like atte Brook, Profete and Futur survived the demographic collapse and are to be found in the account rolls before and after 1348–1349. New names fill up the gaps left by the families who can not be traced any more after 1350, these included Martyn, Dagheman, Crisp and atte Water.

Appendix 3: Labour and Harvest on the Manor of Gnatingdon

The manor of Gnatingdon has been chosen to illustrate the supply of labour available for the harvesting process. Before the accounting reform in 1354–1355 the number of the core labour force taking their meals at the lord's table is given in the *autumpnus* account. It includes the manorial staff and with it the *famuli* (of which not all were employed directly in the harvest) and the reapers or cutters. Usually the number of the reapers comprized three quarters of that of the manorial staff. The work of the core group was assisted by hired people and labourers from the village. The exact number of these and the number of the works performed by each group is not given. Unsurprisingly the size of the core harvester group correlates highly with the acreages of demesne land under plough at 0.83. Therefore in the period 1256–1349 the size of the core group oscillates between 23 and 37, in the decades immediately preceding the Black Death it stands in the mid-30s.

The impact of the Great Pestilence is visible in 1349–1350, 1352–1353 and 1353–1354. Whereas the number of the manorial staff halved, the number of the reapers was down by 70–85%, a very low amount of labour per acre was available in 1350. In Gnatingdon and Sedgeford the sown acreage of demesne land was greatly diminished as a consequence of the plague, but reached its lowest point not in 1349–1350, but in 1352–1353 and 1353–1354 (Fig. 2.3). The fall in cultivated acreage may not only have been the result of the post-plague socio-economic chaos, but may also have been connected to the drought conditions of the early 1350s, to which the villages Sedgeford and Gnatingdon, being situated on freely draining sandy soils, were particularly vulnerable.[12] Since the number of harvesters is not fluctuating much in the accounts of those years, the amount of labour spent by the core harvester group on each harvested acre increased in fact to pre-plague levels. However, in those years the relation between harvest date and length collapsed in Gnatingdon: harvests were longer than would be expected considering the harvest date and independent documentary evidence on summer and harvest season weather. The labour input per acre from hired workers probably shrank in comparison to pre-plague levels and/or works were performed unwillingly.[13] The demographic crisis and economic changes necessitated adaptive strategies and as a compensation for the shrunken labour supply, the quicker mowing of grain was introduced in Gnatingdon. The first mower is mentioned for the harvest of 1350. For another aspect of the crisis in the early 1350s, the frequently changing administrative personnel – certainly not the usual policy for Sedgeford and Gnatingdon – see Appendix 2.

After the accounting reform 1354–1355 very detailed information on the harvesting process is available in the works account. Included are the number of labourers as well as works and days performed by each harvesting group:[14] *famuli*, labour hired for the whole of the harvest, labour hired for individual days (cockers, probably wandering harvest workers), and the workers rendering the various labour dues. The labour dues included those performed by the customary tenants, the *opera* (day work) and *precariae* (boon works). The use of all these works is also detailed, apparently the optimum aimed at were about two works per acre for cutting corn and one work per acre for mowing corn, the latter generally replacing two and a half cutting works. This demonstrates the time and labour saving capacity of mowing corn. The last time customary labour dues were used for harvesting in Gnatingdon is the harvest of 1388. Over the period from 1360–1361, when the first account for Gnatingdon after the accounting reform is available, until 1387–1388 the number of customary tenants participating in the harvest dwindled, especially in the 1360s. At the same time the use of labourers hired for the whole harvest was raised from three in 1361 to 15 in 1388. Greater numbers of works were also done by cockers. Whereas in the 1360s c.20–40 works were performed by cockers at the harvest in Gnatingdon, this number increased to c.90–150 in the 1380s. Obviously Gnatingdon complied well to the national picture in a time when the enforcement of labour dues became increasingly difficult, whereas the use of hired labour augmented considerably.[15] Over this period the amount of acres of demesne land sown and the number of harvesters including the manorial staff, customary tenants and the labourers hired for the whole harvest (excluding mowers) correlates merely at 0.47. This is not

---

[12] For the drought and agricultural problems in the early 1350s, see Sect. 6.5.

[13] This would result in a slower harvesting process. For example the lower productivity of customary labour as compared to hired labour, leads to lower hay yields, slower corn harvesting and slower weeding on the manor of Wisbech Barton in Cambridgeshire, see Stone, Wisbech Barton, 648–649, 652–654.

[14] One work unit equals one man's work per day.

[15] In Cuxham, Oxfordshire, the development was even much faster, labour services almost ceased until 1359 and were replaced with hired labour, see Harvey (ed.), Medieval Oxfordshire village, 85–86.

surprising in a period of recurrent demographic crises, experimentation in harvesting methods and the annually varying contribution of the cockers. Plagues or other diseases hit in 1361, 1365, 1369, 1375, 1383, 1387 and 1390–1391 (Chap. 10).

By 1390–1391 the situation in Gnatingdon was – from a managerial point of view – changed profoundly. All labour dues apart from the *precaria* had disappeared. However, from 1394 onwards the harvest *precaria*-service on the fields of Gnatingdon was performed by the tenants of Sedgeford, and for the Gnatingdon tenants no harvest boon work is listed. Apart from the boon works performed by the Sedgeford tenants, the fields of Gnatingdon were harvested with the work of the manorial staff and hired labour alone. Harvesters hired for the whole harvest absorbed the share formerly performed by the customary tenants. After 1397 the number of works performed by cockers dwindled and they were rarely employed except in wet and difficult harvests such as 1406. Over time the size of both groups, the manorial staff and the labourers employed for the whole harvest, decreased and on the shrinking sown acreage mowers became more and more important. Whereas in the 1360s c.35% of the sown acreage were mown,[16] this increased to c.55% in the 1390s and the first decade of the fifteenth century and jumped to potentially c.70% afterwards.[17] The sensitivity of the harvest process in relation to weather and labour is also underlined by the fact that after the early 1380s the *compoti* explain the reason for additional works, and the need for activities such as turning, opening and rebinding the sheaves. The cause was usually abundant rainfall. Between 1390 and 1431 demesne size and number of manorial staff and labourers hired for the whole harvest (excluding mowers) correlates well again at 0.79, although local and occasionally extra-regional epidemics still occurred in England (Chap. 10).

[16] Several times a 'Flemyng' (a person from Flanders) was performing this task in Gnatingdon, thereby indicating the roots and ways of spreading the method; after 1373 no mower was referred to as 'Flemyng' any more.

[17] The last accounts, 1408–1409 to 1429–1430 NRO, LEST/IC 34–42, give the works performed by the mowers, not the actual acreage. Usually one mowing work would be used for one acre, but it is not sure, if all works performed by the mowers were really mowing works. If they were, then as much as 70 per cent of the crops was scythed, and some winter crops must have been harvested in this way, too. The occasional mowing of winter crops also happened in Hinderclay in Suffolk, Stone, Medieval agriculture, 250. Stone specifies, that in Hinderclay mowing was used in times of crisis, as 1348–1353, and when grain prices were low 1389–1402 and 1411–1429. This coincides with the adoption of mowing in Gnatingdon, where, however, it was much more readily embraced. The number of mowers climbed from one in the early 1350s to two after the Second Plague 1361 and temporarily even to four in the late 1360s. In Gnatingdon mowing corn was never abandoned until 1430, and at the beginning of the phases of especially low grain prices mentioned by Stone, shortly after 1390 and around 1410, scything intensified considerably in Gnatingdon and cockers almost totally disappeared. In the 1390s often five to six mowers were employed, and after 1409 it was usually four, but then the number of sown acres had fallen. In the last account, 1429–1430 NRO, LEST/IB 42, c.83% of the acreage were mown. Obviously labour costs were too high in relation to the grain price, so high indeed, that during harvest time it was hardly possible any more to offset this problem with the labour-saving mowing. Under such conditions it was unsustainable for Norwich Cathedral Priory to keep the manor of Gnatingdon under direct management. It must have been farmed out in the early 1430s.

Appendix 4: Reconstructed East Anglian April–July Mean Temperatures

Appendix 4.1 Reconstructed Norfolk April-July (AMJJ) mean temperatures 1256–1431

Year	AMJJ temperature	Year	AMJJ temperature	Year	AMJJ temperature	Year	AMJJ temperature	Year	AMJJ temperature
1251		1291		1331	13.01	1371	13.32	1411	11.78
1252		1292	12.93	1332	13.52	1372	12.26	1412	12.34
1253		1293	13.10	1333	13.85	1373	11.92	1413	12.51
1254		1294	12.00	1334	13.01	1374	11.50	1414	12.00
1255		1295	13.18	1335	11.43	1375	12.76	1415	12.08
1256	12.68	1296	12.17	1336	13.01	1376	12.26	1416	12.76
1257		1297	13.68	1337	12.93	1377	12.71	1417	12.17
1258	12.27	1298	13.60	1338	12.69	1378	12.84	1418	13.06
1259		1299	13.09	1339	12.34	1379		1419	12.51
1260		1300	13.01	1340	12.59	1380	12.07	1420	12.79
1261		1301	12.68	1341	11.92	1381	12.26	1421	11.58
1262	12.97	1302	12.64	1342	12.26	1382	11.75	1422	12.84
1263		1303	12.41	1343	12.68	1383	11.92	1423	11.75
1264	13.10	1304	13.43	1344	12.71	1384		1424	12.42
1265	13.26	1305	12.34	1345	12.34	1385	13.10	1425	12.51
1266		1306	13.43	1346	12.64	1386		1426	12.68
1267	13.77	1307	13.55	1347	11.96	1387	12.26	1427	
1268		1308		1348	11.64	1388	12.00	1428	10.91
1269	12.60	1309	12.84	1349		1389	11.92	1429	12.17
1270	13.05	1310	12.34	1350	12.26	1390	13.18	1430	12.76
1271		1311		1351	12.80	1391	12.51	1431	13.18
1272		1312	13.01	1352	13.18	1392	12.51		
1273	13.18	1313	12.34	1353	13.10	1393	12.84	N	147
1274	12.76	1314	11.92	1354	13.43	1394	12.93		
1275	12.10	1315	11.79	1355	12.18	1395	11.84		
1276		1316	12.98	1356	12.34	1396	12.93		
1277		1317		1357	12.68	1397	12.76		
1278	13.18	1318	13.68	1358		1398	12.84		
1279	13.18	1319	11.92	1359	13.06	1399	11.75		
1280	12.26	1320	12.84	1360	12.81	1400	13.26		
1281	13.26	1321	13.01	1361	14.53	1401	11.84		
1282		1322		1362	12.26	1402	11.84		
1283	11.97	1323	11.92	1363	12.68	1403	12.68		
1284		1324		1364	11.58	1404			
1285		1325	12.84	1365	13.59	1405	12.26		
1286	12.34	1326	13.94	1366	12.34	1406	11.84		
1287		1327	12.34	1367	11.92	1407	12.51		
1288	13.10	1328	13.43	1368	12.26	1408	12.59		
1289	12.72	1329	12.35	1369	13.10	1409	13.35		
1290	–	1330	11.76	1370	11.67	1410	12.84		

Appendix 5: East Anglian July–September Precipitation Index

Appendix 5.1 East Anglian July-September precipitation index 1256–1431

Year	JAS temperature	Year	JAS temperature	Year	JAS temperature	Year	JAS temperature	Year	JAS temperature
1251		1291		1331	−0.58	1371	0.61	1411	−1.00
1252		1292	0.51	1332	−1.00	1372	−0.18	1412	−0.87
1253		1293	−0.53	1333	−0.73	1373	0.72	1413	−1.34
1254		1294	0.98	1334	−0.62	1374	−0.11	1414	−1.24
1255		1295	0.26	1335	0.34	1375	−0.18	1415	−0.55
1256	1.00	1296	0.01	1336	−0.51	1376	−0.24	1416	−1.40
1257		1297	−0.05	1337	−1.31	1377	0.51	1417	−1.11
1258	2.18	1298	−0.65	1338	0.29	1378	0.86	1418	0.90
1259		1299	0.84	1339	−0.70	1379	0.77	1419	−0.39
1260		1300	0.06	1340	−1.11	1380	1.68	1420	−0.52
1261		1301		1341	−0.16	1381	0.70	1421	1.93
1262	0.08	1302	1.06	1342	−0.12	1382	0.83	1422	−0.33
1263		1303	0.71	1343	−0.32	1383	−0.54	1423	1.54
1264	0.71	1304	−0.72	1344	−1.14	1384	−1.07	1424	−0.34
1265	0.15	1305	−0.70	1345	0.86	1385	−0.33	1425	−0.07
1266		1306	−0.95	1346	0.81	1386	−0.54	1426	−1.29
1267	0.80	1307	2.07	1347	−1.09	1387	1.57	1427	−1.01
1268		1308	−0.43	1348	−0.78	1388	−0.34	1428	1.66
1269	−0.94	1309	0.29	1349	1.73	1389	−0.37	1429	−1.26
1270	0.25	1310	−0.21	1350	0.79	1390	−0.56	1430	−1.63
1271		1311	−0.59	1351	0.28	1391	−0.50	1431	−1.63
1272		1312	−0.83	1352	0.13	1392	−0.43	1432	
1273	0.17	1313	0.11	1353	0.71	1393	−0.51	1433	
1274	0.60	1314	0.78	1354	0.13	1394	0.23	1434	
1275	0.73	1315	1.62	1355	0.82	1395	−0.53	1435	1.39
1276		1316	−0.97	1356	0.79	1396	−0.49	1436	0.44
1277		1317		1357	0.90	1397	−0.19	1437	
1278	0.03	1318	−0.61	1358	−0.25	1398	−0.56	1438	
1279	0.79	1319	0.55	1359	1.48	1399	−0.34	1439	
1280	0.22	1320	0.81	1360	0.51	1400	−0.28	1440	−1.46
1281	0.72	1321	0.10	1361	0.44	1401	0.08	1441	
1282		1322	1.81	1362	0.99	1402	0.55	1442	0.44
1283	0.32	1323	−0.10	1363	0.80	1403	0.47	1443	
1284		1324	0.79	1364	0.88	1404	−0.72	1444	
1285		1325	−1.38	1365	1.60	1405	0.91	1445	−0.51
1286	−1.26	1326	−1.01	1366	2.12	1406	0.72	1446	−1.46
1287		1327	−0.49	1367	0.71	1407	−0.34	1447	
1288	−0.65	1328	−1.18	1368	0.12	1408	1.12	1448	0.44
1289	0.92	1329	−0.65	1369	−0.46	1409	−1.19		
1290	–	1330	2.21	1370	−0.16	1410	−1.30	N	164

Appendix 6: Late Medieval English Plague Waves and Weather Conditions, 1348–1500

Appendix 6.1 Late medieval English plague waves and weather conditions, 1348–1500

Category	Year	Summer index	First-difference	Plague occurrence	Comments
PP	**1420**	9	4	Supraregional plague in Norfolk, Essex and Kent	
PP	1422	9	4		Within five years after last supraregional plague
PP	1473	9	4	Plague in Hull, Dysentery in southern England	Within five years after last supraregional plague
PP	**1471**	8	5	Supraregional plague	
PP	**1393**	8	4	Regional plague in Essex	
PP	1424	8	4		Within five years after last supraregional plague
PP	**1390**	8	3	Supraregional plague; Fifth Pestilence	
PP	**1400**	8	3	Supraregional plague	
PP	**1383**	8	2	Regional plague in Norfolk	
PP	**1434**	8	2	Supraregional plague	
PP	1442	8	2	Plague in London	Within five years after last supraregional plague
PP	1495	8	2		Conditions for plague fulfilled by no plague
P	**1361**	8	1	Supraregional plague; Second Pestilence	First difference low, but plague probably started before
P	1385	8	1		First difference low, third warm summer in sequence
P	1447	8		Plague in Canterbury	First difference unknown
PP	1360	7	6	Minor plague or other disease	Follows on cold summer
PP	1492	7	6		Follows on cold summer
PP	**1464**	7	4	Supraregional plague	
PP	1371	7	3		Within five years after last supraregional plague
PP	**1375**	7	3	Supraregional plague; Fourth Pestilence	

(continued)

Appendix 6.1 (continued)

Category	Year	Summer index	First-difference	Plague occurrence	Comments
PP	**<u>1457</u>**	7	3	Supraregional plague	
P	1352	7	1		First difference low; within five years after last supraregional plague
P	**1426**	7	1	Regional plague in London, Sussex and potentially Great Yarmouth	First difference low
P	1461	7	1		First difference low; within five years after last supraregional plague
P	**<u>1479</u>**	7	1	Supraregional plague	First difference low, but plague started before
P	1384	7	−1		First difference low, second warm summer in sequence
P	**1452**	6	3	Regional plague in London, Reading and Suffolk	Summer conditions slightly above average
P	**1466**	6	3	Regional plague in southern Lincolnshire, northern Cambridgeshire, eastern Norfolk	Summer conditions slightly above average; within five years after last supraregional plague
P	1482	6	3		Summer conditions slightly above average; within five years after last supraregional plague
P	**1478**	6	3	Regional plague in London and Oxford	Summer conditions slightly above average; is the start of the supraregional of the year following
	1351	6	2		Summer conditions slightly above average; within five years after last supraregional plague
	1410	6	2		Summer conditions slightly above average; within five years after last supraregional plague

(continued)

Appendix 6.1 (continued)

Category	Year	Summer index	First-difference	Plague occurrence	Comments
	1412	6	2	Plague in Canterbury	Summer conditions slightly above average; within five years after last supraregional plague
	1417	6	2		Summer conditions slightly above average; within five years after last supraregional plague
	1441	6	2		Summer conditions slightly above average; within five years after last supraregional plague
	1469	6	2		Summer conditions slightly above average; within five years after last supraregional plague
	1490	6	2		Summer conditions slightly above average
	1494	6	2		Summer conditions slightly above average
	1498	6	2		Summer conditions slightly above average
	1500	6	1	Supraregional plague	Summer conditions slightly above average; first difference low, but plague started before
	1458	6	−1	Supraregional plague	Summer conditions slightly above average, first difference low, but plague started before
	1486	5	4		Summer moderate; follows on cold summer; one year after English Sweat
	1429	5	3		Summer moderate
	1367	5	3		Summer moderate; follows on cold summer; two years after 'Pokkes'
	1407	5	3	Supraregional plague	Summer moderate
	1357	5	2		Summer moderate

(continued)

Appendix 6.1 (continued)

Category	Year	Summer index	First-difference	Plague occurrence	Comments
	1395	5	2		Summer moderate; within five years after last supraregional plague
	1449	5	2	Plague in London	Summer moderate
	1348	5	1	Supraregional plague linked to cool-wet conditions and possibly malnutrition; First Pestilence	Summer moderate
	1439	5	1	Supraregional plague linked to cool-wet conditions and malnutrition, famine disease present	Summer moderate
	1368	5	0	Supraregional plague linked to cool-wet conditions and possibly malnutrition; Third Pestilence	Summer moderate
	1413	5	−1	Regional plague in the southeast maybe also in the east	Summer moderate; first difference low, but plague generally present between 1412 and 1415
	1499	5	−1	Supraregional plague	Summer moderate; first difference low, but plague started before
	1391	5	−3	Plague and famine disease in Norfolk, northern and western England	Summer moderate; first difference low, but plague started before
	1437	4	3	Plague in London	Summer conditions slightly below average; follows on cold summer; within five years after last supraregional plague
	1489	4	3		Summer conditions slightly below average; follows on cold summer
	1350	4	0	Plague still in some places in England; First Pestilence	Summer moderate
	1405	4	0	Low level epidemic (plague? linked to cool-wet conditions)	Summer conditions slightly below average

(continued)

Appendix 6.1 (continued)

Category	Year	Summer index	First-difference	Plague occurrence	Comments
	1438	4	0	Supraregional plague linked to cool-wet conditions and malnutrition, famine disease present	Summer conditions slightly below average
	1349	4	−1	Supraregional plague linked to cool-wet conditions and possibly malnutrition; First Pestilence	Summer conditions slightly below average
	1369	4	−1	Supraregional plague linked to cool-wet conditions and possibly malnutrition; Third Pestilence	Summer conditions slightly below average
	1406	2	−2	Low level epidemic (plague? linked to cool-wet conditions)	Summer cold

Given are supraregional plague waves (*bold* and *underlined*), regional outbreaks (*bold*) and local outbreaks outside London (*underlined*). Displayed are all summer seasons indexed as warm (7–9), all summer seasons in which the first difference equalled or exceeded 2 and all plague years in England, excepting those for northern England, because meteorological conditions in northern England are not under the geographical coverage of the climate indices used here. The summer temperature index is provided by van Engelen et al., A millennium of weather, winds and water in the Low Countries. First difference of summer conditions (compared to previous summer) is based on the summer index from van Engelen et al. For information on sources of plague outbreaks see Chap. 10. Years with summer conditions raising the risk for large-scale plague epidemics, high temperatures (7–9) and a first difference equalling or exceeding 2, are marked with PP. Years with summer conditions, that still increased the chances for large-scale plague epidemics, but to a lesser degree, high temperatures (7–9) with a first difference <2, or temperatures slightly above average (6) combined with a first difference of 3, are market with P.

Bibliography

Archival Material

Bodleian Library, Oxford (BLO)

MS Digby 147, fols. 125–138	Merle, William, Tractatus de pronosticacione aeris, c.1340
MS Rolls, Norfolk 20–45	Eaton, bailiffs' accounts

Chicago University Library (CUL)

Bacon 325–374	Redgrave, *compotus* accounts
Bacon 405–510	Hinderclay, *compotus* accounts

Climatic Research Unit, Norwich (CRU)

Minister's account rolls of Christ Church Canterbury, extracts, 1305-c.1386, extracted by J. Burgon Bickersteth, 1969 (copy given by Bickersteth to Hubert Lamb).

The National Archives, London (TNA)

SC 6/935/20-38	Framingham Parva
SC 6/942/15-16	Walsoken Popenhoe, reeves' and serjeants' accounts

Norfolk Record Office, Norwich (NRO)

Calendar of Documents Relating to Kempstone (WIS)

WIS 02-06, 8-10, 12-15, 17-19, 21-37	Kempstone, reeves' and bailiffs' accounts

Dean and Chapter Records (DCN)

DCN 60/02/01	Aldeby, bailiff's account
DCN 60/04/01-102	Catton, bailiffs' accounts
DCN 60/07/01-7	Denham, bailiffs' accounts
DCN 60/08/01-28	Eaton, bailiffs' accounts
DCN 60/10/01-35,	North Elmham, bailiffs' accounts; tithe accounts (DCN 60/10/41-6)
DCN 60/13/01-31	Gateley, bailiffs' accounts
DCN 60/14/01-25	Gnatingdon, bailiffs' accounts
DCN 60/15/01-16	Hemsby, bailiffs' accounts
DCN 60/16/01-6	Henley, bailiffs' accounts
DCN 60/18/01-71	Hindolveston, bailiffs' accounts
DCN 60/20/01-41	Hindringham, bailiffs' accounts
DCN 60/23/01-25	Martham, bailiffs' accounts
DCN 60/25/01-5	Melton, bailiffs' accounts
DCN 60/26/01-25	Monks' Grange, bailiffs' accounts

DCN 60/28/01-10	Newton, bailiffs' accounts
DCN 60/29/01-46A	Plumstead, bailiffs' accounts
DCN 60/30/01-12	Scratby, bailiffs' accounts
DCN 60/33/01-32	Sedgeford, bailiffs' accounts
DCN 60/35/01-52	Taverham, bailiffs' accounts
DCN 60/37/01-21	Thornham, bailiffs' accounts
DCN 60/38/01-26	Worstead, bailiffs' accounts
DCN 61/06	Arminghall, bailiff's account
DCN 61/10-11	Lakenham, bailiffs' accounts
DCN 61/14	Le Gannoc, fragmentary servant's account roll
DCN 61/15-16	Attlebridge, bailiff's accounts
DCN 61/18-24	Bawburgh, bailiffs' accounts
DCN 61/29	Fordham, custodian's account
DCN 61/31	Hardingham, tithe account
DCN 61/33-34	Heythe, custodians' accounts
DCN 61/39-41	Ormesby, bailiffs' accounts
DCN 61/42-43	Postwick, bailiff's, messor's and farmer's account
DCN 61/45	Wicklewood, servant's account
DCN 61/46	Wiggenhall, account of manor
DCN 61/47-48	Witton, tithe account and bailiff's account
DCN 61/51-52	Chalk, Kent, tithe account and servant's account
DCN 61/53	Unidentified manor, bailiff's account
DCN 61/56	Non cathedral, Stoke, bailiff's account
DCN 62/01	Grouped manor accounts: 1326–1327
DCN 62/02	Grouped manor accounts: 1333–1334
DCN 95/04-14	Various manors, bailiffs' and castleward's accounts (Episcopal and Diocesan, including St Benet's Abbey)

Hare of Stow Bardolph Collection (HARE)

HARE 780-790	Fincham, reeves' and bailiffs' accounts

Le Strange of Hunstanton (LEST)

LEST/BG 01-19	Hunstanton, bailiffs' accounts
LEST/DG 01-5	Heacham, bailiffs' accounts
LEST/DF 21	Sedgeford, bailiff's account, 1429–1430
LEST/EG 01-12	Ringstead, bailiffs' accounts
LEST/IB 15-71	Sedgeford, bailiffs' accounts
LEST/IB 72-74	Sedgeford, farmers' accounts
LEST/IB 79-80	Sedgeford, tithe accounts
LEST/IC 01-42	Gnatingdon, bailiffs' accounts
LEST/LA 14-28	Personal journals, 1809–1827 (Nicholas Styleman of Snettisham)

Minor Collection (MC)

MC 120/01-44	Farming diaries, 1768–1816 (Stephen Frost of Langham)
MC 120/45-57	Farming diaries, 1817–1831 (Thomas Rippingall)
MC 120/58	Farming diary, 1832 (Th. Rippingall, Stephen Frost Rippingall)
MC 120/61-85	Farming diaries, 1833–1858 (Rev. Stephen Frost Rippingall)
MC 120/87	Farming notebook and diary, 1847–1861/1867 (Stephen Frost Rippingall)
MC/120/121-123	Personal diaries, 1859–1861 (William Rippingall)
MC 150/52, 2 vols.	Farming diaries for Fritton estate, 1802–1827 (Thomas Howes)
MC 212/01-17	Great Cressingham, bailiffs' accounts

Norfolk Record Society (NRS)

NRS 5889, 20 D1, 5922, 20 D3	Martham, bailiffs' accounts 1323–1324–1437–1438

Records of the City of Norwich (NCR)

Bailiffs' accounts of St Giles's Hospital (Great Hospital), Norwich

NCR Press G Case 24 Shelves c-i	Calthorpe, Costessey, Cringleford, Hardley, South Walsham, bailiffs' accounts

Records of the Estates of the Bishop of Norwich, I: Records of St Benet's Abbey (DN/EST)

DN/EST 01/01-22	Various manors, bailiffs' accounts, grange accounts
DN/EST 09/01-2	Ashby, bailiffs' accounts
DN/EST 09/03-21	Flegg, bailiffs' and messors' accounts
DN/EST 10/01-5	North Walsham, bailiffs' accounts
DN/EST 11/01-12	Heigham, Potter Heigham, Scottow, Shotesham, bailiffs' accounts

Raynham Hall, Norfolk

Akenham manorial records 1278–1397

Published Primary Sources

Annales Monasterii de Bermundeseia, AD 1042–1432, in: Annales Monastici 3, ed. by Luard, Henry Richards (Rerum Britannicarum Medii Aevi Scriptores 36), London 1866.

Annales Monasterii de Oseneia, AD 1016–1347, in: Annales Monastici 4, ed. by Luard, Henry Richards (Rerum Britannicarum Medii Aevi Scriptores 36), London 1869.

Annales Monasterii de Wintonia, AD 519–1277, in: Annales Monastici 2, ed. by Luard, Henry Richards (Rerum Britannicarum Medii Aevi Scriptores 36), London 1865.

Annals of Loch Cé. A chronicle of Irish affairs from AD 1014 to AD 1590, ed. by Hennessy, William M., vol. 1, London 1871.

Annales of Ulster, otherwise Annals of Senat. A chronicle of Irish affairs AD 431–1131, 1155–1541, ed. by MacCarthy, B., vol. 3 AD 1379–1541, Dublin 1895.

Annales Paulini, in: Chronicles of the reign of Edward I and Edward II, ed. by Stubbs, William (Rerum Britannicarum Medii Aevi Scriptores 76.1), London 1882.

Annales Prioratus de Dunstaplia, AD 1–1297, in: Annales Monastici 3, ed. by Luard, Henry Richards (Rerum Britannicarum Medii Aevi Scriptores 36), London 1866.

Annales Prioratus de Wigornia, AD 1–1377, in: Annales Monastici 4, ed. by Luard, Henry Richards (Rerum Britannicarum Medii Aevi Scriptores 36), London 1869.

Annales Ricardi Secundi et Henrici Quarti, in: Chronica Monasterii S. Albani, ed. by Riley, Henry Thomas (Rerum Britannicarum Medii Aevi Scriptores 28.3), London 1866.

Arnold, Richard, The customs of London, otherwise called Arnold's Chronicle, London 1811.

Ashenden, John, Summa astrologiae judicialis de accidentibus mundi quae anglicana vulgo nuncupatur, Venice 1489.

Avesbury, Robert, De Gestis Mirabilibus Regis Edwardi Tertii, ed. by Thompson, Edward Maunde (Rerum Britannicarum Medii Aevi Scriptores 93.2), London 1889.

Brief Latin chronicle, in: Three fifteenth-century chronicles. With historical memoranda by John Stowe, the antiquary, and contemporary notes of occurrences written by him in the reign of Queen Elizabeth, ed. by Gairdner, James, London 1880.

Chronica Buriensis, 1212–1301/ The Chronicle of Bury St Edmunds, 1212–1301, ed. and transl. by Gransden, Antonia (Medieval Texts), London and Edinburgh 1964.

Chronicle of London from 1089 to 1493, ed. by Nicolas, London 1827.

Chronicle of the Grey Friars in London, ed. by Nichols, J.G. (Camden Society, Original series 53) 1852.

Chronicon Angliae, ab Anno Domini 1328 usque ad Annum 1388. Auctore monacho quodam Sancti Albani, ed. by Thompson, Edward Maunde (Rerum Britannicarum Medii Aevi Scriptores 64), London 1874.

Chronicon Anonymi Cantuariensis, in: Chronica Johannis de Reading et Anonymi Cantuariensis, 1346–1367, ed. by Tait, James (Historical Series 20), Manchester 1914.

Chronicon Rerum Gestarum in Monasterii Sancti Albani, in: Chronica Monasterii S. Albani, ed. by Riley, Henry Thomas (Rerum Britannicarum Medii Aevi Scriptores 28.5), vol. 1, London 1870.

Chronicon Westmonasteriense, 1381–1394. The Westminster Chronicle, 1381–1394, ed. and transl. by Hector, L.C., Harvey, Barbara F., Oxford 1982.

Clyn, John, Annalium Hiberniae Chronicon ad annum MCCCXLIX/ The Annals of Ireland. By friar John Clyn, of the convent of Friars Minors, Kilkenny; and Thady Dowling, Chancellor of Leighlin. Together with the Annals of Ross, ed. by Butler, Richard (Irish Archaeological Society), Dublin 1849, 1–39.

Continuatio Adami Murimuthensis, in: Adami Murimuthensis chronica sui temporis, nunc primum per decem annos aucta, (MCCCIII-MCCCXLVI) cum eorundem continuatione (ad MCCCLXXX) a quodam anonymo, ed. by Hog, Thomas, London 1846.

Continuatio chronici Florentii Wigorniensis, in: Florentii Wigorniensis monachi chronicon ex chronicis, [...], cui accesserunt continuationes duae quarum una ad annum MCXLI, altera, nunc primum typis vulgata, ad annum MCCXCV perducta, ed. by Benjamin Thorpe, vol. 2, London 1849.

Continuatio Eulogii, in: Eulogium (historiarum sive temporis). Chronicon ab orbe condito usque ad annum Domini M.CCC.LXVI., a monacho quodam Malmesburiensi exaratum ; accedunt continuationes duæ, quarum una ad annum M.CCCC.XIII., altera ad annum M.CCCC.XC. perducta est, ed. by Haydon, Frank Scott (Rerum Britannicarum Medii Aevi Scriptores 9), vol. 3, London 1863.

Continuation of Higden, in: Polychronicon Ranulphi Higden Monachi Cestrensis, together with the English translations of John Trevisa and of an unknown writer of the fifteenth century, ed. by Lumby, Joseph Rawson (Rerum Britannicarum Medii Aevi Scriptores 41), vol. 8, London 1882.

Cornelius Stovin, Journals of a Methodist farmer 1871–1875, ed. by Stovin, Jean, London 1982.

Cozens-Hardy, B. (ed.), Mary Hardy's Diary (Norfolk Record Society 37), Cambridge 1968.

Fabyan, Robert, The new chronicles of England and France (The concordance of histories). Reprinted from Pynson's edition of 1516, the first part collated with the editions of 1533, 1542, and 1559; and the second with a manuscript of the author's own time, as well as the subsequent editions, ed. by Ellis, Henry, London 1811.

Flores Historiarum, ed. by Henry Richards Luard, (Rerum Britannicarum Medii Aevi Scriptores 95.3), London 1890.

Grace of Kilkenny, James, Annales Hiberniae, ed. by Butler, Richard (Irish Archaeological Society), Dublin 1842.

Grafton's Chronicle, or, History of England, to which is added his table of the bailiffs, sheriffs, and mayors of the City of London, from the year 1189 to 1558, vol. 2, London 1809.

Gransden, A (1957) A fourteenth-century chronicle from the Grey Friars at Lynn. Engl Hist Rev 72(283):270–278.

Gregory, William, The historical collections of a citizen of London in the fifteenth century. III: William Gregory's Chronicle of London, ed. by Gairdner, James (Camden Society, Original scrics) 1876.

Griffiths, Elizabeth (ed.), William Windham's Green Book, 1673–1688 (Norfolk Record Society 66), Cambridge 2002.

Guisborough, Walter of, Chronicle of Walter of Guisborough, ed. by Rothwell, H. (Camden Society, Third Series 89), London 1957.

Hanawalt, Barbara (ed.), Crime in East Anglia in the fourteenth century. Norfolk goal delivery rolls, 1307–1316 (Norfolk Record Society 44), Norwich 1976.

Harvey, P.D.A. (ed.), Manorial Records of Cuxham, Oxfordshire, circa 1200–1359 (Historical Manuscript Commissio Joint Publications 23, Oxfordshire Record Society I), London 1976.

Herryson, John, Abbreviata Cronica ab anno 1377 usque ad annum 1469, ed. by Smith, J.J. (Publications of the Cambridge Antiquarian Society 2), Cambridge 1840.

Higden, Ranulf, Polychronicon Ranulphi Higden Monachi Cestrensis, together with the English translations of John Trevisa and of an unknown writer of the fifteenth century, ed. by Lumby, Joseph Rawson (Rerum Britannicarum Medii Aevi Scriptores 41), vol. 8, London 1882.

Holinshed's chronicles of England, Scotland and Ireland, vol. 2 England, London 1807.

Ingulph's chronicle of the Abbey of Croyland. With the continuations by Peter of Blois and anonymous writers, ed. and transl. by Riley, Henry T., London 1908.

Kershaw, Ian, The Bolton Priory 'Compotus', 1286–1325: together with a priory account roll for 1377–1378 (Yorkshire Archaeological Society record series), Leeds: Yorkshire Archaeological Society, 2000.

Knighton, Henry, Chronicon Henrici Knighton vel Cnitthon, Monachi Leycestrensis, ed. by Lumby, Joseph Rawson (Rerum Britannicarum Medii Aevi Scriptores 92.2), London 1895.

Langland, William, The vision of Piers Plowman. A critical edition of the B-text, ed. by Schmidt, Aubrey Vincent Carlyle, London and New York 1978.

Lynam, Charles (ed.), Translation of chronicle as relating to the Abbey, in: The Abbey of St. Mary Croxden, Staffordshire, London 1911.

McLaren, Mary-Rose (ed.), London Chronicle, Bradford, West Yorkshire Archives MS 32D86/42, in: The London chronicles of the fifteenth century. A revolution in English writing, Woodbridge (UK) and Rochester (USA) 2002.

Medieval Framlingham. Select documents 1270–1524, ed. by Ridgard, John (Suffolk Records Society 27), Exeter 1985.
Merle, William, Merle's MS: Consideraciones temperiei pro 7 annis, per Magistrum Willelmum Merle, socium domus de Merton, 1337–1344, ed by Symons, G.J., London 1891.
Murimuth, Adam, Continuatio Chronicarum, in: Adæ Murimuth Continuatio Chronicarum, Robert de Avesbury, De Gestis Mirabilibus Regis Edwardi Tertii, ed. by Thompson, Edward Maunde (Rerum Britannicarum Medii Aevi Scriptores 93), London 1889.
Norwich Cathedral Priory gardener's accounts, 1329–1530: in Farming and gardening in late medieval Norfolk, ed. by Noble, Claire, Moreton, Charles and Rutledge, Paul (Norfolk Record Society 61), Norwich 1997.
Oschinsky, Dorothea (ed.), Walter of Henley and other treatises on estate management and accounting, Oxford 1971.
Postles, David, Stubbington manorial accounts (Winchester College Muniments 15376–15384, 15386–15391), 2003, transcription available at: www.historicalresources.myzen.co.uk/STUBB/0prelim.html
Reading, John, Chronica, in: Chronica Johannis de Reading et Anonymi Cantuariensis, 1346–1367, ed. by Tait, James (Historical Series 20), Manchester 1914.
Short English chronicle, in: Three fifteenth-century chronicles. With historical memoranda by John Stowe, the antiquary, and contemporary notes of occurrences written by him in the reign of Queen Elizabeth, ed. by Gairdner, James, London 1880.
The Brut, or the chronicles of England, ed. by. Brie, Friedrich W.D., vol. 2, London and Oxford 1908.
The Pipe Roll of the Bishopric of Winchester, 1208–1209, ed. by Hall, Hubert, London 1903.
The Pipe Roll of the Bishopric of Winchester, 1210–1211 (PRO Eccl. 2-22-159270B), ed. by Holt, Neville Richard, Manchester 1964.
The Pipe Roll of the Bishopric of Winchester, 1301–1302, ed. by Page, M., (Hampshire Record Series 14), 1996.
The Pipe Roll of the Bishopric of Winchester, 1409–1410, ed. by Page, M. (Hampshire Record Series 16), 1999.
The St. Albans Chronicle, 1406–1420, ed. by Galbraith, V.H., Oxford 1937.
Trokelowe, John de, Johannis de Trokelowe annales, in: Chronica Monasterii S. Albani, ed. by Riley, Henry Thomas (Rerum Britannicarum Medii Aevi Scriptores 28.3), London 1866.
Tusser, Thomas, A hundreth good pointes of husbandrie, 1557. Renascence Editions text transcr. by Bear, Risa from the Dobell edition 1909., available at: https://scholarsbank.uoregon.edu/xmlui/bitstream/handle/1794/855/hundreth.pdf
Vitellus A XVI, in: Chronicles of London, ed. by Lethbridge Kingsford, Charles, Oxford 1905.
Wade Martins, Susanne and Williamson, Tom (eds.), The Farming Journal of Randall Burroughes, 1794–1799 (Norfolk Record Society 58), Cambridge 1995.
Walsingham, Thomas, Chronica Maiora, 1376–1394 (The St Albans Chronicle), ed. and transl. by Taylor, John, Childs, Wendy R., Watkiss, Leslie, vol. 1, Oxford 2003.
Walsingham, Thomas, Historia Anglicana, 1272–1381, 1382–1422: in: Chronica Monasterii S. Albani, ed. by Riley, Henry Thomas (Rerum Britannicarum Medii Aevi Scriptores 28,1), 2 vols., London 1863 and 1864.
Wriothesley, Charles, A chronicle of England during the reigns of the Tudors, from A.D. 1485 to 1559, ed. by Hamilton, William Douglas (Camden Society, New series 11), vol.1, London 1875.

Secondary Literature

Abel, Wilhelm, Massenarmut und Hungerkrisen im vorindustriellen Deutschland, Göttingen 1977 (1974).
Afton, Bethanie, Investigating agricultural production and land productivity. Methodology and opportunities using English farm records, in: Histoire et Mesure, vol. 15 3/4, 2000, 233–245.

Alexander, L.V., Jones, P.D., Updated precipitation series for the UK and discussion of recent extremes, in: Atmospheric Science Letters, vol. 1, 2000, 142–150.
Alexandre, Pierre, Le climat en Europe au Moyen Âge: contribution à l'histoire des variations climatiques de 1000 à 1425, d'après les sources narratives de l'Europe occidentale (École des Hautes Études en Sciences Sociales), Paris 1987.
Anonymous, Meußthurn, von wunderbarlicher Natur, Art, und Eygenschafft. Auch häuffigem Uffkommen, unnd endlichem Abnemmen, deß schädlichen landverderblichen Meußungeziffers, Frankfurt am Main 1618.
Anonymous, The practical Norfolk farmer. Describing the management of a farm throughout the year, Norwich 1808.
Anonymus (1823) Landwirtschaftliche Berichte. Oekonomische Neuigkeiten und Verhandlungen 25(1):369–372.
Appleby, Andrew B., Grain prices and subsistence crises in England and France, 1590–1740, in: The Journal of Economic History, vol. 39/4, 1979, 865–887.
Audoin-Rouzeau, Frédérique, Les chemins de la peste. Le rat, la puce et l'homme, Rennes 2003.
Ault, W.O., Open-field farming in medieval England. A study of village by-laws (Historical problems: Studies and documents, vol. 16), London and New York 1972.
Bailey, Mark, Demographic decline in late medieval England. Some thoughts on recent research, in: The Economic History Review, 2nd ser. vol. 49/1, 1996, 1–19.
Bailey, Mark, A marginal economy? East Anglian Breckland in the Later Middle Ages, Cambridge 1989.
Bailey, Mark, Medieval Suffolk. An economic and social history, 1200–1500 (History of Suffolk, vol. 1), Woodbridge (UK) and Rochester (USA) 2010.
Bailey, Mark, *Per impetum maris*. Natural disaster and economic decline in eastern England, 1275–1350, in: Campbell, Bruce M. S. (ed.), Before the Black Death. Studies in the 'crisis' of the early fourteenth century, Manchester and New York 1991, 184–208.
Bauernfeind, Walter, Woitek, Ulrich, The influence of climatic change on price fluctuations in Germany during the 16th century price revolution, in: Climatic Change, vol. 43/1, 1999, 303–321.
Bean, J. M. W., Plague, population and economic decline in England in the Later Middle Ages, in: The Economic History Review, vol. 15/3, 1963, 423–437.
Bell, W.T., Ogilvie, A.E.J., Weather compilations as a source of data for the reconstruction of European climate during the medieval period, in: Climatic Change, vol. 1, 1978, 331–348.
Ben Ari, Tamara, Neerinck, Simon, Gage, Kenneth L. et al., Plague and climate. Scales matter, in: PLoS Pathogens, vol. 7/9, 2011, e1002160.
Benedictow, Ole Jørgen. The Black Death, 1346–1353. The complete history. Woodbridge (UK) 2004.
Bennett, H.S., Life on the English manor. A study of peasant conditions, 1150–1400, Cambridge 1937.
Beveridge, W.H., The Winchester rolls and their dating, in: The Economic History Review, vol. 2/3, 1929, 93–113.
Biraben, Jean Noël, Les hommes et la peste en France et dans les pays européens et méditerranéens, vol. 1. La peste dans l'histoire (Civilisations et Sociétés, ed. by École des Hautes Études en Sciences Sociales, vol. 35), Paris 1975.
Black, John, Black, Dorothy, Plague in East Suffolk 1906–1918, in: Journal of the Royal Society of Medicin, vol. 93, 2000, 540–543.
Blake, P.W., Bull, J., Cartwright, A.R. et al., The Norfolk we live in, 4th edition, Norwich 1975.
Blomefield, Francis, An essay towards a topographical history of the county of Norfolk, containing a description of the towns, villages, and hamlets, with the foundations of monasteries, churches, chapels, chantries, and other religious buildings. Also an account of the ancient and present state of all the rectories, vicarages, donatives, and impropriations, their former and present patrons and incumbents, with their several valuations in the King's books, whether discharged or not; likewise, an historical account of the castles, seats, and manors, their present and ancient owners; together with the epitaphs, inscriptions, and arms in all the parish

churches, and chapels; with several draughts of churches, monuments, arms, ancient ruins, and other relicks of antiquity. Collected out of ledger books, registers, records, evidences, deeds, court-rolls, and other authentick memorials, 11 vols., London 1805–10 (reprint).

Bolton, J.L., Looking for *Yersinia pestis*. Scientists, historians and the Black Death, in: Clark, Linda, Rawcliffe, Carole (eds.), Society in an age of plague (The fifteenth century, vol. 12), Woodbridge (UK) 2013, 15–38.

Brandon, P.F., Late-medieval weather in Sussex and its agricultural significance, in: Transactions of the Institute of British Geographers, vol. 54, 1971, 1–17.

Brázdil, Rudolf, Dobrovolný, Petr, Luterbacher, Jürg et al., European climate of the past 500 years. New challenges for historical climatology, in: Climatic Change, vol. 101, 2010, 7–40.

Brázdil, Rudolf, Durdáková, Michaela, The effect of weather factors on fluctuations of grain prices in the Czech Lands in the 16th-18th centuries, in: Prace Geograficzne, vol. 108, 2000 19–25.

Brázdil, Rudolf, Kotyza, Oldrich, History of weather and climate in the Czech Lands I. Period 1000–1500 (Zürcher Geographische Schriften, vol. 62), Zürich 1995.

Brázdil, Rudolf, Kotyza, Oldrich, History of weather and climate in the Czech Lands IV. Utilisation of economic sources for the study of climate fluctuation at Louny and surroundings in the fifteenth-seventeenth centuries, Brno 2000.

Brázdil, Rudolf, Pfister, Christian, Wanner, Heinz, et al. Historical climatology in Europe – the state of the art, in: Climatic Change, vol. 70, 2005, 363–430.

Bridbury, A.R., Before the Black Death, in: The Economic History Review, 2nd ser., vol. 30/3, 1977, 393–410.

Bridbury, A.R., The Black Death, in: The Economic History Review, 2nd ser., vol. 26/4, 1973, 577–592.

Britnell, Richard H., Britain and Ireland 1050–1530, in: Daunton, Martin (ed.), Economy and society (Economic and Social History of Britain), Oxford et al. 2004.

Britnell, Richard H., Commercialisation and economic development in England, 1000–1300, in: Britnell, R. H., Campbell, Bruce M.S. (eds.), A commercialising economy. England 1086 to c. 1300, Manchester and New York 1995.

Britnell, Richard H., The commercialisation of English society 1000–1500, Cambridge 1993.

Britnell, Richard H., The Pastons and their Norfolk, in: Agricultural History Review, vol. 36/2, 1988, 132–144.

Britnell, Richard H., The Winchester Pipe Rolls and medieval English society, Woodbridge (UK) 2003.

Britton, C.E., A meteorological chronology to AD 1450 (Meteorological Office Geophysical Memoirs, vol. 70), London 1937.

Brooks, C.E.P., Glasspoole, J., British floods and droughts, London 1928.

Buisman, Jan, Duizend Jaar Weer, Wind en Water in de Lage Landen, Franeker, vol. 1 (to 1300) 1995, vol. 2 (1300–1450) 1996, vol. 3 (1450–1575) 1998.

Büntgen, Ulf, Trouet, Valerie, Frank, David et al., Tree-ring indicators of German summer drought over the last millennium, in: Quaternary Science Reviews, vol. 29, 2010, 1005–1016.

Camenisch, Chantal, Endless cold. A seasonal reconstruction of temperature and precipitation in the Burgundian Low Countries during the 15th century based on documentary evidence, in: Climate of the Past, vol. 11/8, 2015, 1049–1066.

Camenisch, Chantal, Keller, Kathrin, Salvisberg, Melanie et al., The early Spörer Minimum. A period of extraordinary climate and socio-economic changes in western and central Europe, in: Climate of the Past, Discussions, in review 2016.

Campbell, Bruce M.S., 'A fair field once full of folk.' Agrarian change in an era of population decline, 1348–1500, in: Agricultural History Review, vol.41/1, 1993, 60–70.

Campbell, Bruce M.S., Agricultural progress in medieval England: some evidence from eastern Norfolk, in: The Economic History Review, 2nd ser. vol. 36, 1983, 26–46.

Campbell, Bruce M.S., Arable productivity in medieval England. Some evidence from Norfolk, in: The Journal of Economic History, vol. 43/2, 1983, 379–404.

Campbell, Bruce M.S., Ecology versus economics in late thirteenth- and early fourteenth century English agriculture, in: Sweeny, Del (ed.), Agriculture in the Middle Ages. Technology, practice, and representation, Philadelphia 1995, 76–108.

Campbell, Bruce M.S., English seigniorial agriculture 1250–1450 (Cambridge Studies in Historical Geography), Cambridge 2000.
Campbell, Bruce M.S., Four famines and a pestilence. Harvest, price, and wage variations in England, 13th to 19th centuries, in: Liljewall, Britt et al., Agrarhistoria på många sätt, 28 studier om människan och jorden. Festschrift to Janken Myrdal (Skogs- och lantbrukshistoriska meddelanden nr. 47), Gävle 2009.
Campbell, Bruce M.S., The great transition. Climate, disease and society in the late-medieval world, Cambridge 2016.
Campbell, Bruce M. S., Ó Gráda, Cormac, Harvest shortfalls, grain prices, and famines in pre-industrial England, in: The Journal of Economic History, vol. 71/4, 2011, 859–886.
Campbell, Bruce M.S., Land, labour, livestock, and productivity trends in English seigniorial agriculture, 1208–1450, in: Campbell, Bruce M.S., Overton, Mark (eds.), Land, labour and livestock. Historical studies in European agricultural productivity, Manchester 1991, 144–182.
Campbell, Bruce M.S., Nature as historical protagonist: environment and society in pre-industrial England, in: The Economic History Review, 2nd ser. vol. 63, 2010a, 281–314.
Campbell, Bruce M.S., Overton, Mark, A new perspective on medieval and early modern agriculture. Six centuries of Norfolk farming c. 1250 – c. 1850, in: Past and Present, vol. 141, 1993, 38–105.
Campbell, Bruce M.S., Physical shocks, biological hazards and human impacts. The crisis of the fourteenth century revisited, in: Cavaciocchi, Simonetta (ed.), Le interazioni fra economia e ambiente biologico nell'Europa preindustriale secc. XIII–XVIII – Economic and biological interactions in pre-industrial Europe from the 13th to the 18th centuries (Serie 2, Atti delle 'Settimane de Studi' e altri Convegni vol. 41), Florence 2010b, 13–32.
Campbell, Bruce M.S., The regional uniqueness of English field systems? Some evidence from eastern Norfolk, in: Agricultural History Review, vol. 29, 1981, 16–28.
Campbell, James, Norwich before 1300, in: Rawcliffe, Carole, Wilson, Richard (eds.), Medieval Norwich, London and New York 2004, 29–49.
Chance, Eleanor, Colvin, Christina, Cooper, Janet et al., Medieval Oxford, in: Crossley, Alan, Elrington, C. R. (eds.), A history of the county of Oxford, vol. 4, the City of Oxford (Victoria County History), London 1979, 3–73.
Cavanaugh, Dan C., Marshall, John D., The influence of climate on the seasonal prevalence of plague in the Republic of Vietnam, in: Journal of Wildlife Diseases, vol. 8/1, 1972, 85–94.
Cheney, C.R., Jones, Michael, A handbooks of dates. For students of British history (Royal Historical Society. Guides and Handbooks No. 4), Cambridge et al. 2000.
Cheney, C.R., Norwich Cathedral Priory in the fourteenth century, in: Bulletin of the John Rylands Library, vol. 20, 1962, 93–120.
Chuine, I., Yiou, P., Viovy, N. et al., Grape ripening as a past climate indicator, in: Nature, vol. 432, 2004, 289–90.
Clitomachus [Döderlein, Johann Alexander], Curieuses Gespräch unter einigen guten Freunden von Mäusen, deren natürlichen Beschaffenheiten, Unterschiedlichen Gattungen, mancherley Eigenschafften, und dahero von den Gelehrten in der Sitten-Lehr beliebten Tugend und Lehr-Sprüchen; ingleichen dem durch dieselben zum öfftern verursachten mercklichen Schaden, Verheerungen ganzer Landschafften, auch Bestraffungen einzeler Personen. Schwabach and Leipzig 1743.
Cohn, Norman, Europe's inner demons. The demonization of Christians in medieval Christendom, London 2005 (1975).
Cohn, Samuel, The Black Death transformed. Disease and culture in early Renaissance Europe, London 2002.
Cohn, Samuel, Changing pathology of plague, in: Cavaciocchi, Simonetta (ed.), Le interazioni fra economia e ambiente biologico nell'Europa preindustriale secc. XIII–XVIII – Economic and biological interactions in pre-industrial Europe from the 13th to the 18th centuries (Serie 2, Atti delle 'Settimane de Studi' e altri Convegni vol. 41), Florence 2010, 33–56.
Cook, Charles Henry, The curiosities of ale and beer. An entertaining history by John Bickerdyke [pseudonym], London 1889.

Collins, J.T., Harvest technology and labour supply in Britain, 1790–1870, in: Economic History Review, New Series, vol. 22/3, 1969, 453–473.

Cooper, J. Richard, Melvin, Thomas M., Tyers, Ian et al., A tree-ring reconstruction of East Anglian (UK) hydroclimate variability over the last millennium, Climate Dynamics, vol. 40/3, 2013, 1019–1039.

Creighton, Charles. A history of epidemics in Britain. From AD 664 to the extinction of plague, Cambridge 1891.

Davenport, Frances Gardiner, The economic development of a Norfolk Manor, 1086–1565, Cambridge 1906.

Dawson, A.G., Hickey, K., Mayewski, P.A. et al., Greenland (GISP2) ice core and historical indicators of complex North Atlantic climate changes during the fourteenth century, in: The Holocene, vol. 17/4, 2007, 427–434.

Denholm-Young, N., Seigniorial Administration in England, London 1937.

Desai, Meghnad, The agrarian crisis in medieval England: A Malthusian tragedy or a failure of entitlements?, in: Bulleting of Economic Research vol. 43/ 3, 1991, 223–258.

Dickel, Horst, Der Begin der Mähdruschernte des Roggens als phänologische Phase, in: Wetter und Leben. Zeitschrift für angewandte Meteorologie, vol. 24/1–2, 1972, 74–8.

Dodds, B., Estimating arable output using Durham Priory tithe receipts, 1341–1450, in: The Economic History Review, 2nd ser. vol. 57/ 2, 2004, 245–285.

Dodwell, Barbara, The monastic community, in: Atherton, Ian, Fernie, Eric, Harper-Bill, Christopher et al. (eds.), Norwich Cathedral. Church, city, and diocese, 1096–1996, London 1996, 231–254.

Dodwell, Barbara, The Muniments and the library, in: Atherton, Ian, Fernie, Eric, Harper-Bill, Christopher et al. (eds.), Norwich Cathedral. Church, city, and diocese, 1096–1996, London 1996, 325–338.

Drew, J.S., Manorial accounts of St Swithun's Priory, Winchester, in: Carus-Wilson, E. M. (ed.) Essays in Economic History, vol. 2, London 1962,12–30.

Dunn, Penelope, Trade, in: Rawcliffe, Carole, Wilson, Richard (eds.), Medieval Norwich, London and New York 2004, 213–234.

Dyer, Christopher, Everyday life in medieval England, London 1994.

Dyer, Christopher, Seasonal patterns in food consumption in the Later Middle Ages, in: Woolgar, C. M., Serjeantson, D., Waldron, T. (eds.), Food in medieval England. Diet and nutrition, Oxford et al. 2006, 201–14.

Dyer, Christopher, Standards of living in the Later Middle Ages. Social change in England c. 1200–1520, Cambridge Medieval Textbooks, Cambridge et al. 1989.

Edwards, J.F., Hindle, B. P., Comment: inland water transportation in medieval England, in: Journal of Historical Geography, vol. 19, 1993, 12–14.

Elton, Charles S., Periodic fluctations in the numbers of animals. Their causes and effects, in: British Journal of Experimental Biology, vol. 2, 1924, 119–163.

Elton, Charles S., Voles, mice and lemmings. Problems in population dynamics, Oxford 1942.

Ernle, English farming. Past and present, London, Melbourne and Toronto 1961 (1912).

Esther, Alexandra, Imholt, Christian, Perner, Jorg et al., Correlations between weather conditions and common vole densities identified by regression tree analysis, in: Basic and Applied Ecology, vol. 15/1, 2014, 75–84.

Evans, George Ewart, The farm and the village, London 1969.

Fagan, Brian, The Little Ice Age. How climate made history 1300-1850, New York 2000.

Farmer, David L., Prices and wages, 1350–1500, in: Miller, Edward (ed.), The Agricultural History of England and Wales, vol. 3 (1348–1500), Cambridge et al. 1991, 431–525.

Farmer, David L., Two Wiltshire manors and their markets, in: Agricultural History Review, vol. 37/1, 1989, 1–11.

Farrell, Julianne, Mouse plagues – when, where and why, in: Wright, E.J., Webb, M.C., Highley, E. (eds.), Stored grain in Australia (Australian postharvest technical conference, Canberra, 25–27 June 2003), Canberra 2003.

Folland, Chris K., Knight, Jeff, Linderholm, Hans W. et al, The summer North Atlantic Oscillation. Past, present, and future, in: Journal of Climate, vol. 22/5, 2009, 1082–10103.

Frank, Robert Worth, The 'Hungry Gap', crop failure, and famine. The fourteenth century agricultural crisis and Piers Plowman, in: Sweeny, Del (ed.), Agriculture in the Middle Ages. Technology, practice, and representation, Philadelphia 1995, 227–243.

Gatterer, Christoph Wilhelm Jakob, Ueber die Verminderung der Feldmäuse, in: Forst-Archiv zur Erweiterung der Forst-und Jagd-Wissenschaft und der Forst- und Jagd-Literatur, vol. 27, Ulm 1803, 52–84.

Gérard, Charles, Essai d'une faune historique des mammifères sauvages de l'Alsace. Paris 1871.

Glaser, Rüdiger, Klimageschichte Mitteleuropas. 1200 Jahre Wetter, Klima, Katastrophen. Darmstadt 2013 (2001).

Goosse, H., Arzel, O., Luterbacher, J., et. al., The origin of the European 'Medieval Warm Period', in: Climate of the Past, vol. 2, 2006, 99–113.

Gottfried, Robert S., Epidemic disease in fifteenth century England. The medical response and the demographic consequences, Leicester 1978.

Gottfried, Robert S., The Black Death. Natural and human disaster in medieval Europe, London 1983.

Gottschalk, M.K. Elisabeth, Stormvloeden en rivieroverstromingen in Nederland (Sociaal Geografische Studies), vol. 1 de periode vóór 1400, Assen 1971, vol. 2, 1400–1600, Assen 1975.

Gransden, A., Historical writing in England, vol. 1 c. 550 to c. 1307, Ithaca 1974.

Gransden, A., Historical writing in England, vol. 2 c. 1307 to the early sixteenth century, London and Henley 1982.

Grant, Alexander, Independence and nationhood. Scotland, 1306–1469, London 1984

Grotefend, Hermann, Zeitrechnung des Deutschen Mittelalters und der Neuzeit, 2 vols., Hannover 1891–1898.

Hacket-Pain, Andrew J., Friend, Andrew D., Lageard, Jonathan G. A. et al., The influence of masting phenomenon on growth-climate relationships in trees. Explaining the influence of previous summers' climate on ring width, in: Tree physiology, vol. 35/3, 2015, 319–330.

Haensch S, Bianucci R, Signoli M et al (2010) Distinct clones of Yersinia pestis caused the Black Death. PLOS Pathog 6(10):e1001134. doi:10.1371/journal.ppat.1001134.

Halcrow, E.M., The decline of demesne farming on the estates of Durham Cathedral Priory, in: Economic History Review, New Series, vol. 7/3, 1955, 345–356.

Hallam, H.E., The climate of eastern England 1250–1350, in: The Agricultural History Review, vol. 32, 1984, 124–132.

Hallam, H.E., Farming techniques. Eastern England, in: Hallam., H.E. (ed.), The agrarian history of England and Wales, vol. 2 1042–1350, Cambridge et al. 1988, 272–312.

Hallam, H.E., Rural England and Wales, 1042–1350, in: Hallam., H.E. (ed.), The agrarian history of England and Wales, vol. 2 1042–1350, Cambridge et al. 1988, 966–1008.

Harper-Bill, Christopher, The Medieval church and the wider world, in: Atherton, Ian, Fernie, Eric, Harper-Bill, Christopher et al. (eds.), Norwich Cathedral. Church, city, and diocese, 1096–1996, London 1996, 281–313.

Harrison, C.J., Grain price analysis and harvest qualities, 1465–1634, in: The Agricultural History Review, vol. 19/2, 1971, 135–155.

Harvey, Barbara, Living and dying in England, 1100–1540. The monastic experience, Oxford 1995.

Harvey, B. F, Westminster Abbey and its estates in the Middle Ages, Oxford 1977.

Harvey, P.D.A., A medieval Oxfordshire village: Cuxham 1240–1400 (Oxford Historical Series), Oxford 1965.

Harvey, P.D.A., The pipe rolls and the adoption of demesne farming in England, in: The Economic History Review, 2nd ser., vol. 27, 1974, 345–359.

Hatcher, John, Mortality in the fifteenth century. Some new evidence, in: The Economic History Review, vol. 39/1, 1986, 19–38.

Hatcher, John, Plague, population and the English economy 1348–1530 (Studies in economic and social history), London and Basingstoke 1984 (1977).
Henisch, Bridget Ann, Fast and feast. Food in medieval society, University Park (US) 1999 (1976).
Henisch, Bridget Ann, The Medieval calendar year, University Park (US) 1999.
Heyman, Paul, Thoma, Bryan R., Marié, Jean-Lou et al., In search for factors that drive hantavirus epidemics, in: Frontiers in physiology, vol. 3/237, 2012, 1–23.
Hooker, Reginald H., The weather and the crops in eastern England, 1885–1921, in: Quarterly Journal of the Royal Meteorological Society, vol. 48/202, 1922, 115–138.
Horrox, Rosemary (ed. and transl.), The Black Death (Manchester Medieval Sources series), Manchester and New York 1994.
Hoskins, W.G., Harvest fluctuations and English economic history, 1480–1619, in: Agricultural History Review, vol. 12/1, 1964, 28–46.
Hoskins, W.G., Harvest fluctuations and English economic history, 1620–1759, in: Agricultural History Review, vol. 16/1, 1968, 15–31.
Hudson, William, Cottingham Tingey, John, Revised catalogue of the records of the City of Norwich. As arranged in the muniment room, in the castle museum, Norwich 1898.
Hyams, E., Vineyards in England. A practical handbook for the restoration of vine cultivation and wine making to southern Britain, London 1953.
Hybel, N., The grain trade in northern Europe before 1350, in: The Economic History Review, 2nd ser. vol. 55/2, 2002, 219–247.
Imholt, Christian, Esther, A., Perner, J. et al., Identification of weather parameters related to regional population outbreak risk of common voles (*Microtus arvalis*) in eastern Germany, in: Wildlife Research, vol. 38/7, 2011, 551–559.
Imholt, Christian., Essbauer, Sandra Simone, Jacob, Jens et al., Klima, Nagetiere und Nagetier-assoziierte Krankheitserreger, in: Lozán, J.L., Grassl, H., Karbe, L. et al. (eds.), Warnsignal Klima. Gefahren für Pflanzen, Tiere und Menschen (2nd electronic edition, Ch. 3.2.19), 2014, 1–10, available at: www.warnsignale.uni-hamburg.de
Inwards, Richard, Weather lore. A collection of proverbs, sayings and rules concerning the weather, London 1994 (1893).
Jacob, Jens, Tkadlec, Emil, Rodent outbreaks in Europe. Dynamics and damage, in: Singleton, Grant, Belmain, Steve, Brown, Peter et al. (eds.), Rodent outbreaks. Ecology and impacts, International Rice Research Institute 2010, 207–223.
James, Margery Kirkbride, Studies in the medieval wine trade, London 1971.
Jäschke, Kurt-Ulrich, Englands Weinwirtschaft in Antike und Mittelalter, in: Schrenk, Christhard, Weckbach, Hubert (eds.), Weinwirtschaft und Mittelalter. Zur Verbreitung, Regionalisierung und wirtschaftlichen Nutzung einer Sonderkultur aus der Römerzeit (Vorträge des gleichnamigen Symposiums vom 21. bis 24. März 1996 in Heilbronn) (Quellen und Forschungen zur Geschichte der Stadt Heilbronn, ed. by Christhard Schrenk, vol. 9), Heilbronn 1997.
Jenks, Stuart, Astrometeorology in the Middle Ages, in: Isis, vol. 74/2, 1983, 185–210.
Jones, E.L., Seasons and prices. The role of the weather on English agricultural history, London 1965.
Jones, P.D., Briffa, K.R., Barnett, T.P. et al., High-resolution palaeoclimatic records for the last millennium. Interpretation, integration and comparison with General Circulation Model control-run temperatures, in: The Holocene, vol. 8/4 1998, 455–471.
Jones, P.D., Briffa, K.R., Osborn, T.J. et al., High-resolution palaeoclimatology of the last millennium: a review of current status and future prospects, in: The Holocene, vol. 19/1 2009, 3–49.
Jones, Phil, Hulme, Mike, The changing temperature of 'Central England', in: Hulme, Mike, Barrow, Elaine (eds.), Climates of the British Isles. Present, past and future, London 1997, 173–96.
Jordan, William Chester, The Great Famine. Northern Europe in the early fourteenth century, Princeton und Chichester 1996.
Karlsson, Gunnar, Plague without rats. The case of fifteenth-century Iceland, in: Journal of Medieval History, vol. 22/3, 1996, 263–284.

Kausrud, Kyrre Linné, Viljugrein, Hildegunn, Frigessi, Arnoldo et al., Climatically driven synchrony of gerbil populations allows large-scale plague outbreaks, in: Proceedings of the Royal Society B: Biological Sciences, vol. 274/1621, 2007, 1963–1969.
Kausrud, Kyrre Linné, Yang, Meixue, Xu, Lei et al., Modeling the epidemiological history of plague in Central Asia. Palaeoclimatic forcing on a disease system over the past millennium, in: BMC Biology, vol. 8/112, 2010, 1–14.
Kent, Nathaniel, General view of the agriculture of the county of Norfolk, with observations for the means of its improvement (Board of Agriculture), London 1796.
Kermode, Jennifer, Medieval merchants. York, Beverley and Hull in the Later Middle Ages (Cambridge Studies in Medieval Life and Thought, Fourth series), Cambridge 1998.
Kershaw, Ian, The Great Famine and agrarian crisis in England, 1315–1322, in: Hilton, R.H. (ed.), Peasants, knights and heretics. Studies in medieval English social history, Cambridge, London and New York 1976, 85–132 (reprint of paper in: Past and Present vol. 59, 1973, 3–50).
King, E., Estate management and reform movement, in: Ormrod, W. M.(ed.), England in the thirteenth century. Proceedings of the 1989 Harlaxton Symposium, Stamford 1991.
Kiss, Andrea, Floods and weather in 1342 and 1343 in the Carpathian Basin, in press, 2011.
Kiss, Andrea, Wilson, R. and Bariska, I., An experimental 392-year documentary- based multi-proxy (vine and grain) reconstruction for May-July temperatures for Közseg, West Hungary, in: International Journal of Biometeorology, 2010.
Körner, Martin, Geschichte und Zoologie interdisziplinär: Feld- und Schermäuse in Solothurn 1538–1643. Ein Beitrag zur historischen Demographie wild lebender Tierarten, in: Jahrbuch für solothurnische Geschichte, vol. 66, 1993, 441–454.
Lamb, H.H., Climate. Present, past and future, vol. 2. Climatic history and the future, London and New York 1977.
Langdon, John, The economics of horses and oxen in medieval England, in: The Agricultural History Review, vol. 30/1, 1982, 31–40.
Langdon, John, Was England a technological backwater in the Middle Ages?, in: Astill, Grenville, Langdon, John (eds.), Medieval farming and technology. The impact of agricultural change in northwest Europe, Leiden, New York and Köln 1997, 275–291.
Lavigne, Franck, Degeai, Jean-Philippe, Komorowski, Jean-Christophe et al., Source of the great A.D. 1257 mystery eruption unveiled, Samalas volcano, Rinjani Volcanic Complex, Indonesia, in: Proceedings of the National Academy of Sciences, vol. 110/42, 2013, 16742–16747.
Lawrence, E.N., The earliest known journal of the weather, in: Weather, vol. 27, 1972, 494–501.
Legislation.gov.uk, delivered by The National Archives, available at: www.legislation.gov.uk/changes/chron-tables/private/25#f2
Le Roy Ladurie, Emmanuel, Histoire humaine et comparée du climat, vol. 1: Canicules et glaciers XIIIe–XVIIIe siècles, Paris 2004
Lisle, Edward, Observations in husbandry, London 1757.
Long, C., The oldest European weather diary?, in: Weather, vol. 29, 1974, 233–7.
Löpmeier, F.-J., Trampf, W., Die agrarmeteorologische Situation im Jahr 2007, in: Müller-Westermeier, Gerhard, Willing, Patricia (eds.), Klimastatusbericht 2007. Aktuelle Ergebnisse des Klimamonitorings, DWD, Offenbach 2008, 50–60.
Lucas, Henry S., The Great European Famine of 1315, 1316 and 1317, in: Speculum, vol. 5, 1930, 343–77.
Luterbacher, Jürg, Werner, J.P., Smerdon, J.E. et al., European summer temperatures since Roman times, in: Environmental Research Letters, vol. 11/1, 2016, 024001.
Manley, G., The mean temperature of central England, 1698–1952, in: The Quarterly Journal of the Royal Meteorological Society, vol. 79, 1953, 242–261.
Manley, G. Central England temperatures: monthly means 1659–1973, in: The Quarterly Journal of the Royal Meteorological Society, 100, 1974, 389–405.
Mann, M., Zhang, Z. Rutherford, S. et al. Global signatures and dynamical origins of the Little Ice Age and the Medieval Climate Anomaly, in: Science, 326, 2009, 1256–60.

Marshall, William, A review of the reports to the Board of Agriculture. From the eastern department of England, comprizing Lincolnshire, Norfolk, Suffolk and northeast Essex, London 1811.
Marvin, Julia, Cannibalism as an aspect of famine in two English chronicles, in: Carlin, Martha, Rosenthal, Joel T. (eds.), Food and eating in medieval Europe, London and Rio Grande 1998, 73–86.
Masschaele, James, Peasants, merchants, and markets. Inland trade in medieval England, 1150–1350, Basingstoke and London 1997.
Mate, Mavis, Agrarian economy after the Black Death: The manors of Canterbury Cathedral Priory, 1348–1391, in: The Economic History Review, 2nd ser., vol. 37/3, 1984, 341–354.
Mate, Mavis, Agricultural technology in southeast England, 1348–1530, in: Astill, Grenville, Langdon, John (eds.), Medieval farming and technology. The impact of agricultural change in northwest Europe, Leiden, New York and Köln 1997, 251–274.
Matthews, J. D., The influence of weather on the frequency of beech mast years in England, in: Forestry, vol. 28/2, 1955, 107–116.
Mayhew, N. J., Money and prices in England from Henry II to Edward III, in: Agricultural History Review, vol. 35/2, 1987, 121–132.
Meaden, G.T., Merle's weather diary and its motivation, in: Weather, 28, 1973, 210–1.
Meeres, Frank, Guide to the records of Norwich Cathedral, Norwich 1998.
Messmer, M., Gómez-Navarro, J. J., Raible, C. C., Climatology of Vb-cyclones, physical mechanisms and their impact on extreme precipitation over Central Europe, in: Earth System Dynamics Discussions, vol. 6/1, 2015, 541–553.
Miller, E., Introduction. Land and people, in: Miller, Edward, Thirsk, Joan (eds.), The agrarian history of England and Wales, vol. 3 1348–1500, Cambridge et al. 1991. 1–33.
Miller, E., The Abbey and Bishopric of Ely: The social history of an ecclesiastical estate from the tenth century to the early fourteenth century, Cambridge, 1951.
Ministry of Agriculture, Fisheries and Food (MAFF), Agricultural land classification of England and Wales, Sheet 124 King's Lynn, 1972.
Ministry of Agriculture, Fisheries and Food (MAFF), Agricultural land classification of England and Wales, Sheet 126 Norwich, 1972.
Ministry of Agriculture, Fisheries and Food (MAFF), Agricultural land classification of England and Wales, Sheet 125 Fakenham, 1972.
Mortimer, R., William Merle's weather diary and the reliability of historical evidence for medieval climate, in: Climate Monitor, vol. 10/2, 1981, 42–45.
Možný, Martin, Brázdil, Rudolf, Dobrovolný, Petr et al., Cereal harvest dates in the Czech Republic between 1501 and 2008 as a proxy for March–June temperature reconstruction, in: Climatic Change, vol. 110/3–4, 2012, 801–821.
Müller-Haubold, Hilmar, Hertel, Dietrich, Leuschner, Christoph, Climatic drivers of mast fruiting in European beech and resulting C and N allocation shifts, in: Ecosystems, vol. 18/6, 2015, 1083–100.
Munro, John H., Revisions of the Phelps Brown and Hoskins 'Basket of consumables' commodity price series, 1264–1700, available at: www.economics.utoronto.ca/munro5/Prices%20and%20 Wages.htm
Neal, Robert A., Phillips, I.D., Summer daily precipitation variability over the East Anglian region of Great Britain, in: International Journal of Climatology, vol. 29, 2009, 1661–1679.
Noble, Claire, Aspects of life at Norwich Cathedral Priory in the late medieval period, University of East Anglia, 2001 (unpubl. Ph.D. thesis University of East Anglia).
Nordli, P.Ø., Reconstruction of nineteenth century summer temperatures in Norway by proxy data from farmer's diaries, in: Climatic Change, vol. 48, 2001, 201–208.
North, John David, Cosmos. An illustrated history of astronomy and cosmology, Chicago and Bristol 2008.
Ogilvie, Astrid, Farmer, Graham, Documenting the medieval climate, in: Hulme, Mike, Barrow, Elaine (eds.), Climates of the British Isles. Present, past and future, London 1997, 112–133.

Oliver, J., Problems of agro-climatic relationships in Wales in the eighteenth century, in: Taylor, James A. (ed.), Weather and agriculture, Oxford et al. 1967, 187–200.

Ollerenshaw, C.B., Climatic factors and liver fluke disease, in: Taylor, James A. (ed.), Weather and agriculture, Oxford et al., 1967, 129–135.

Oman, Charles William Chadwick, The great revolt of 1381, Oxford 1906.

Ordnance Survey: 132, North West Norfolk. King's Lynn and Fakenham 1:50000 (Landranger map), Southampton 2006.

Ordnance Survey: 133, North East Norfolk. Cromer and Wroxham 1:50000 (Landranger map), Southampton 2006.

Page, William (ed.), A history of the county of Norfolk (Victoria County History), 2 vols., London 1906.

Parry, Martin L., Secular climatic change and marginal agriculture, in: Transactions of the Institute of British Geographers, vol. 64, 1975, 1–13.

Pestell, Tim, Landscapes of monastic foundations. The establishment of religious houses in East Anglia c.650–1200, Woodbridge (UK) and Rochester (USA) 2004.

Pfister, Christian, Luterbacher, Jürg, Wanner, Heinz et al., Documentary evidence as climate proxies. Proxy-specific white paper produced from the PAGES/CLIVAR workshop, Trieste 2008, PAGES (Past global changes), 2009, available at: http://pages-142.unibe.ch/products/scientific_foci/pdfs/Trieste_documentary.pdf

Pfister, Christian, Schwarz-Zanetti, Gabriela, Wegmann, Milene, Winter severity in Europe: The fourteenth century, in: Climatic Change, vol. 34, 1996, 91–108.

Pfister, Christian, Variations in the spring-summer climate of Central Europe from the High Middle Ages to 1850, in: Wanner, Heinz, Siegenthaler, U. (eds.), Lecture notes in earth sciences, vol. 16: Long term and short term variability of climate, Berlin 1988, 57–82.

Pfister, Christian, Agrarkonjunkur und Witterungsverlauf im westlichen Schweizer Mittelland zur Zeit der Ökonomischen Patrioten, 1755–1797, Bern 1975.

Pfister, Christian, Fluctuations climatiques et prix céréaliers en Europe du XVIe au XXe siècle, in: Annales. Histoire, Sciences Sociales, vol. 43/1, 1988, 25–53.

Pfister, Christian, Getreide-Erntebeginn und Frühsommertemperaturen im schweizerischen Mittelland seit dem 17. Jahrhundert, in: Geographica Helvetica, vol. 1, 1979, 23–35.

Pinot, Adrien, Gauffre, Bertrand, Bretagnolle, Vincent, The interplay between seasonality and density. Consequences for female breeding decisions in a small cyclic herbivore, in: BMC Ecology, vol. 14/17, 2014, 1–13.

Postles, David A., Heads of religious houses as administrators, in: Ormrod, W. M. (ed.), England in the thirteenth century. Proceedings of the 1989 Harlaxton Symposium, Stamford 1991, 37–50.

Postles, David, Cleaning the medieval arable, in: Agricultural History Review, vol. 37/2, 1989, 130–143.

Pribyl, Kathleen, Cornes, Richard C., Pfister, Christian, Reconstructing medieval April–July mean temperatures in East Anglia, 1256–1431, in: Climatic Change, 2012, 393–412.

Pribyl, Kathleen, The study of the climate of medieval England. A review of historical climatology's past achievements and future potential, in: Weather, vol. 69/5, 2014, 116–120.

Proctor, C.J., Baker A., Barnes, W.L. et al., A thousand year speleothem proxy record of North Atlantic climate from Scotland, in: Climate Dynamics, vol. 16/10, 2000, 815–820.

Raftis, J. Ambrose, The estates of Ramsey Abbey. A study in economic growth and organization (Pontifical Institute of Medieval Studies, Studies and Texts 3), Toronto 1957.

Rawcliffe, Carole, Medicine for the soul. The life, death and resurrection of an English medieval hospital, St Giles's Norwich, c. 1249–1550, Stroud 1999.

Rawcliffe, Carole, Urban bodies. Communal health in late medieval English towns and cities. Woodbridge (UK) 2013.

Reijniers, Jonas, Begon, Mike, Ageyev, Vladimir S. et al., Plague epizootic cycles in Central Asia, in: Biology letters, vol. 10, 2014, 20140302.

Richardson, H.G., The exchequer year, in: Transactions of the Royal Historical Society, 4th ser. vol. 8, 1925, 171–190.

Rogers, A.G.L., Was rye ever the ordinary food of the English?, in: The Economic Journal, vol. 32/125, 1922, 119–124.
Rohr, Christian, Extreme Naturereignisse im Ostalpenraum. Naturerfahrung im Spätmittelalter und am Beginn der Neuzeit, Köln 2007.
Rösener, Werner, Bauern im Mittelalter, München 1991 (1985).
Roth, Werner, Mäuse wieder auf dem Vormarsch, delivered by Dienstleistungszentrum Ländlicher Raum Eifel, available at: www.dlr.rlp.de/Internet/global/themen.nsf/0/516c21aad03d1edfc125794a00554e8c/$FILE/Roth_Maeuse_2014.pdf, 2014
Rutledge, Elizabeth, Economic life, medieval Norwich, ed. by Rawcliffe, Carole and Wilson, Richard, London and New York 2004, 158–188.
Samia, Noelle I., Kausrud, Kyrre Linné, Hans Heesterbeek et al., Dynamics of the plague–wildlife–human system in Central Asia are controlled by two epidemiological thresholds, in: Proceedings of the National Academy of Sciences of the United States of America, vol. 108/35, 2011, 14527–14532.
Saunders, Herbert Washington, An introduction to the obedientiary and manor rolls of Norwich Cathedral Priory, Norwich 1930.
Saunders, G.R., Giles, J.R., A relationship between plagues of the house mouse, *Mus Musculus* (*Rodentia Muridae*) and prolonged periods of dry weather in south-eastern Australia, in: Australian Wildlife Research, vol. 4, 1977, 241–247.
Schmid, Boris V., Büntgen, Ulf, Easterday, W. Ryan et al., Climate-driven introduction of the Black Death and successive plague reintroductions into Europe, in: Proceedings of the National Academy of Sciences, vol. 112/10, 2015, 3020–3025.
Scott, Susan, Duncan, S.R., Duncan, C.J., The origins, interactions and causes of the cycles in grain prices in England, 1450–1812, in: Agricultural History Review, vol. 46/1, 1998, 1–14.
Seifert, Lisa, Wiechmann, Ingrid, Harbeck, Michaela et al., Genotyping *Yersinia pestis* in historical plague. Evidence for long-term persistence of *Y. Pestis* in Europe from the 14th to the 17th century, in: PLoS One, vol. 11/1, 2016, 1–8.
Semenza, Jan C., Menne, Bettina, Climate change and infectious diseases in Europe, in: Lancet Infectious Diseases, vol. 9/6, 2009, 365–375.
Sharp, Buchanan, Famine and scarcity in late medieval and early modern England. The regulation of grain marketing, 1256–1631, Cambridge 2016.
Short, Thomas, A general chronological history of the air, weather, seasons meteors etc., London 1749.
Shrewsbury, John Findlay Drew, A history of bubonic plague in the British Isles, Cambridge 1970.
Slavin, Philip, Bread and ale for the brethren. The provisioning of Norwich Cathedral Priory 1260–1536 (Studies in regional and local history, ed. by Nigel Goose, vol.11), Hatfield (UK) 2012.
Slavin, Philip, Communities of famine. A fourteenth-century environmental shock in the British Isles, 2016 in press.
Slavin, Philip, Feeding the brethren. Grain provisioning of Norwich Cathedral Priory, c. 1280–1370, 2008, (unpubl. Ph.D. thesis University of Toronto).
Slavin, Philip, The fifth rider of the apocalypse: The great cattle plague in England and Wales and its economic consequences, 1319–1350, in: Cavaciocchi, Simonetta (ed.), Le interazioni fra economia e ambiente biologico nell'Europa preindustriale secc. XIII–XVIII – Economic and biological interactions in pre-industrial Europe from the 13th to the 18th centuries (Serie 2, Atti delle 'Settimane de Studi' e altri Convegni vol. 41), Florence 2010, 165–79.
Slicher van Bath, B.H., The agrarian history of western Europe, A.D. 500–1850, London 1963.
Smith, R.A.L., Canterbury Cathedral Priory. A study in monastic administration, Cambridge 1943.
Snedegar, Keith V., Between scholasticism and folk wisdom. The weather lore of William Merle, in: Southern African Journal of Medieval and Renaissance Studies, vol. 5, 1997, 29–36.
Soils of England and Wales: Sheet 4 Eastern England 1:250000 (Soil Survey of England and Wales), Rothamstead Experimental Station, Harpenden 1983.
Stenseth, Nils Chr., Ageyev, Vladimir S., Klassovskiy, Nikolay L. et al., Plague dynamics are driven by climate variation, in: Proceedings of the National Academy of Sciences of the United States of America, vol. 103/35, 2006, 13110–13115.

Stern, D.V., A Hertfordshire manor of Westminster Abbey. Profits, productivity and weather, ed. by Christopher Thornton (Studies in regional and local history, ed. by Nigel Goose, vol.1), Hatfield (UK) 2000 (1978).

Sticker, Georg, Abhandlungen aus der Seuchengeschichte und Seuchenlehren, vol. 1. Die Pest, Giessen 1908–10.

Stone, David, Decision-making in medieval agriculture, Oxford et al. 2005.

Stone, David, The consumption of field crops in late medieval England, in: Woolgar, C. M., Serjeantson, D., Waldron, T. (eds.), Food in medieval England. Diet and Nutrition, Oxford et al. 2006, 11–26.

Stone, David, The productivity of hired and customary labour: Evidence from Wisbech Barton in the fourteenth century, in: The Economic History Review, 2nd ser. vol. 50/4, 1997, 640–656.

Stone, Eric, Profit-and-loss accountancy at Norwich Cathedral Priory, in: Transactions of the Royal Historical Society, 5th ser. vol. 12, 1962, 25–48.

Stone, Eric, The estates of Norwich Cathedral Priory, 1100–1300, 1956, (unpubl. Ph.D. thesis Oxford University).

Symons, G.J., ed. Merle's MS: consideraciones temperiei pro 7 annis, 1337–1344, London 1891.

Tanner, Norman, The cathedral and the city, in: Atherton, Ian, Fernie, Eric, Harper-Bill, Christopher et al. (eds.), Norwich Cathedral. Church, city, and diocese, 1096–1996, London 1996, 255–280.

Tarand, A., Kuiv, P., The beginning of the rye harvest – a proxy indicator of the summer climate in the Baltic area, in: Frenzel, Burkhard, Pfister, Christian, Gläser, Birgit (eds.), Climatic trends and anomalies in Europe 1675–1715. High resolution spatio-temporal reconstructions from direct meteorological observations and proxy data. Methods and results, Stuttgart, Jena and New York 1994, 61–72.

Thorndike, Lynn, A history of magic and experimental science, vol. 3 fourteenth and fifteenth century, New York 1934.

Tickell, John, The history of the town and county of Kingston Upon Hull. From its foundation in the reign of Edward the First to the present time, Hull 1796.

Timmreck, Claudia, Lorenz, Stephan J., Crowley, Thomas J. et al., Limited temperature response to the very large AD 1258 volcanic eruption, Geophysical Research Letters, vol. 36 , 2009, L 21708.

Titow, Jan, Evidence of weather in the account rolls of the Bishopric of Winchester 1209–1350, in: The Economic History Review, vol. 12, 1960, 360–407.

Titow, Jan, Le climat à travers les rôles de comptabilité de l'évêché de Winchester (1350–1450), in: Annales. Économies, Sociétés, Civilisations, vol. 25/2, 1970, 312–350.

Titow, Jan, Winchester yields. A study in medieval agricultural productivity, Cambridge 1972.

Trouet, Valerie, Esper, Jan, Graham, Nicholas E. et al., Persistant positive North Atlantic Oscillation mode dominated the Medieval Climate Anomaly, in: Science, vol. 324, 78–80.

Turner, William H., Coxe, H.O., Calendar of charters and rolls preserved in the Bodleian library, Oxford 1878.

Unwin, P.T.H., Wine and the vine. An historical geography of viticulture and the wine trade, London and New York 1991.

van Engelen, Aryan F.V., Buisman, Jan and IJnsen, Folkert, A millennium of weather, winds and water in the Low Countries, in: Jones, P.D., Ogilvie, A.E.J., Davies, T.D. et al. (eds.), History and climate: memories of the future?, New York et al. 2001, 101–24.

van Zwanenberg, D., The last epidemic of plague in England? Suffolk 1906–1918, in: Medical history, vol. 14/1, 1970, 63–74.

Virgoe, Roger, The estates of Norwich Cathedral Priory, 1101–1538, in: Atherton, Ian, Fernie, Eric, Harper-Bill, Christopher et al. (eds.), Norwich Cathedral. Church, city, and diocese, 1096–1996, London 1996, 339–359.

Wade Martins, Susanna, Williamson, Tom, Roots of change. Farming and the landscape in East Anglia, c. 1700–1870, The Agricultural History Review Supplement 2nd ser., Exeter 1999.

Wetter, Oliver, Pfister, Christian, Spring-summer temperatures reconstructed for northern Switzerland and southwestern Germany from winter rye harvest dates, 1454–1970, in: Climate of the Past, vol. 7/4, 2011, 1307–1326.

Williamson, Tom, Explaining regional landscapes: East Anglia and the Midlands in the Middle Ages, in: Harper-Bill, Christopher (ed.), Medieval East Anglia, Woodbridge (UK) and Rochester (USA) 2005, 11–32.

Wilson, Rob, Miles, Dan, Loader, Neil J. et al., A millennial long March–July precipitation reconstruction for southern-central England, in: Climate Dynamics, vol. 40/3, 2013, 997–1017.

Wolf, Hinrich, Über die Feldmäuse insonderheit in Nord-Dithmarschen samt einem Anhange über die bekanntesten Arten des Unkrauts, Hamburg 1786.

Woolgar, C.M., The great household in late medieval England, New Haven and London 1999.

Xu, Lei, Liu, Qiyong, Stige, Leif Chr. et al., Nonlinear effect of climate on plague during the third pandemic in China, in: Proceedings of the National Academy of Sciences of the United States of America, vol. 108/25, 2011, 10214–19.

Yaxley, David, The documentary evidence, in: Wade-Martins, Peter (ed.), Excavations in North Elmham Park 1967–1972 (East Anglian Archaeology, Report No. 9 Norfolk, ed. by Norfolk Archaeological Unit, Norfolk Museum Service), Gressenhall 1980, 517–635.

Yaxley, David, The prior's manor-houses. Inventories of eleven of the manor-houses of the Prior of Norwich, Dereham 1988.

Young, Arthur, General view of the agriculture of the county of Norfolk (Board of Agriculture), London 1804.

Zwolak, Rafał, Bogdziewicz, Michał, Wróbel, Aleksandra et al., Advantages of masting in European beech. Timing of granivore satiation and benefits of seed caching support the predator dispersal hypothesis, in: Oecologia, vol. 180/3, 2016, 749–758.

Index

K. Pribyl, *Farming, Famine and Plague*, DOI 10.1007/978-3-319-55953-7

FSC
www.fsc.org
MIX
Papier | Fördert gute Waldnutzung
FSC® C083411

Kolibri 360 GmbH
Ferdinand-Jühlke-Straße 7
99095 Erfurt, Deutschland
produktsicherheit@kolibri360.de